Electronic Devices on Discrete Components for Industrial and Power Engineering

Electronic Devices on Discrete Components for Industrial and Power Engineering

Vladimir Gurevich
Israel Electric Corp, Haifa

CRC Press
Taylor & Francis Group
Boca Raton London New York

CRC Press is an imprint of the
Taylor & Francis Group, an **informa** business

CRC Press
Taylor & Francis Group
6000 Broken Sound Parkway NW, Suite 300
Boca Raton, FL 33487-2742

Printed in the United States of America on acid-free paper
10 9 8 7 6 5 4 3 2 1

International Standard Book Number-13: 978-1-4200-6982-2 (Hardcover)

Visit the Taylor & Francis Web site at
http://www.taylorandfrancis.com

and the CRC Press Web site at
http://www.crcpress.com

Preface

Integral microchips and microprocessors have come into our lives so swiftly and completely that sometimes it seems that modern equipment simply cannot exist without them, which is true. However, dependence of modern equipment on microelectronics and microprocessors does not mean that there are no problems in this area. The integrity of many functions distributed earlier among separate devices of a complex system in a single microprocessor leads to the reduction of system reliability because damage to the microprocessor or to any number of peripheral elements serving the microprocessor leads to failure of the whole system but not of its separate functions as it was in pre-microprocessor time. Added to this is the extra sensitivity of microelectronic and microprocessor-based equipment to electromagnetic interferences (EMI) and the possibility of intentional remote actions breaking the normal operation of the microprocessor-based devices (electromagnetic weapons, electromagnetic terrorism). Intensive investigations into the electromagnetic weapons field are being carried out in Russia, the U.S., England, Germany, China, and India. Many world-leading companies work intensively in this sphere creating new devices of these weapon systems functioning at a distance of several dozens of meters to several kilometers, which while specialized in their use are still available to everybody (as they are freely sold on the market).

The need for specialized power supplies of microprocessors, different types of memory, special input and output circuits, special software – in short, all of the above-mentioned – has led to the situation where documentation and manufacturing of automation devices has become available only to serious companies having all the necessary resources for this. Development tendencies of this area of technique make it more and more unavailable to individual engineers and technicians wishing to apply their knowledge and ingenuity to improve production or technological processes to their companies. At the same time, lately in the market a number of new types of small-size, discrete electronic components with previously inaccessible parameters appeared. They are miniature transistors meant for currents of dozens of amperes and voltages of 1200 – 1600 V; miniature vacuum reed switches with operational speeds of milliseconds capable of sustaining voltages of 1,000 – 2,000 V; and other no less interesting elements. These new discrete components serve as the basis for creating industrial automation and control devices that are fed directly from networks of 220 – 250 V and work directly with input and output signals of the same voltage level. Hybrid devices combining advantages of semiconductor (transistors, thyristors) and electromechanical (reed switches) elements are of particular interest.

This book is concerned with the description of different functional units and automation devices for industry and electric power engineering implemented by modern discrete electronic elements without using microelectronics and microprocessor-based technologies. The devices described in this book turn out to be much simpler and cheaper; they may be produced not only by large companies, but even by independent amateurs. This book presents for the readers' judgment dozens of unusual but very simple realizable devices, which may be easily created by any engineer or technician wishing to improve automation systems. Some of the technical decisions presented by the author may serve as the basis for the creation of new types of devices of relay protection and automation free from disadvantages of complex microelectronic systems.

The book consists of seven chapters and appendices containing reference data. The first three chapters are devoted to the theory and operating principles of modern discrete components designed for automation devices: transistors, thyristors, dinistors, reed switches, and high-voltage reed switch relays. The fourth chapter describes dozens of different functional modules of automation systems incorporating discrete elements with direct supply from 220 – 250 V networks: switching devices, generators and multivibrators, timers, logic elements, elements sensitive to overcurrents and overvoltages, voltage regulators and stabilizers, pulse expanders, etc. The fifth, sixth, and seventh chapters are devoted to the description of concrete examples of automation devices for industry and electric power engineering based on discrete electronic components and also hybrid ones: semiconductors and reed switches.

The book makes a smooth transition from theory and the properties of modern electronic elements by means of examination of operating principles and examples of realization of separate functional units of automation devices to the description of concrete examples of those that are finished and ready for use. The author thinks that this approach to the material will make it possible for the readers not only to repeat the constructions that are described, but to understand and master the general principles of automation devices on discrete elements and to apply them in the future for creation of new necessary constructions. As an aid to complete understanding, voluminous reference material has been included containing information about the most modern components specially selected by the author and classified in the appendices.

Author

Vladimir Gurevich was born in Kharkov, Ukraine, in 1956. He received an M.S.E.E. degree (1978) at the Kharkov Technical University, named after P. Vasilenko, and a Ph.D. degree (1986) at Kharkov National Polytechnic University. His employment experience includes: teacher, assistant professor and associate professor at Kharkov Technical University, and chief engineer and director of Inventor, Ltd. In 1994, he arrived in Israel and works today at Israel Electric Corp. as a specialist of the Central Electric Laboratory. He is the author of more than 140 professional papers and 4 books and holder of nearly 120 patents in the field of electrical engineering and power electronics. In 2006 he was Honorable Professor with the Kharkov Technical University, and since 2007 he has served as an expert with the TC-94 Committee of International Electrotechnical Commission.

Contents

1

Solid-State Electronic Elements

1.1 SEMICONDUCTING MATERIALS AND *P-N*-JUNCTION

As is known, all substances depending on their electro-conductivity are divided into three groups: conductors (usually metals) with a resistance of 10^{-6}-10^{-3} Ohm·cm, dielectrics with a resistance of 10^{9}-10^{20} Ohm·cm, and semiconductors (many native-grown and artificial crystals) covering an enormous intermediate range of values of specific electrical resistance.

The main peculiarity of crystal substances is typical, well-ordered atomic packing into peculiar blocks – crystals. Each crystal has several flat symmetric surfaces and its internal structure is determined by the regular positional relationship of its atoms, which is called the lattice. Both in appearance and in structure, any crystal is like any other crystal of the same given substance. Crystals of various substances are different. For example, a crystal of table salt has the form of a cube. A single crystal may be quite large in size or so small that it can only be seen with the help of a microscope. Substances having no crystal structure are called amorphous. For example, glass is amorphous in contrast to quartz, which has a crystal structure.

Among the semiconductors that are now used in electronics, one should point out germanium, silicon, selenium, copper-oxide, copper sulfide, cadmium sulfide, gallium arsenide, and carborundum. To produce semiconductors two elements are mostly used: germanium and silicon.

In order to understand the processes taking place in semiconductors, it is necessary to consider phenomena in the crystal structure of semiconductor materials, which occur when their atoms are held in a strictly determined relative position to each other due to weakly bound electrons on their external shells. Such electrons, together with electrons of neighboring atoms, form *valence bonds* between the atoms. Electrons taking part in such bonds are called *valence electrons*. In absolutely pure germanium or silicon at very low temperatures there are no free electrons capable of creating electric current, because under such circumstances all four valence electrons of the external shells of each atom that can take part in the process of charge transfer are too strongly held by the valence bounds. That's why that substance is an insulator (dielectric) in the full sense of the word – it does not let electric current pass at all.

When the temperature is increased, due to the thermal motion some valence electrons detach from their bonds and can move along the crystal lattice. Such electrons are called *free electrons*. The valence bond from which the electron is detached is called a *hole*. It possesses properties of a positive electric charge, in contrast to the electron, which has a negative electric charge. The more the temperature is, the more the number of free electrons capable of moving along the lattice, and the higher the conductivity of the substance is.

Moving along the crystal lattice, free electrons may run across holes – valence bonds missing some electrons – and fill up these bonds. Such a phenomenon is called *recombination*. At normal temperatures in the semiconductor material, free electrons occur constantly, and recombination of electrons and holes takes place.

If a piece of semiconductor material is put into an electric field by applying a positive or negative terminal to its ends, for instance, electrons will move through the lattice towards the positive electrode and holes – to the negative one. The conductivity of a semiconductor can be enhanced considerably by putting specially selected admixtures to it – metal or non-metal ones. In the lattice the atoms of these admixtures will replace some of the atoms of the semiconductors. Let us remind ourselves that external shells of atoms of germanium and silicon contain four valence electrons, and that electrons can only be taken from the external shell of the atom. In their turn the electrons can be added only to the external shell, and the maximum number of electrons on the external shell is eight.

When an atom of the admixture that has more valence electrons than required for valence bonds with neighboring atoms of the semiconductor, additional free electrons capable of moving along the lattice occur on it. As a result the electro-conductivity of the semiconductor increases. As germanium and silicon belong to the fourth group of the periodic table of chemical elements, donors for them may be elements of the fifth group, which have five electrons on the external shell of atoms. Phosphorus arsenic, and stibium belong to such donors (*donor admixture*).

If admixture atoms have fewer electrons than needed for valence bonds with surrounding semiconductor atoms, some of these bonds turn out to be vacant and holes will occur in them. Admixtures of this kind are called *p*-type ones because they absorb (accept) free electrons. For germanium and silicon, p-type admixtures are elements from the third group of the periodic table of chemical elements, external shells of atoms of which contain three valence electrons. Boron, aluminum, gallium, and indium can be considered *p*-type admixtures (*accepter admixture*).

In the crystal structure of a pure semiconductor all valence bonds of neighboring atoms turn out to be fully filled, and occurrence of free electrons and holes can be caused only by deformation of lattice, arising from thermal or other radiation. Because of this, conductivity of a pure semiconductor is quite low under normal conditions.

If some donor admixture is injected, the four electrons of the admixture, together with the same number in the filled valence, bond with the latter. The fifth electron of each admixture atom appears to be "excessive" or "redundant," and therefore can freely move along the lattice.

When an accepter admixture is injected, only three filled valence bonds are formed between each atom of the admixture and neighboring atoms of the semiconductor. To fill up the fourth, one electron is lacking. This valence bond appears to be vacant. As a result, a hole occurs. Holes can move along the lattice like positive charges, but instead of an admixture atom, which has a fixed and permanent position in the crystal structure, the vacant valence bond moves. It goes like this. An electron is known to be an elementary carrier of an electric charge. Affected by different causes, the electron can escape from the filled valence bond, having left a hole which is a vacant valence bond and which *behaves like a positive charge equaling numerically the negative charge of the electron.* Affected by the attracting force of its positive charge, the electron of another atom near the hole may "jump" to the hole. At that point recombination of the hole and the electron occurs, their charges are mutually neutralized and the valence bond is filled. The hole in this place of the lattice of the semiconductor disappears. In its turn a new hole, which has arisen in the valence bond from which the electron has escaped, may be filled with some other electron which has left a hole. Thus, moving of electrons in the lattice of the semiconductor with a *p*-type admixture and recombination of them with holes can be regarded as moving of holes. For better understanding one may imagine a concert hall in which for some reason some seats in the first row turn out to be vacant. As spectators from the second row move to the vacant seats in the first row, their seats are taken by spectators of the third row, etc. One can say that in some sense vacant seats "move" to the last rows of the concert halls, although in fact all the stalls remain screwed to the floor. "Moving" of holes in the crystal is very much like "moving" of such vacant seats.

Semiconductors with electro-conductivity enhanced, due to an excess of free electrons caused by admixture injection, are called semiconductors with *electron-conductivity* or in short, *n-type semiconductors.* Semiconductors with electro-conductivity influenced mostly by moving of holes are called *semiconductors with p-type conductivity* or just *p-type semiconductors.*

There are practically no semiconductors with only electronic or only p-type conductivity. In a semiconductor of *n*-type, electric current is partially caused by moving of holes arising in its lattice because of an escaping of electrons from some valence bonds, and in semiconductors of *p*-type current is partially created by the moving of electrons. Because of this it is better to define semiconductors of the *n*-type as semiconductors in which *the main current carriers are electrons* and semiconductors of the *p*-type as semiconductors in which *holes are the main current carriers.* Thus a semiconductor belongs to this or that type depending on what type of current carrier predominates in it. According to this, the other opposite charge carrier for any semiconductor of a given type is a *minor carrier.*

One should take into account that any semiconductor can be made a semiconductor of *n*- or *p*-type by putting certain admixtures into it. In order to obtain the required conductivity it is enough to put in a very small amount of the admixture, about one atom of the admixture for 10 millions of atoms of the semiconductor. All of this imposes special requirements for the purification of the original semiconductor material, and accuracy in dosage of admixture injection. One should also take into consideration that the speed of current carriers in a semiconductor is lower than in a metal conductor or in a vacuum. Moving of electrons is slowed down by obstacles on their way in the form of inhomogeneities in the crystal. Moving of holes is half as slow because they move due to jumping of electrons to vacant valence bounds. Mobility of electrons and holes in a semiconductor is increased when the temperature goes up. This leads to an increase of conductivity of the semiconductor.

The functioning of most semiconductors is based on the processes taking place in an intermediate layer formed in the semiconductor, at the boundary of the two zones with the conductivities of the two different types: " *p* " and " *n.* " The boundary is usually called the *p-n junction* or the *electron-hole junction,* in accordance with the main characteristics of the type of main charge carriers in the two adjoining zones of the semiconductor.

There are two types of p-n junctions: *planar* and *point junctions,* which are illustrated schematically in Fig. 1.1. A planar junction is formed by moving a piece of the admixture – for instance indium, to the surface of the germanium – of n-type, and further heating until the admixture is melted.

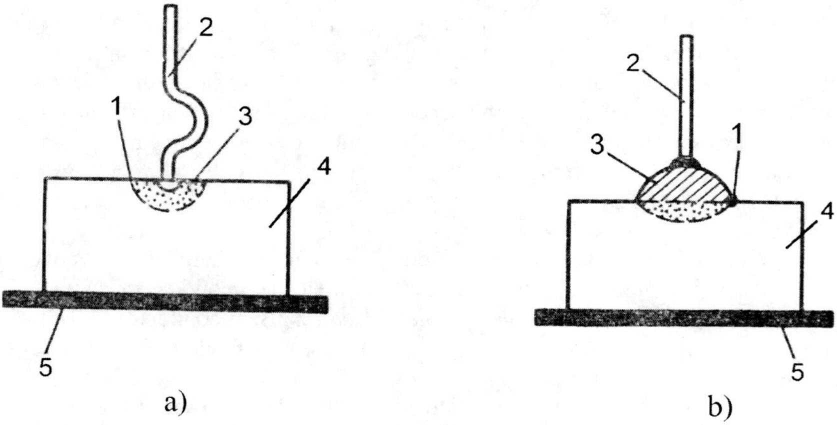

FIGURE 1.1 Construction of point (a) and planar (b) p-n junctions of the diode 1 – p-n junction; 2 – wire terminal; 3 – p-area; 4 – crystal of n-type; 5 – metal heel piece.

When a certain temperature is maintained for a certain period of time, there is diffusion of some admixture atoms to the plate of the semiconductor, to a small depth, and a zone with conductivity opposite to that of the original semiconductor is formed. In the above case it is *p*-type, for *n*-germanium.

Point junction results from tight electric contact of the thin metal conductor (wire), which is known to have electric conductivity, with the surface of the p-type semiconductor. This was the basic principle on which the first crystal detectors operated. To decrease dependence of diode properties on the position of the pointed end of the wire on the surface of the semiconductor, and the clearance of its momentary surface point, junctions are formed by fusing the end of the thin metal wire to the surface of a semiconductor of the n-type. Fusion is carried the moment a short-term powerful pulse of electric current is applied. Affected by the heat formed for this short period of time, some electrons escape from atoms of the semiconductor, which are near the contact point, and leave holes. As a result of this some small part of the n-type semiconductor in the immediate vicinity of the contact turns into a semiconductor of the p-type (area 3 on Fig.1.1a).

Each part of semiconductor material, taken separately (that is before contacting), was neutral, since there was a balance of free and bound charges (Fig. 1.2a). In the n-type area, concentration of free electrons is quite high and that of holes quite low. In the p-type area on the contrary, concentration of holes is high, and that of electrons low. Joining of semiconductors with different concentrations of main current carriers causes diffusion of these carriers through the junction layer of these materials: the main carriers of the p-type semiconductor – holes – diffuse to the n-type area because the concentration of holes in it is very low. And vice versa, electrons from the n-type semiconductor, with a high concentration of them, diffuse to the n-type area, where there are few of them (Fig. 1.2b).

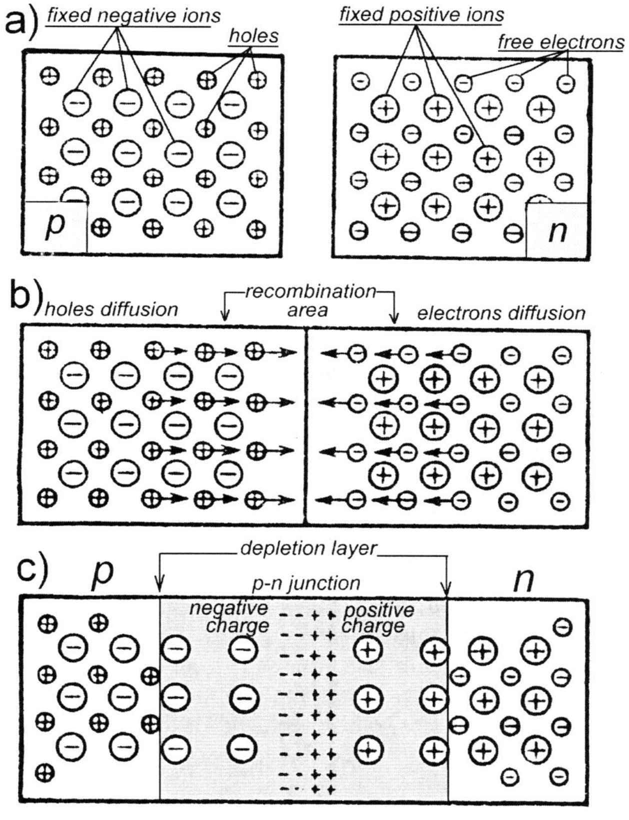

FIGURE 1.2 Formation of a blocking layer when semiconductors of different conductivity are connected.

On the boundary of the division of the two semiconductors, from each side a thin zone with conductivity opposite to that of the original semiconductor is formed. As a result, on the boundary (which is called a p-n junction) a space charge arises (the so called potential barrier), which creates a diffusive electric field and prevents the main current carriers from flowing after balance has been achieved (Fig. 1.2c).

Strongly pronounced dependence of electric conductivity of a p-n junction, from polarity of external voltage applied to it, is typical of the p-n junction. This can never be noticed in a semiconductor with the same conductivity. If voltage applied from the outside creates an electric field coinciding with a diffusive electric field, the junction will be blocked and current will not pass through it (Fig. 1.3).

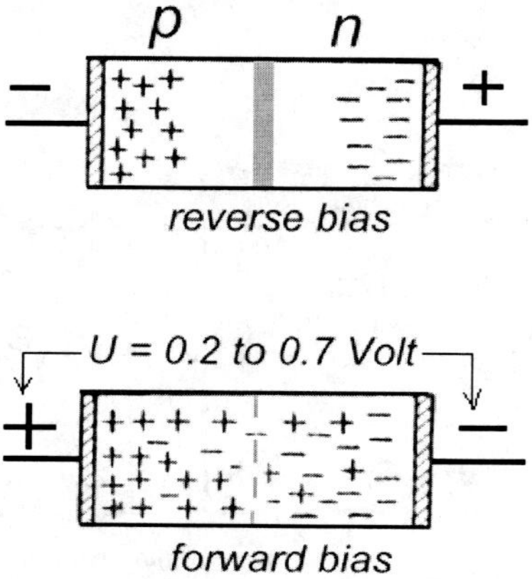

FIGURE 1.3 *p-n-* junction with reverse and forward bias.

Moreover, moving of minor carriers becomes more intense, which causes enlargement of the blocking layer and lifting of the barrier for main carriers. In this case it is usually said that the junction *is reversely bias.* Moving of minor carriers causes a small current to pass through the blocked junction. This is the so-called *reverse current* of the diode, or *leakage current.* The smaller it is, the better the diode is.

When the polarity of the voltage applied to the junction is changed, the number of main charge carriers in the junction zone increases. They neutralize the space charge of the blocking layer by reducing its width and lowering the potential barrier that prevented the main carriers from mobbing through the junction. It is usually said that the junction is *forward biased.* The voltage required for overcoming of the potential barrier in the forward direction is about 0.2V for germanium diodes, and 0.6-0.7V for silicon ones.

To overcome the potential barrier in the reverse direction, tens and sometimes even thousands of Volts are required.

If the barrier is overpassed, irreversible destruction of the junction and its breakdown takes place, which is why threshold values of reverse voltage and forward current are indicated for junctions of different appliances.

Fig. 1.4 illustrates an approximate volt-ampere characteristic of a single junction, which is dependence of current passing through it on the polarity and external voltage applied to the junction. Currents of forward and reverse direction (up to the breakdown area) may differ by tens and hundreds of times. As a rule, planar junctions withstand higher voltages and currents than point ones, but do not work properly with high frequency currents.

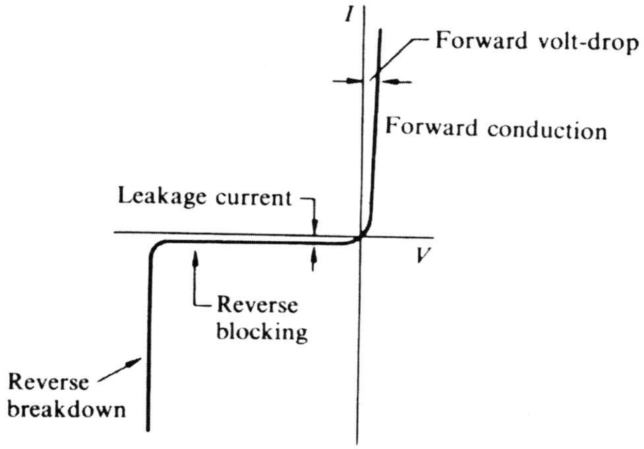

FIGURE 1.4 Volt-ampere characteristic of a single p-n junction (diode).

1.2 THE TRANSISTOR'S PRINCIPLE

The idea of somehow using semiconductors had been tossed about before World War II, but knowledge about how they worked was scant, and manufacturing semiconductors was difficult.

In 1945, however, the vice president for research at Bell Laboratories established a research group to look into the problem. The group was led by William Shockley and included Walter Brattain, John Bardeen, and others, physicists who had worked with quantum theory, especially in solids. The team was talented and worked well together.

In 1947 John Bardeen and Walter Brattain, with colleagues, created the first successful amplifying semiconductor device. They called it a transistor (from "transfer" and "resistor"). In 1950 Shockley made improvements to it that made it easier to manufacture. His original idea eventually led to the development of the silicon chip. Shockley, Bardeen, and Brattain won the 1956 Nobel Prize for the development of the transistor. It allowed electronic devices to be built smaller, lighter, and even cheaper.

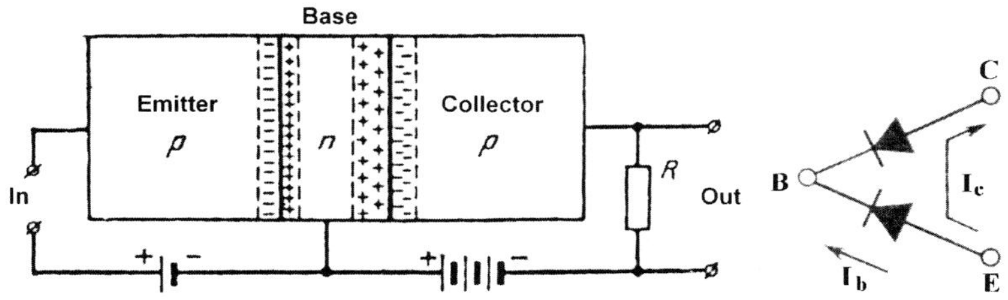

FIGURE 1.5 Circuit and the principle of operation of a transistor.

It can be seen in Fig. 1.5 that a transistor contains two semiconductor diodes, connected together, and having a common area. Two utmost layers of the semiconductor (one of them is called an "emitter" and the other, a "collector") have *p*-type conductivity with a high concentration of holes, and the intermediate layer (called a "base") has *n*-type conductivity with a low concentration of electrons. In electric circuits low voltage is applied to the first (the emitter) *p-n* junction because the junction is connected in the forward (carrying) direction, and much higher voltage is applied to the second (the collector) junction, in the reverse (cut-off) direction. In other words, emitter junction is a forward biased and collector junction is a reverse biased. The collector junction remains blocked until there is no current in the emitter-base circuit. The resistance of the whole crystal (from the emitter to the collector) is very high. As soon as the input circuit (Fig. 1.5) is closed, holes from the emitter seem to be injected (emitted) to the base and quickly saturate it (including the area adjacent to the collector). As the concentration of holes in the emitter is much higher than the concentration of electrons in the base, after recombination there are still many vacant holes in the base area, which is affected by the high voltage (a few or tens of Volts) applied between the base and the collector, easily overpassing the barrier layer between the base and the collector.

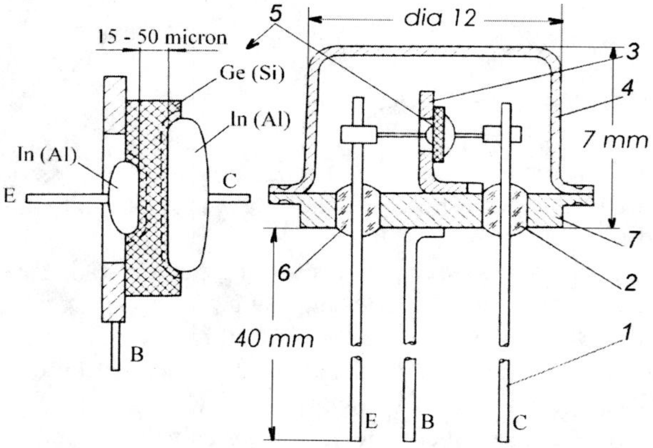

a)

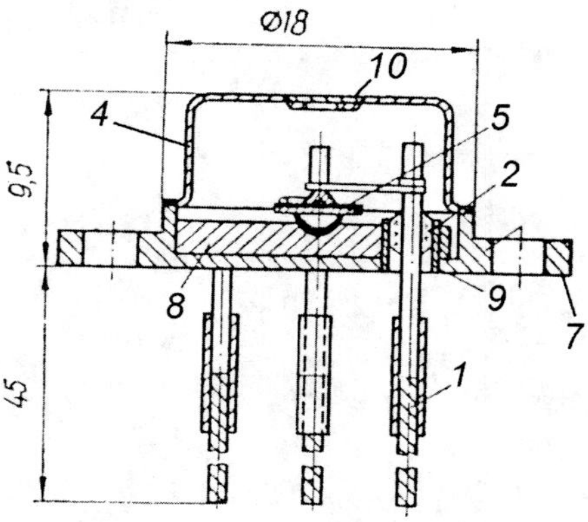

b)

FIGURE 1.6 Transistors produced in the '70's: a) low power transistor; b) power transistor.

1 – outlets; 2 and 6 – glass insulators; 3 – crystal holder; 4 – protection cover; 5 – silicon (germanium) crystal; 7 – flange; 8 – copper heat sink; 9 – Kovar bushing; 10 – hole for gas removal after case welding and disk for sealing-in.

Increased concentration of holes in the cutoff collector junction causes the resistance of this junction to fall rapidly, and it begins to *conduct current in the reverse direction.* The high strength of the electric field in the "base-collector" junction results in a very high sensitivity of the resistance of this junction in the reverse (cutoff) state to a concentration of the holes in it.

That's why even a small number of holes injected from the emitter under the effect of weak input current can lead to sharp changes of conductivity of the whole structure, and considerable current in the collector circuit.

The ratio of collector current to base current is called the *"current amplification factor."* In low power transistors this amplification factor has values of tens and hundreds, and in power transistors – tens.

1.3 SOME TRANSISTOR KINDS

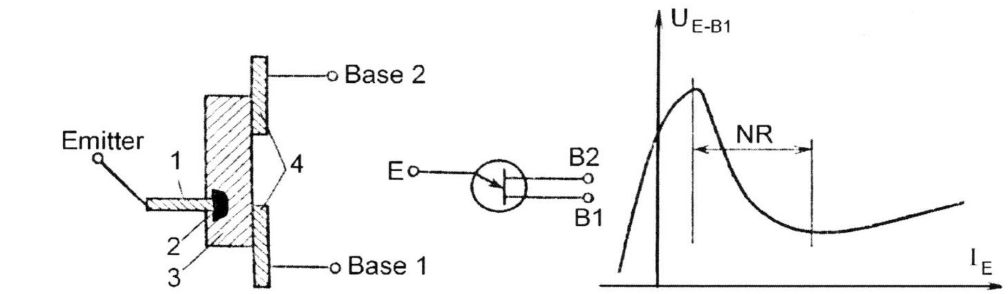

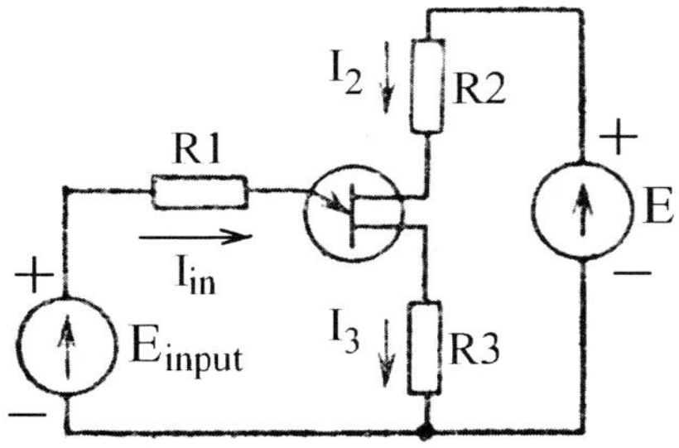

FIGURE 1.7 A unijunction transistor (or two-base diode) and its circuit.
1 – p-type core; 2 – p-n junction; 3 – n-type plate; 4 – ohmic contacts; NR- negative resistance area.

In the 1970s, transistor engineering developed very rapidly. Hundreds of types of transistors and new variants of them appeared (Fig. 1.6). Among them appeared transistors with reverse conductivity or n-p-n transistors, and also unijunction transistors (as it contains only one junction such a transistor is sometimes called a two-base diode) (Fig. 1.7). This transistor contains one junction formed by welding a core made from p-material to a single-crystal wafer made from n-type material (silicon). The two outlets, serving as bases, are attached to the wafer. The core, placed asymmetrically with regard to the base, is called an emitter. Resistance between the bases is about a few thousands Ohms. Usually the base B_2 is biased in a positive direction from the base B_1. Application of positive voltage to the emitter causes strong current of the emitter (with insignificant voltage drop between the emitter E and the base B_1). One can observe the area of negative resistance (NR – see Fig. 1.7) on the emitter characteristic of the transistor where the transistor is very rapidly enabled, operating like a switch.

a

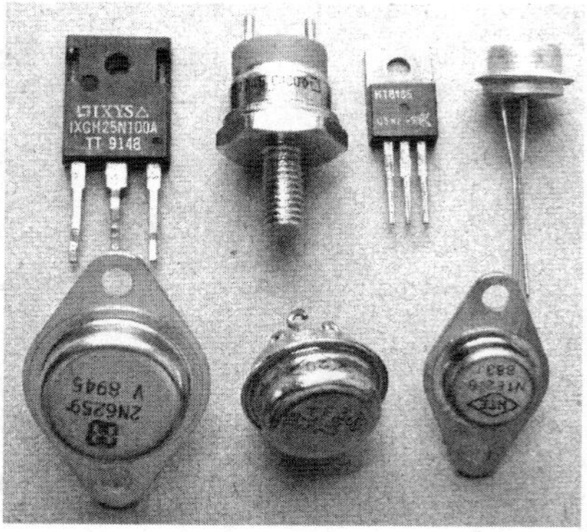

b

c

FIGURE 1.8 This is how modern low (a), power (b) and high power (c) transistors look.

In fact modern transistors (Fig. 1.8) are characterized by such a diversity of types that it is simply impossible to describe all of them in this book, therefore only a brief description of the most popular types of modern semiconductor devices, and the relays based on them, are presented here.

Besides the transistors described above, which are called *Bipolar Junction Transistors* or just "bipolar transistors" (Fig. 1.9), so called Field Effect Transistors (FET, – Fig. 1.10) have become very popular recently. The first person to attempt to construct a field effect transistor in 1948 was again William Shockley. But it took many years of additional experiments to create a working FET with a control p-n junction called a "unitron" (Unipolar Transistor), in 1952.

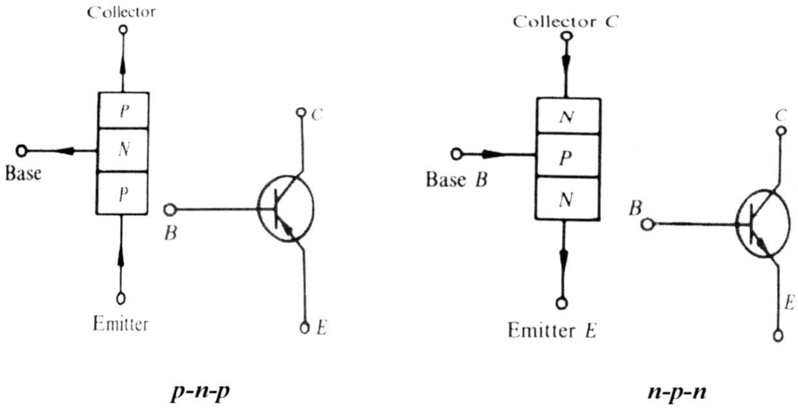

FIGURE 1.9 Structure and symbolic notation on the schemes of bipolar transistors of *p-n-p* and *n-p-n* types.

Such a transistor was a semiconductor three-electrode device in which control of the current caused by the ordered motion of charge carriers of the same sign between two electrodes was carried out with help of an electric field (that is why it is called "field") applied to the ford electrode.

Electrodes between which working currents pass are called *source* and *drain* electrodes. The source electrode is the one through which carriers flow into the device. The third electrode is called a "*gate*." Change of value of the working current in a unipolar transistor is carried out by changing the effective resistance of the current conducting area, the semiconductor material between the source and the drain called the "*channel.*" That change is made by increasing or decreasing area 5 (Fig. 1.10). Increase of voltage of the initial junction bias leads to expansion of the depletion layer. As a result, the rest area of the section of the conductive channel in the silicon decreases and the transistor is blocked, and vice versa, when the value of the blocking voltage on the gate decreases, the area (5) depleted by current carriers contracts and turns into a pointed wedge. At the same time the section of the conductive channel increases and the transistor is enabled.

Depending on the type of the conductivity of semiconductor material of the channels, there are unipolar transistors with *p* and *n* channels. Because of the fact that control of the working current of unipolar transistors is carried out with the help of a channel, they are also called "*channel transistors.*" The third name of the same semiconductor device – a "*field transistor*" or "*field effect transistor*" *(FET)* points out that working current control is carried out by an electric field (voltage) instead of electric current as in a bipolar transistor. The latter peculiarity of unipolar transistors, which allows them to obtain very high input resistances, estimated in tens and hundreds of megohms, determined their most popular name: field transistors.

It should be noted that apart from field transistors with *p-n* junctions between the gate and the channel (FET), there are also field transistors with an insulated gate: Metal-Oxide-Semiconductor FET Transistors (MOSFET). The latter were suggested by S. Hofstein and F. Heiman in 1963.

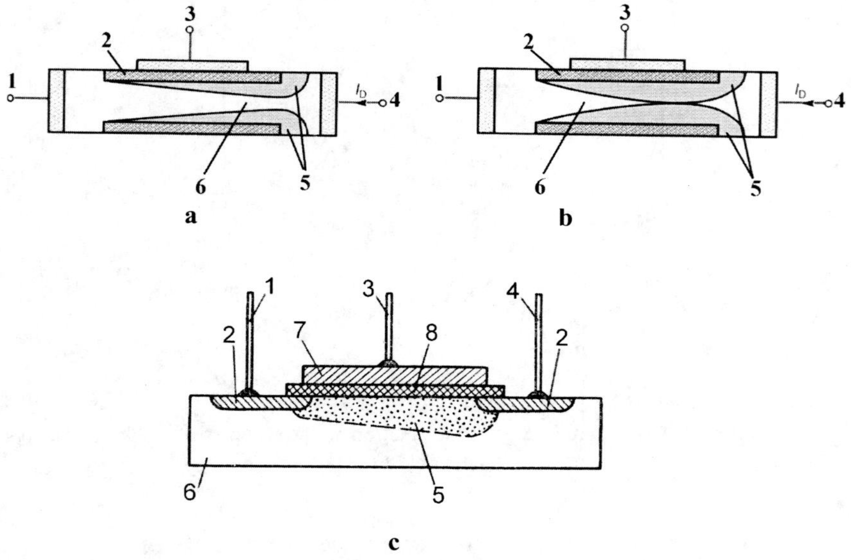

FIGURE 1.10 Simplified structure of FET (a, b) and MOSFET (c) transistors
1 – «source»; 2 – *n*-type admixture; 3 – «gate»; 4 – «drain»; 5 – area consolidated by current carriers (depletion layer); 6 – conductive channel in silicon of *p*-type; 7 – metal; 8 – silicon dioxide.

Field transistors with an insulated gate appeared as a result of searching for methods to further increase input resistance and frequency range extensions of field transistors with *p-n* junctions. The distinguishing feature of such field transistors is that the junction biased in a reverse direction is replaced with a control structure "metal – oxide – semiconductor," or a MOSFET-structure in abbreviated form. As shown in Fig. 1.10 this device is based on a silicon mono crystal, in this case of p-type. The source and drain areas have conductivity opposite to the rest of the crystal, that is of the n-type. The distance between the source and the drain is very small, usually about 1 micron. The semiconductor area between the source and the drain, which is capable of conducting current under certain conditions, is called a channel, as in the previous case.

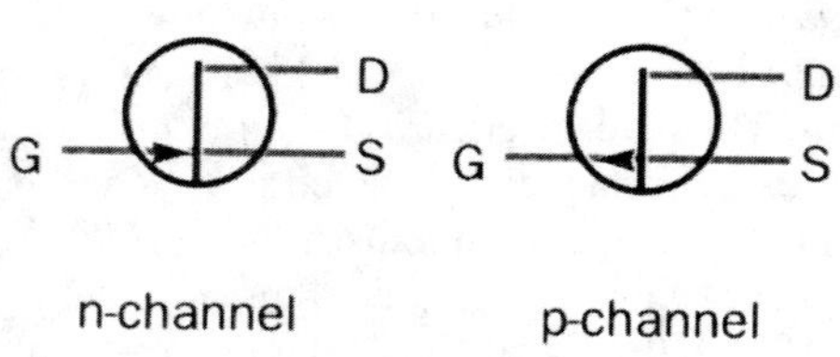

FIGURE 1.11 Symbolic notation of FET transistors with *n*- and *p*-channels:
G – gate; S – source; D – drain.

In fact the channel is an n-type area formed by diffusion of a small amount of the donor admixture to the crystal with p-type conductivity. The gate is a metal plate covering source and drain zones. It is isolated from the mono crystal by a dielectric layer only 0.1 micron thick. The film of silicon dioxide formed at this high temperature is used as a dielectric. Such film allows us to adjust the concentration of the main carriers in the channel area by changing both value and polarity of the gate voltage.

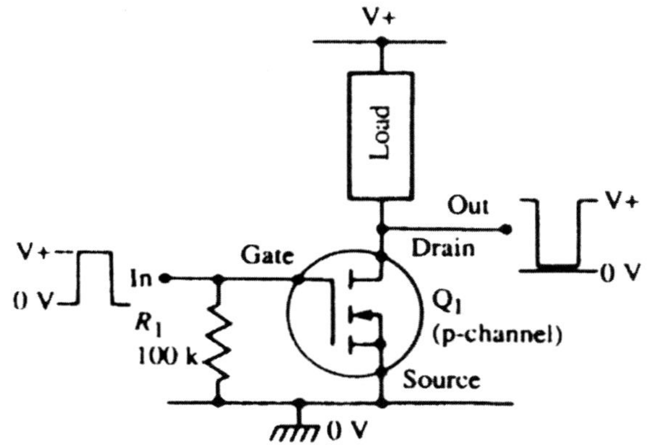

FIGURE 1.12 Symbolic notation and the circuit of a MOSFET-transistor.

This is the major difference of MOSFET-transistors, as opposed to field ones with p-n junctions, which can only operate well *with blocking voltage of the gate*. The change of polarity of the bias voltage leads to junction unblocking and to a sharp reduction of the input resistance of the transistor.

The basic advantages of MOSFET-transistors are as follows: first there is an insulated gate allowing an increase in input resistance by at least 1000 times in comparison with the input resistance of a field transistor with a *p-n* junction.

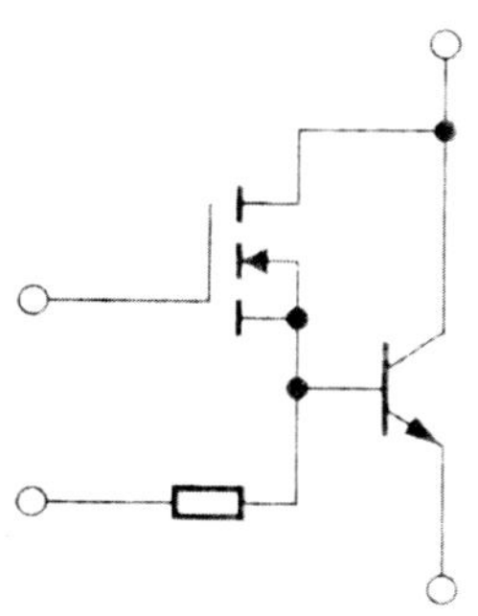

FIGURE 1.13 Compound structure - "*Pobistor.*"

In fact it can reach a billion megohms. Secondly, gate and drain capacities become considerably lower and usually do not exceed 1-2 pF. Thirdly, the limiting frequency of MOSFET – transistors can reach 700 – 1000 MHz, which is at least ten times higher than that of standard field transistors.

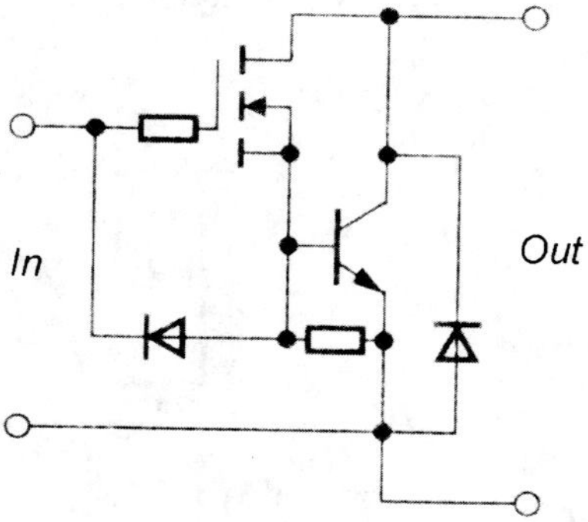

FIGURE 1.14 Scheme of a power switching module CASCADE-CD, with a working voltage of 1000V and currents more than 100A (Mitsubishi Electric).

Attempts to combine in one switching device the advantages of bipolar and field transistors led to the invention of a compound structure in 1978, which was called a *"pobistor"* (Fig. 1.13). The idea of a modular junction of crystals of bipolar and field transistors in the same case was employed by Mitsubishi Electric to create a powerful switching semiconductor module (Fig. 1.14).

Further development of production technology of semiconductor devices allowed development of a single-crystal device with a complex structure with properties of a "pobistor": an *IGBT* transistor. The *Insulated Gate Bipolar Transistor* (IGBT) is a device which combines the fast-acting features and high power capabilities of the bipolar transistor with the voltage control features of the MOSFET gate. In simple terms, the collector-emitter characteristics are similar to those of the bipolar transistor, but the control features are those of the MOSFET. The equivalent circuit and the circuit symbol are illustrated in Fig. 1.15. Such a transistor (Fig. 1.16) has a higher switching power than FET and bipolar transistors and its operation speed is between that of FET and bipolar transistors. Unlike bipolar transistors, the IGBT-transistor doesn't operate well in the amplification mode and is designed for use in the switching (relay) mode as a powerful high-speed switch.

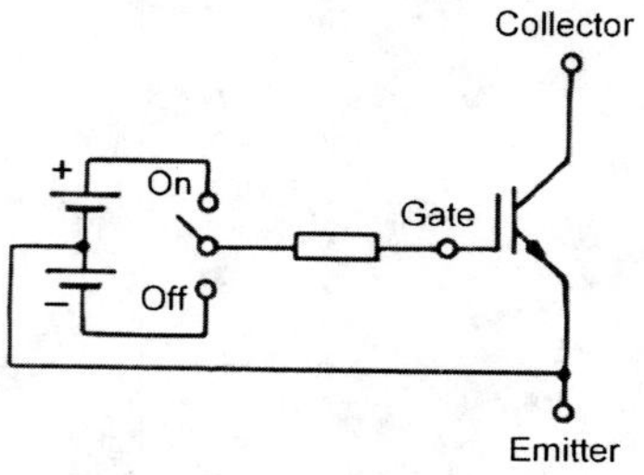

FIGURE 1.15 Insulated Gate Bipolar Transistor (IGBT).

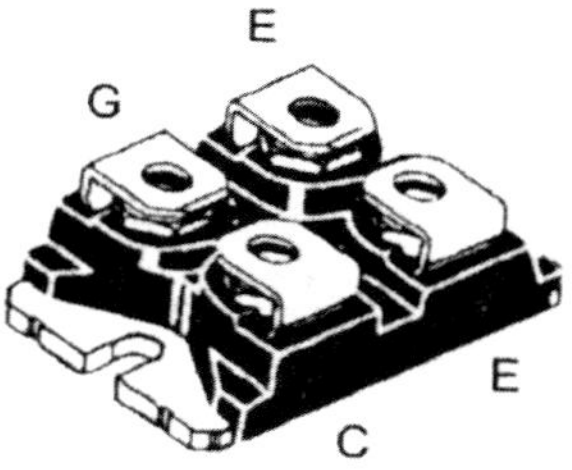

FIGURE 1.16 An IGBT-transistor IXDN75N120A produced by IXYS with a switched current up to 120 A, and maximum voltage of up to 1200 V (dissipated power is 630W). With such high parameters the device is quite small in size: 38 x 25 x 12 mm.

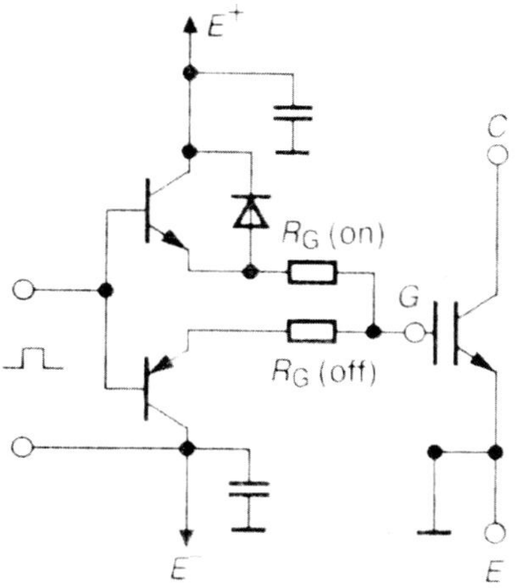

FIGURE 1.17 Model scheme of the IGBT transistor control, providing pulses of opposite polarity on the gate required for reliable blocking and unblocking of the transistor.

The IGBT transistor is enabled by a signal of positive (with regard to the emitter) polarity, with voltage not more than 20V. It can be blocked with zero potential on the gate, however with some types of loads a signal of negative polarity on the gate may be required for reliable blocking (Fig. 1.17).

Many companies produce special devices for IGBT transistor control. They are made as separate integrated circuits or ready-to-use printed circuit cards, so-called *Drivers* (Fig. 1.18). Such drivers are universal as a rule and can be applied to any type of power IGBT transistors. Apart from forming control signals of the required level and form, such devices often protect the IGBT from short circuits. In spite of progress in IGBT transistor development, different firms continue producing standard high power bipolar transistors in capsule packages (Fig. 1.8c). In power devices such transistors, equipped with big aluminum heat sinks and fans, are united to power units which can weigh tens of kilograms in weight. Heat sinks for such transistors are made in the form of two separate halves pulled together with special screw-bolts, with insulation covering between them, inside of which there is a transistor (Fig. 1.19).

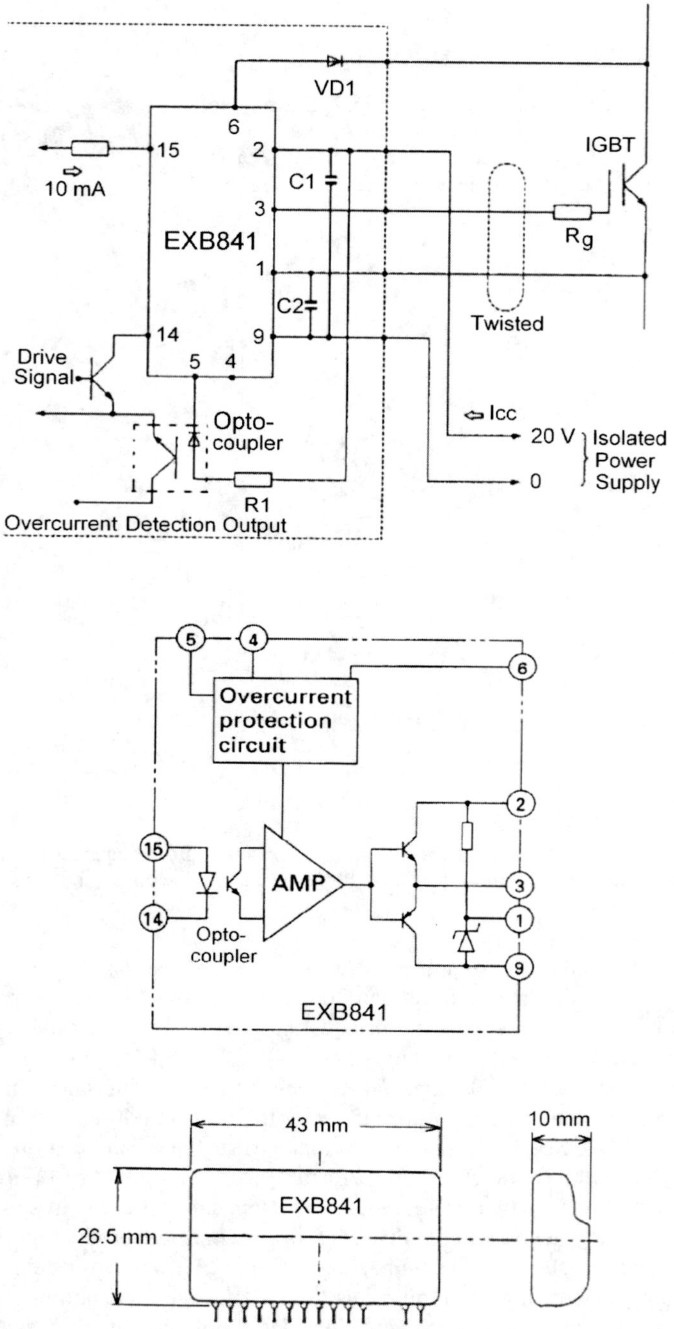

FIGURE 1.18 IGBT-driving hybrid integral circuit EXB841 type (Fuji Electric).

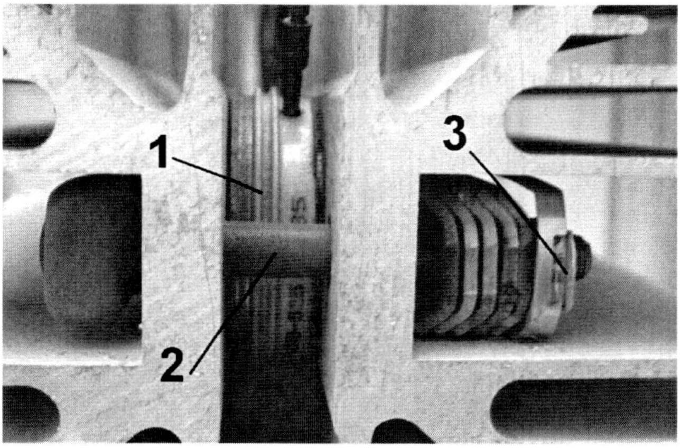

FIGURE 1.19 Attachment point of power wafer transistor in a heat sink. 1 – transistor; 2 – insulated screw bolt; 3 – torque measuring disk with a scale.

To provide a good thermal contact between the transistor and the heat sink the hold-down pressure must be strong enough, but must not exceed the threshold value of the transistor. Special torque spanners or spring disks with a scale are used (Fig .1.19).

To increase switching current, transistors are connected in parallel (Fig. 1.20a). Current grading through appliances connected in parallel is carried out with the help of low-value resistors cut in to a circuit of emitters of the transistors. When there is a great number of parallel connected transistors (Fig. 1.20b), the total current of all base electrodes (control current) becomes commensurable with the working (collector) current, which is why in this case an additional transistor is used on the input side (Fig. 1.20b).

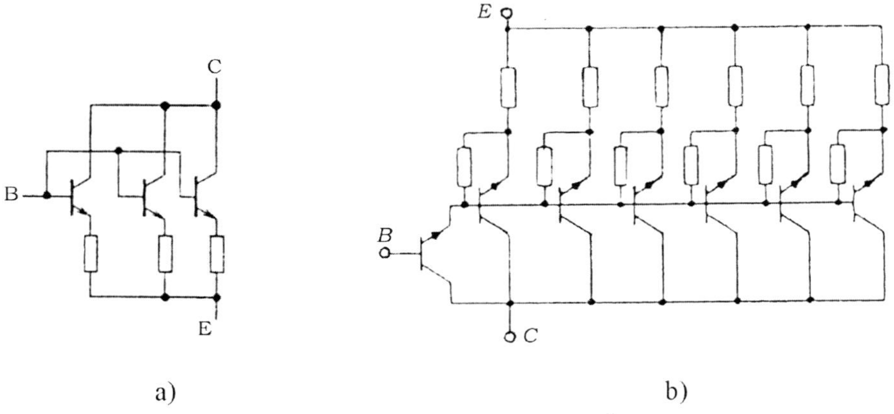

FIGURE 1.20 Parallel connection of bipolar transistors.

1.4 BIPOLAR TRANSISTOR GENERAL MODES

As an element of an electric circuit, the transistor is usually used such that one of its electrodes is input, one is output, and the third is the common with respect to the input and output. A transistor is commonly connected to external circuits in a four-terminal configuration, referred to as a "quadrupole." The source of an input signal requiring amplification is connected to the circuit of the input electrode, and the load on which the amplified signal is dissipated is connected to the circuit of the output electrode. Depending on which electrode is the common for the input and output circuits, transistor connections fall into three basic circuits, as shown in Fig. 1.21.

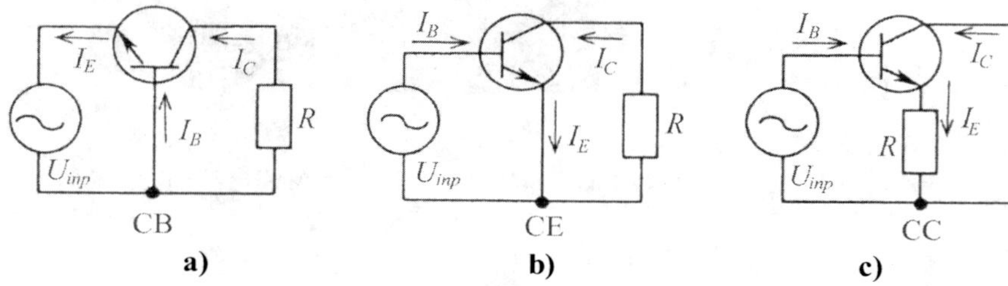

FIGURE 1.21 Transistor basic circuit configurations: a) common base (CB), b) common emitter (CE), c) common collector (CC)
I_E – emitter current; I_B – base current; I_C – collector current; R – load.

In the CB (common base) circuit I_E is the input signal and I_C is the output signal. The current amplification coefficient (also called the "current gain," the ratio of amplifier output current to input current) of a transistor in this configuration is equal to $\alpha = I_C/I_E \approx 1$. A device may have low internal input resistance and high internal output resistance, and for this reason, a change of the load resistance exerts only a minimal influence on the output current (the functional scheme of this mode relates to current source). CB configurations of transistor connections are not commonly used in practice.

The CE (common emitter) circuit is used most often as an amplification stage. Current gain for this circuit is close to the transistor's gain and is equal to $\beta = I_C / I_B \approx 10 - 200$ and more, depending on the type of transistor used. For direct current $\beta = h_{FE}$. (h_{FE} is the DC current gain, parameter specified by manufacturer of transistors). This circuit has rather high input resistance, i.e., it does not shunt and weaken the input signal, and low output resistance.

The CC (common collector) circuit is used quite often in cases where it is needed to stage with very high input resistance. The circuit current amplification coefficient is close to that of the CE circuit; however, the main application of this circuit relates to the functional mode of the voltage, not current amplification, as precisely in this mode one manages to realize the scheme most fully: its very high input resistance. In the voltage amplification mode the scheme has gain close to one.

Regardless of the transistor used in the circuit, it may function in four main modes determined by the polarity of voltage on the emitter and collector junctions. All possible bias modes are illustrated in Fig. 1.22. They are: the forward active mode of operation, the reverse active mode of operation, the saturation mode, and the cut-off mode.

As the base current I_B increases or decreases, the operating point moves up or down the load line, Fig. 1.23. If I_B increases too much, the operating point moves into the saturation region. In the saturation region the transistor is fully turned ON and the value of collector current I_C is determined by the value of the load resistance R_L. The voltage drop across the transistor V_{CE} is near zero.

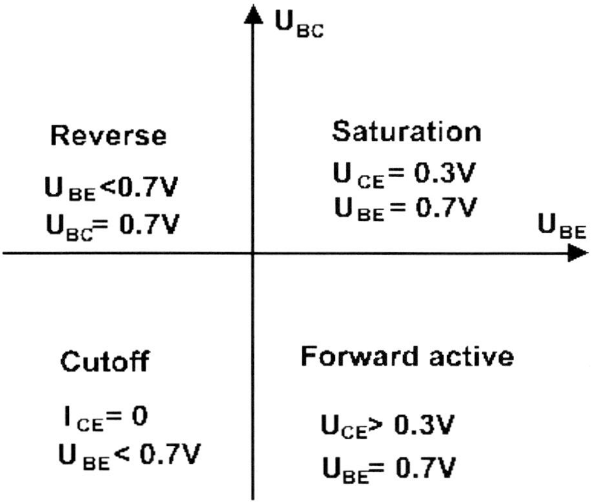

FIGURE 1.22 Possible bias modes of operation of a bipolar junction transistor.

In the cutoff region the transistor is fully turned OFF and the value of the collector current I_C is near zero. Full power supply voltage appears across the transistor. Because there is no current flow through the transistor, there is no voltage drop across the load resistor R_L.

However, bipolar transistors do not have to be restricted to these two of extreme modes operation. As described above, base current "opens a gate" for a limited amount of current through the collector. If this limit for the controlled current is greater than zero but less than the maximum allowed by the power supply and load circuit, the transistor will "throttle" the collector current in a mode somewhere between cutoff and saturation. This mode of operation is called the active mode. A load line is a plot of collector-to-emitter voltage over a range of base currents. The dots marking where the load line intersects the various transistor curves represent the realistic operating conditions for those base currents given.

Let us examine the transistor stage circuit of CE type in detail, Fig. 1.24. It begins to get clearer as to what the load line (Fig. 1.23) is. This line is built on the series of the static volt-ampere characteristics of transistor (these characteristics are parameters set by the manufacturer) on two cross-points with axes corresponding to the idle mode and short-circuit.

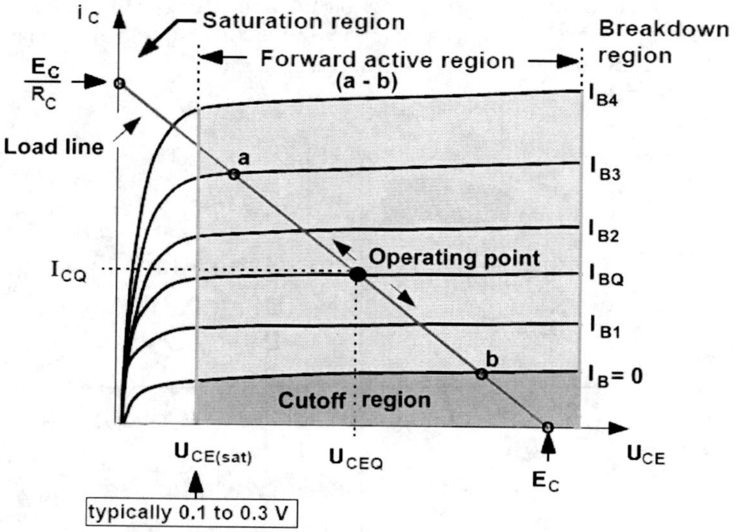

FIGURE 1.23 Output dynamic characteristic of a bipolar junction transistor.

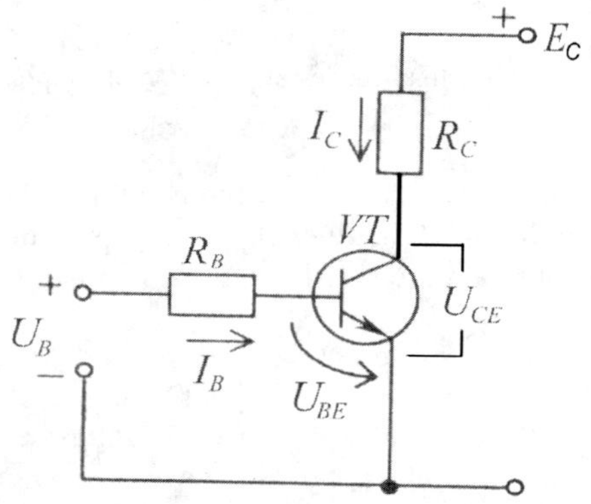

FIGURE 1.24 Transistor stage in CE mode.

Main ratios for this citcuit:

$$I_B = \frac{U_B - U_{BE}}{R_B}, \quad I_C = \beta I_B, \quad U_{CE} = E_C - I_C R_C.$$

For the first of them: $I_C = 0$, $U_{CE} = E_C$ (point on the abscissa axis); for the second: $U_{CE} = 0$, $I_C = E_C / R_C$ (point on the ordinate axis). The intersection points of the load line with any of static characteristics are called the working points corresponding to definite values of output current and output voltage. The transistor functions in the active mode (amplification) when the working point lies within the limits of a-b interval. The functioning of the transistor stage in the amplification mode is characterized by the so-called "quiescent point" or Q-point. The quiescent operating conditions may be shown on the graph in the form of a single point along the load line, Fig. 1.23. For a class A amplifier (widely used in simple automatic devices), the operating point for the quiescent mode (quiescent point) will be in the middle of the load line.

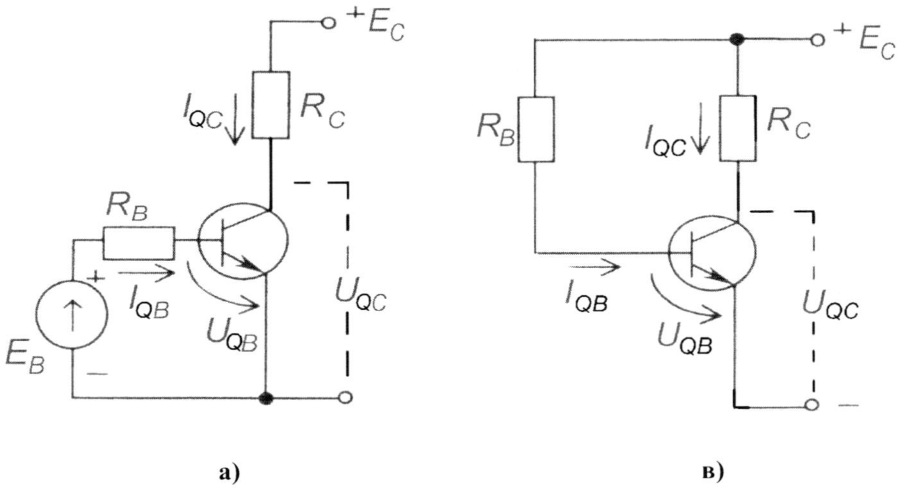

FIGURE 1.25 Circuit diagram of setting of working quiescent point of transistors by fixed current.

Pragmatically, the working Q-point of a transistor chosen according to its characteristic (Fig. 1.23) should be set with the help of the so-called biasing circuit. There are two methods for setting the transistor working point: by fixed current or fixed voltage. The first one is implemented with the help of two circuit diagrams, Fig. 1.25. In the circuit of Fig. 1.25a the biasing circuit is formed by resistor R_B which is calculated as follows:

$$R_B = \frac{E_B - U_{QB}}{I_{QB}},$$

где $U_{BE} \approx 0.7 - 0.9$ V when base-emitter junction is forward biased; I_{QB} – quiescent base current from output dynamic characteristic (Fig. 1.23)

With only one biasing source, as it is shown on Fig. 1.25c, the quiescent mode is ensured by the power supply voltage E_C and by resistor R_B :

$$R_B = \frac{E_C - U_{QB}}{I_{QB}}$$

The method of setting biasing by fixed voltage is shown in Fig. 1.26. It is the most widely used method of setting of transistor quiescent point by means of two resistors *R1* and *R2*.

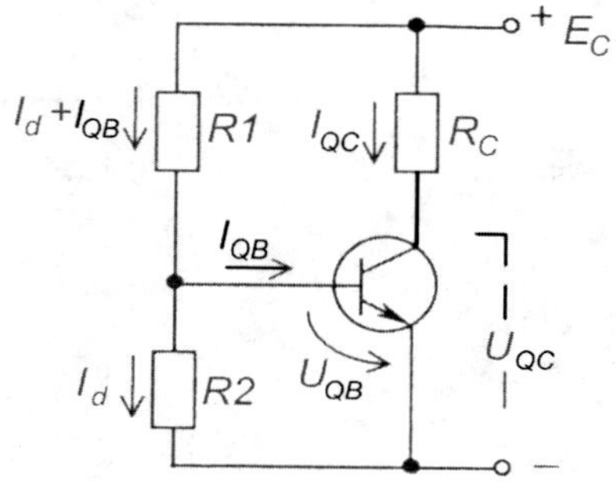

FIGURE 1.26 Circuit diagram of transistor stage with biasing by fixed voltage.

For this circuit the following ratios are appropriate:

$$R_1 = \frac{E_C - U_{QB}}{I_{QB} + I_d}, \quad R_2 = \frac{U_{QB}}{I_d}, \quad I_d = (2-5)I_{QB}$$

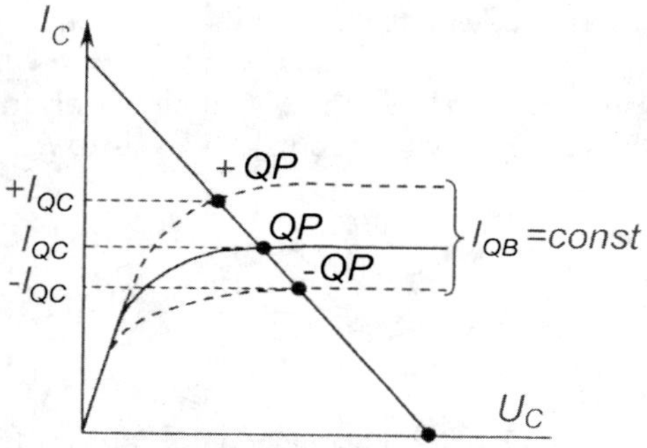

FIGURE 1.27 Changing of transistor quiescent point with displacement direct current I_{QB} temperature-influenced.

As the environment temperature changes, the transistor current gain (β) changes (increases as the temperature increases and decreases with a decrease in temperature), and the quiescent point position will change (Fig. 1.27).

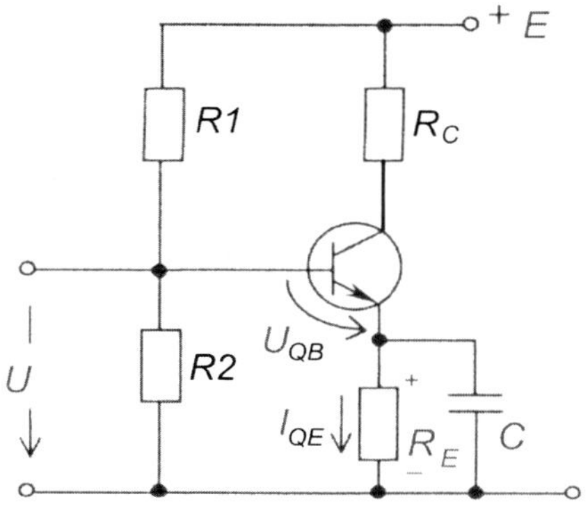

FIGURE 1.28 Amplifying stage with thermostabilization of quiescent point.

In this circuit (Fig. 1.28) for the resistance of the additional resistor R_E in the emitter circuit one chooses based on the equation: $R_E = (0.1 - 0.2)\ R_C$, and the capacitance, C, of capacitor C from the equation:

$$\frac{1}{\omega C_E} << R_E$$

where ω – minimum frequency of enhanced signal.

Thanks to the biasing capability of this capacitor, in AC applications we obtain the amplification stage with common emitter (CE), and in DC applications the amplification stage with negative feedback.

By the biasing of working points higher than point "a" or lower than point "b," the transistor goes into the saturation or cutoff mode, correspondingly. In the saturation mode the transistor is fully turned on and the current $I_{SAT} = E_C / R_C$ flows through it, it does not increase more with an increase of the input signal (that is why this mode is called "saturation"). To force the transistor into the saturation mode, one should make its base current not less than: $I_B = I_{C(SAT)} / \beta$. In many automation devices transistors function in the switch mode, i.e., in two ultimate modes: saturation and cutoff. Conditions for safe cutoff or complete saturation of transistor is ensured, like in the case examined above, by the choice of biasing resistances R_B and R_C.

Equations for voltage in base circuit:

$$U_B = -I_{C0} R_B + U_{BE} \text{ или } U_{BE} = U_B + I_{C0} R_B$$

The safe transistor cutoff is ensured under the condition that $U_{BE} \leq 0$, whence $R_B \leq U_B / I_{C0}$. In this case the value of the base resistance may be calculated as follows:

$$R_B \leq \frac{U_B}{I_{C0\max}}$$

where I_{C0max} is the maximum value of the collector's reverse current (transistor certified value).

In the circuit diagram of Fig. 1.24 the positive input signal of a definite value turns the transistor on (the saturation mode). At this point, currents flowing in the transistor are equal:

$$I_B = \frac{U_B}{R_B}, \quad I_{C(SAT)} = \frac{E_C}{R_C} \leq \beta I_B.$$

Thus one may calculate the value of base resistance for transistor saturation mode:

$$R_B \leq \beta_{\min} R_C \frac{U_B}{E_C}.$$

1.5 TRANSISTOR DEVICES IN SWITCHING MODE

One of the often-used operation modes of a transistor is the switching mode; even a single transistor can work as a high-speed switch (Fig. 1.29).

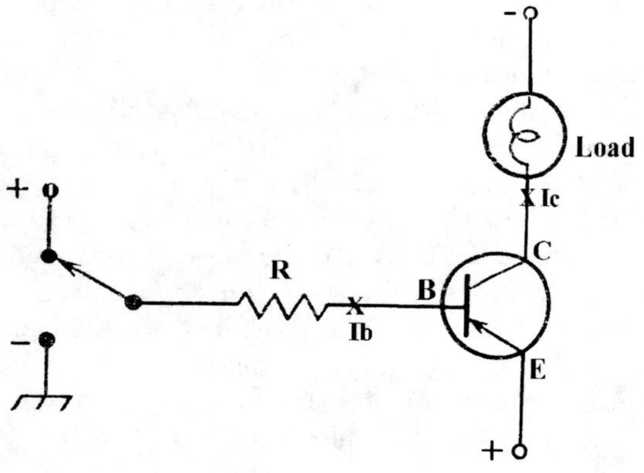

FIGURE 1.29 Electronic switch on a single transistor.

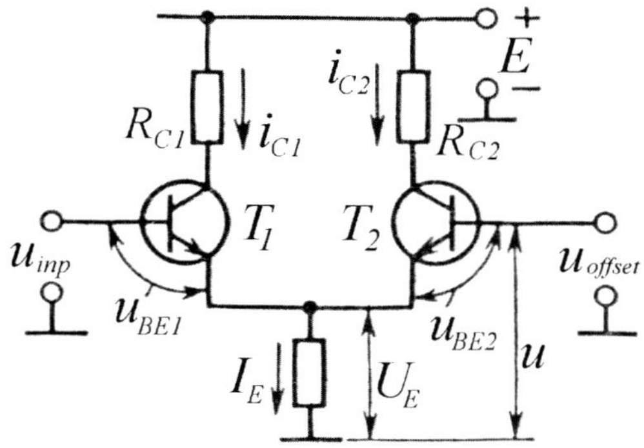

FIGURE 1.30 Transistor switch of two circuits.

For switching current from one circuit to another one, a two-transistor circuit (Fig. 1.30) is used. In this circuit stable offset voltage is applied to the base of the transistor (T_2) and control voltage to base T_1.

When $u_{inp} = u_{offset}$, the currents and voltages in the arms of the circuit are the same. If the input voltage (u_{inp}) exceeds the offset voltage (u_{offset}), transistor T_2 is gradually blocked and the whole current flows only through transistor T_1 and load resistor R_{C1}, and vice versa. When input voltage decreases below the level of the offset voltage ($u_{inp} < u_{offset}$), transistor T_1 is blocked and T_2 is unblocked, switching the sole current to the circuit of the resistor R_{C2}.

As is known, contacts of several electromagnetic relays, connected with each other in a certain way, are widely used in automation systems for carrying out the simplest logical operations with electric signals (Fig. 1.32).

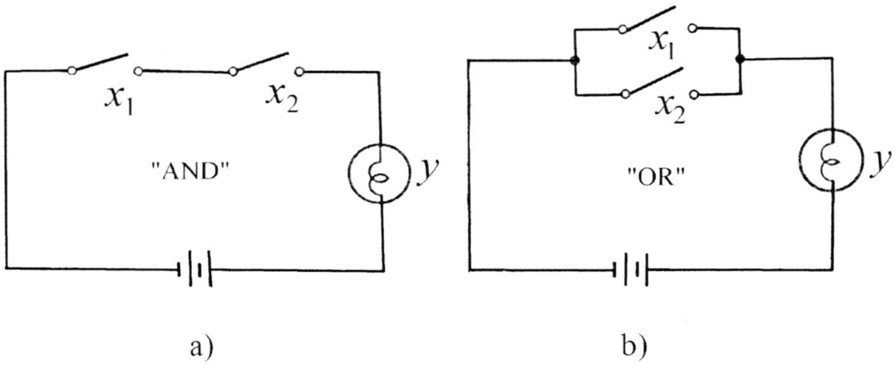

FIGURE 1.31 Implementation of the simplest logical operations, with the help of electromagnetic relay contacts.

For example, the logical operation *AND* is implemented with the help of several contacts connected in series, switched to the load circuit (Fig. 1.31a). The signal Y will be the output of this circuit (that is the bulb will be alight) only if signals on the first input X1 and on the second input X2, operate simultaneously (that is when both contacts are closed). Another simplest logical operation *OR* (Fig. 1.31b) is implemented with the help of several contacts connected in parallel. In this circuit in order to obtain the signal Y on output (that is for switching-on of the bulb), input of signal or on the first input (X1), or on the second input (X2), or on both of the inputs simultaneously, is required. Implementation of logical operations with electric circuits is one of the most important functions of relays. Transistor circuits successfully carry out this task. For example, the function NOT can be implemented on any type of single transistor (Fig. 1.32).

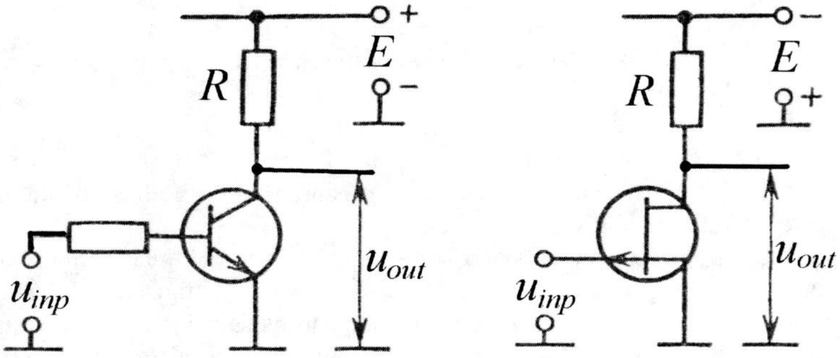

FIGURE 1.32 Logical element *NOT* implemented on a bipolar and field transistors.

In the circuit in Fig. 1.32, when an input signal is missing, the transistor is blocked, that is the whole voltage of the power source E is applied between the emitter and the collector (the drain and the source) of the transistor, and since the output signal is voltage on the collector (the source) of the transistor, that means that if there is no signal at the input, there will be a signal at the output of this circuit. And vice versa, when the signal is applied at the input, the transistor is unblocked and voltage drops to a very small value (fractions of a Volt) and therefore the signal disappears at the output.

The logical element *AND-NOT* can be implemented by different circuit methods. In the simplest case, this is a circuit from transistors connected in series (Fig. 1.33a). When control signals are applied to both inputs X1 and X2 simultaneously, both transistors will be enabled and the voltage drop in the circuit with two transistors connected in series will decrease to a very small value. This means no output signal Y. In the second circuit diagram (Fig. 1.33b) even one signal on any input (X1 or X2) is enough for the voltage on output Y to disappear.

Self-contained logical elements are indicated on circuit diagrams as special signs (Table 1.1).

A signal strong enough for transition of a logical element from one state to another is usually marked as "1." No signal (or a very weak signal incapable of affecting the system state) is usually marked as "0."

Table 1.1 Basic logical elements (according to certain standards logical elements are also indicated as rectangles).

Logical Function	Conventional Symbols	Boolean Identities	Truth Table		
			Inputs		Output
			B	A	Y
AND	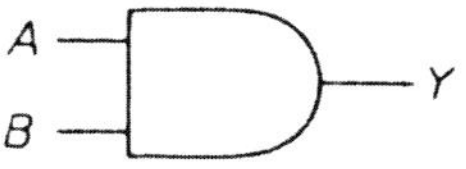	$A \cdot B = Y$	0	0	0
			0	1	0
			1	0	0
			1	1	1
OR	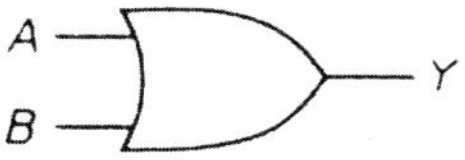	$A + B = Y$	0	0	0
			0	1	1
			1	0	1
			1	1	1
NOT	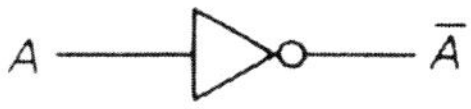	$A = \overline{A}$		0	1
				1	0
AND-NOT (NAND)	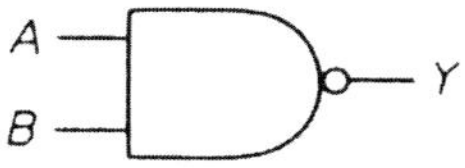	$\overline{A \cdot B} = Y$	0	0	1
			0	1	1
			1	0	1
			1	1	0
OR-NOT (NOR)	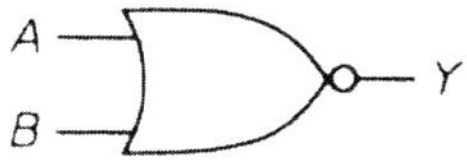	$\overline{A + B} = Y$	0	0	1
			0	1	0
			1	0	0
			1	1	0

The same signs are used for indication of the state of the circuit elements: "1" – switched-on; "0" – switched-off. Such bi-stable (that is having two stable states) devices are called triggers. When supply voltage is applied to such a device (Fig. 1.34) one of the transistors will be immediately enabled and the other one will remain in a blocked state. The process is avalanche-like and is called regenerative. It is impossible to predict which transistor will be enabled because the circuit is absolutely symmetrical and the likelihood of unblocking of both transistors is the same.

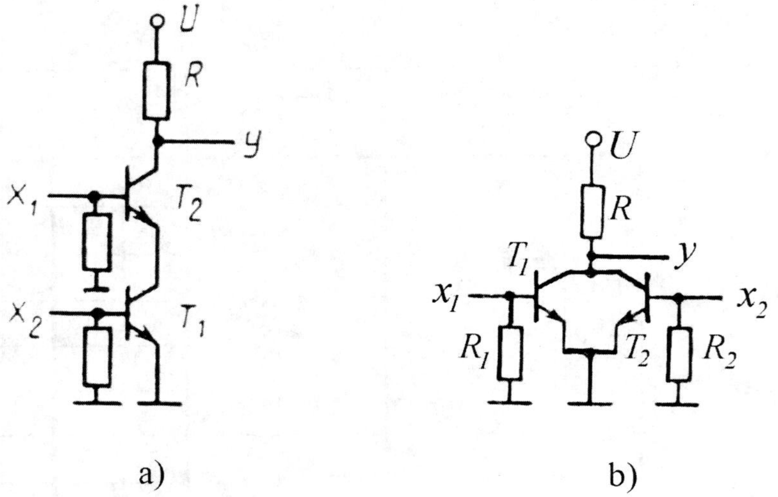

FIGURE 1.33 Transistor logical elements AND-NOT (a) and OR-NOT (b).

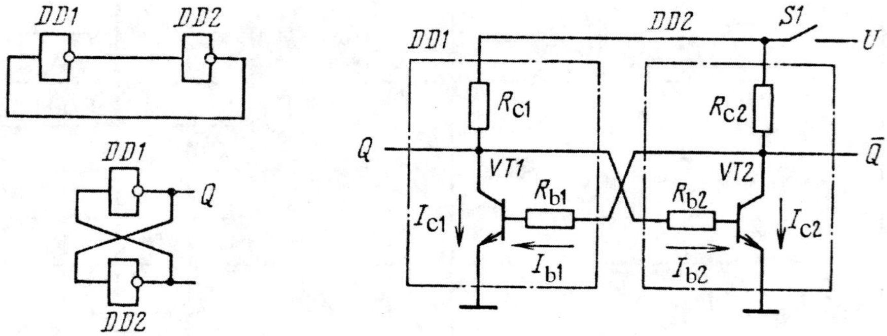

FIGURE 1.34 Bi-stable relay circuit with two logical elements NOT.

This state of the device remains stable just the same. Repeated switching ON and OFF of voltage will cause the circuit to pass into this or that stable state.

The essential disadvantage of such a trigger is no control circuit, which would enable us to control its state at permanent supply voltage.

In practice the so-called *Schmitt-Triggers* are often used as electronic circuits with relay characteristics. There are a lot of variants of such triggers, possessing special qualities. In the simplest variant such a trigger is a symmetrical structure formed by two logical elements connected in a cycle of the type AND-NOT or OR-NOT (Fig.1.35); it is called an *asynchronous RS-trigger*.

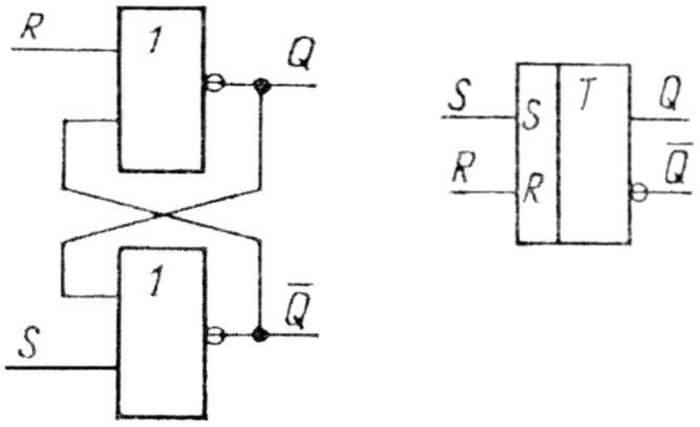

FIGURE 1.35 Asynchronous RS-trigger formed by two logical elements *NOR.*

Table 1.2 Combinations of signals at the inputs and the RS-trigger position.

Input			Output for Logical Element Type			
			AND-NOT		*OR-NOT*	
S (set)	R (reset)	Notes	**Q**	$\overline{Q}$	**Q**	$\overline{Q}$
0	0	forbidden mode for *AND-NOT*	uncertainty		without changes	
1	**0**		**1**	**0**	**1**	**0**
0	**1**		**0**	**1**	**0**	**1**
1	1	forbidden mode for *NOR*	without changes		uncertainty	

One of the trigger outlets is named *direct* (any outlet can be named so as the circuit is symmetrical) and is marked by the letter Q, and the other one is called *inverse* and is marked by the letter $\overline{Q}$ ("Q" under the dash), to signify that in logical sense the signal at this output is opposite to the signal at the direct output. The trigger state is usually identified with the state of the direct output, that is to say that the trigger is in the single (that is switched-on) state when Q = 1, $\overline{Q}$ = 0, and vice versa.

Trigger state transition has a lot of synonyms: "switching," "change-over," "overthrow," "recording," and is carried out with the help of control signals applied at the inputs R and S. The input by which the trigger is set up in the single state is called the S input (from "Set") and the output by which the trigger turns back to the zero position – the R input (from "Reset"). Four combinations of signals are possible at the inputs, each of them corresponding to a certain trigger position (Table 1.2).

As can be seen from the table, when there are no signals on both of the trigger inputs on the elements AND-NOT, or when there are signals on both of the trigger inputs on the elements OR-NOT (NOR), the trigger state will be indefinite, which is why such combinations of signals are prohibited for RS-triggers.

From the time diagram of the asynchronous RS-trigger (Fig. 1.36), it can be seen that after transfer of the trigger to the single state no repeated signals on the triggering input S are capable of changing its state. The return of the trigger to the initial position is possible only after a signal is applied to its "erasing" R input.

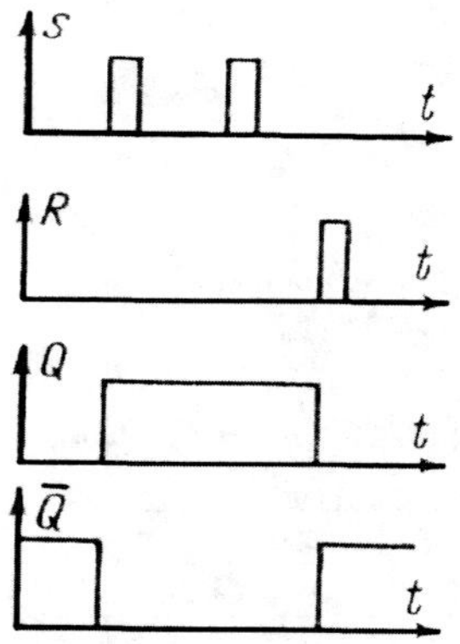

FIGURE 1.36 Time diagram of an asynchronous RS-trigger.

The disadvantage of the asynchronous trigger is its incapacity to distinguish the useful signal of starting from noise occurring in the starting input by chance. Therefore, in practice so-called *synchronous* or D-triggers, distinguished by an additional so-called *synchronizing input,* are frequently used.

Switching of the synchronous trigger to the single state (ON) is carried only with both signals: starting signal at the S input and also with a simultaneous signal on the synchronizing input. Synchronizing (timing) signals can be applied to the trigger (C input, Fig. 1.37) with certain frequencies from an external generator.

A simple amplifier with two transistors, with positive feedback, also has properties of trigger (Fig. 1.38).

In initial position, when there is no voltage (or when voltage is very low) at the input of the circuit, transistor VT1 is closed (locked up). There is voltage on VT1 collector, which opens transistor VT2.

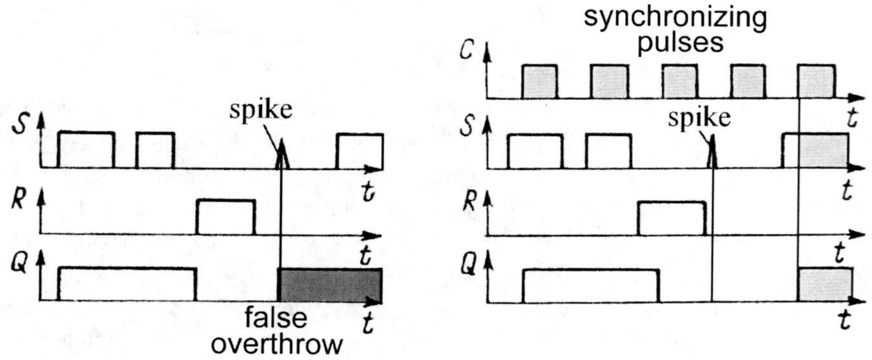

FIGURE 1.37 Time diagrams of operation of an asynchronous trigger (on the left) and synchronous trigger (right) when there is noise.

The emitter current of transistor VT2 causes a voltage drop on the resistor R3, which blocks transistor VT1 and holds it in closed position. If input voltage exceeds the voltage in the emitter on VT1 transistor, it will be opened and will become saturated with very small collector-emitter junction resistance.

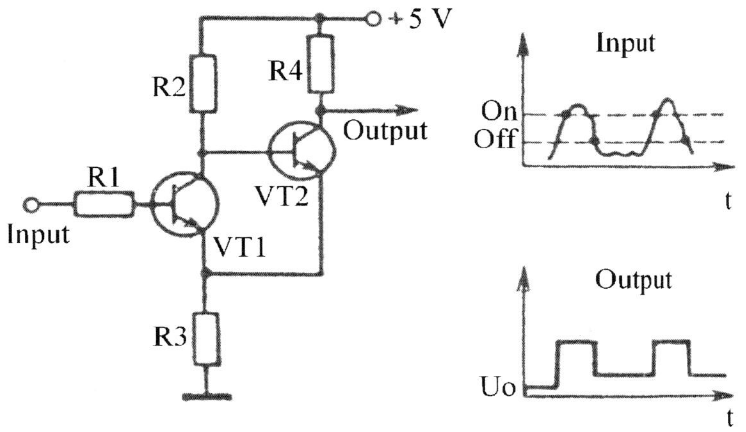

FIGURE 1.38 A simple two-transistor trigger.

As a result, the potentials of the base and the emitter of transistor VT2 will be equal. Transistor VT2 will be blocked. At the output there will be voltage equal to the supply voltage. When input voltage decreases, transistor VT1 leaves the saturation mode, and an avalanche-like process occurs. Emitter current of transistor VT2, causing blocking voltage on resistor R3, accelerates closing of the transistor VT1. As a result, the trigger returns to its initial position.

1.6 THYRISTORS

The history of development of another remarkable semiconductor device with relay characteristics begins with the conception of a "collector with a trap," formulated at the beginning the 1950s by William Shockley, familiar to us from his research on *p-n* junctions. Following Shockley, J. Ebers invented the two-transistor analogy (inter-bounded *n-p-n* and *p-n-p* transistors) of a *p-n-p-n* switch, which became the model of such a device (Fig. 1.39).

The working element of this new semiconductor device with relay characteristics was a four-layer silicon crystal with alternating *p-* and *n-* layers (Fig. 1.40). Such a structure is made by diffusion into the original monocrystal of n_1-silicon (which is a disk 20 – 45 mm in diameter and 0.4 – 0.8 mm thick, or more for high-voltage devices) admixture atoms of aluminum and boron from the direction of its two bases to a depth of about 50 – 80 micron. Injected admixtures form p_1 and p_2 layers in the structure.

The fourth (thinner) layer n_2 (its thickness is about 10-15 micron) is formed by further diffusion of atoms of phosphorus to the layer p_2. The upper layer p_1 is used as an anode in the thyristor, and the lower layer p_2 – as a cathode.

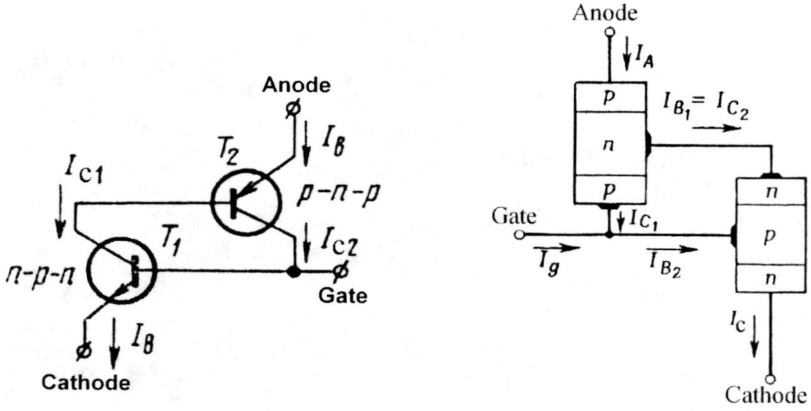

FIGURE 1.39 Two-transistor model of a thyristor.

The power circuit is connected to the main electrodes of the thyristor: the anode and the cathode. The positive terminal of the control circuit is connected through the external electrode to layer p_2, and the negative one to the cathode terminal.

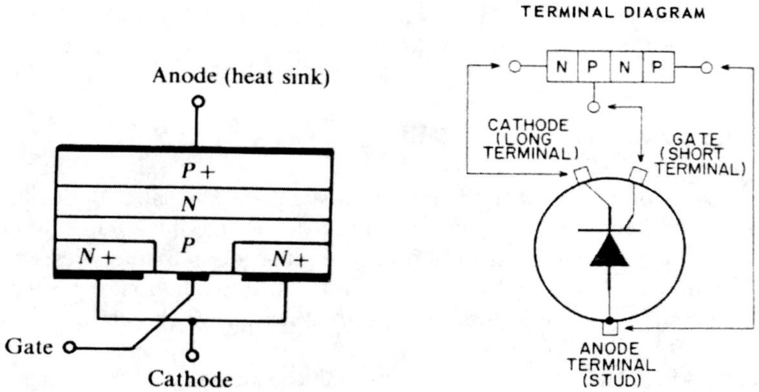

FIGURE 1.40 Structure and symbolic notation of a solid state thyratron – "*thyristor.*"

The volt-ampere characteristic (VAC) of a device with such a structure (Fig. 1.41) much resembles the VAC of a diode by form. As in a diode the VAC of a thyristor has forward and reverse areas. Like a diode, the thyristor is blocked when reverse voltage is applied to it (minus on the anode, plus on the cathode) and when the maximum permissible level of voltage (U_{Rmax}) is exceeded there is a breakdown, causing strong current and irreversible destruction of the structure of the device.

The forward area of the VAC of the thyristor does not remain permanent, as does that of a diode, and can change, being affected by current of the control electrode, called the *Gate*. When there is no current in the circuit of this electrode, the thyristor remains blocked not only in reverse but also in the forward direction, that is it does not conduct current at all (except small leakage current, of course). When the voltage applied in the forward direction between the anode and the cathode is increased to a certain value, the thyristor is quickly (stepwise) enabled and only a small voltage drop (frictions of a Volt) caused by irregularity of the crystal structure remains on it. If low current is applied to the circuit of the gate, the thyristor will be switched ON to much lower voltage between the anode and the cathode. The higher the current, the lower the voltage that is required for unblocking of the thyristor. At a certain current value (from a few milliamperes for low power thyristors, up to hundreds of milliamperes for power ones) the forward branch of the VAC is almost fully rectified and becomes similar to the VAC of a diode. In this mode (that is when control current constantly flows in the gate circuit) the behavior of the thyristor is similar to that of a diode that is fully enabled in the forward direction and fully blocked in the reverse direction. However, it is senseless to use thyristors in this mode: there are simpler and cheaper diodes for this purpose.

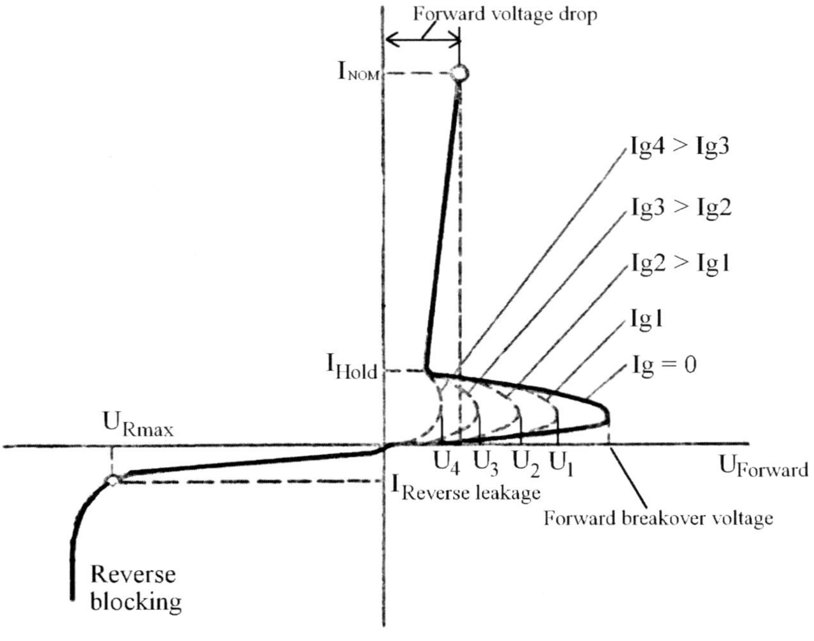

FIGURE 1.41 Volt-ampere characteristic (VAC) of a thyristor.

In fact thyristors are used in modes when the working voltage applied between the anode and the cathode doesn't exceed 50 – 70% of the voltage, causing spontaneous switching ON of the thyristor (when there is no control signal, the thyristor always remains blocked) and control current is applied to the gate circuit only when the thyristor should be unblocked and of such a value that would enable reliable unblocking. In this mode the thyristor functions as a very high-speed relay (unblocking time is a few or tens of microseconds).

Perhaps many have heard that thyristors are used as basic elements for smooth current and voltage adjusting, but if a thyristor is only an electric relay having two stable states like any other relay: a switched ON state and a switched OFF one, how can a thyristor smoothly adjust voltage? The point is that if non-constant alternating sinusoidal voltage is applied, it is possible to adjust unblocking moment of the thyristor by changing the moment of applying a pulse of control current on the gate with regard to the phase of the applied forward sinusoidal voltage. That is, it is as though a part of the sinusoidal current flowing to the load were cut off (Fig. 1.42). The moment of applying a pulse of unblocking control current (such pulses are also called "igniting" by analogy with the control pulses of the thyratron) is usually characterized by the angle α.

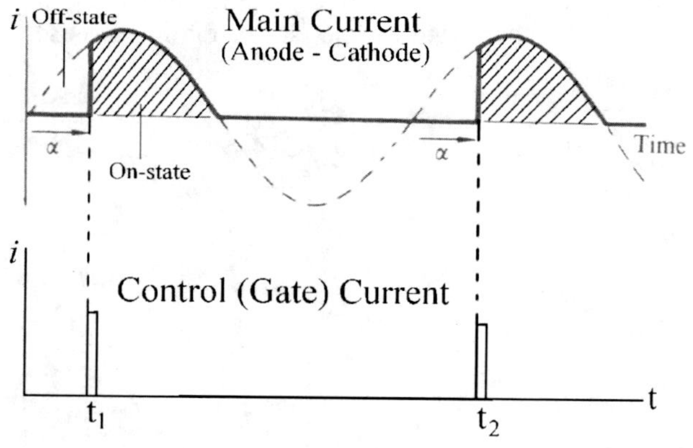

FIGURE 1.42 Principle of operation of a thyristor regulator.

Taking into account that average current value in the load is defined as an integral (that is the area of the rest part of the sinusoid) the principle of operation of a thyristor regulator becomes clear. After unblocking, the thyristor remains in the opened state, even after completion of the control current pulse. It can be switched OFF only by reducing forward current in the anode-cathode circuit to the value less than *hold current* value. In AC circuits the condition for thyristor blocking is created automatically when the sinusoid crosses the zero value. To unblock the thyristor in the second half-wave of the voltage it is necessary to apply a short control pulse through the gate of the thyristor. To control both half-waves of alternating current two thyristors connected anti-parallel are used. Then one of them works on the positive half-wave and the other one – on the negative one.

At present such devices are produced for currents of a few milliamperes to a few thousand amperes, and for blocking voltages up to a few thousands volts. The first industrial examples of power and high-power thyristors produced in different countries had the so-called "pin-like" (*stud* and *flat base* types) construction (Fig. 1.43).

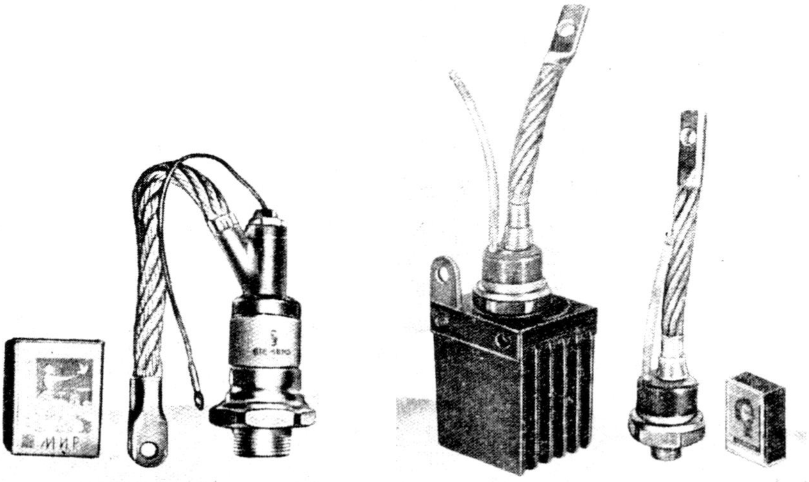

1 2

FIGURE 1.43a Industrial samples of the first pin-like (1, 2 – stud; 3 - flat base) thyristors for currents up to 100-150A, produced in 1960s. 1 – Bst L02 (Siemens); 2 – BTY-16 (AEG); 3 – BKY-100 (Russia).

3

As can be seen from the VAC, a certain voltage drop takes place even on a fully open thyristor because of imperfections in its crystal structure. This voltage is very small in comparison with the working voltage. It totals only fractions of a volt, however when strong working currents pass through the thyristor, such a voltage drop may lead to considerable power dissipation of the thyristor. For example, with voltage on an open thyristor of 1.5V and current of 200A, thermal power equal to 300W is constantly being released. This is very high power and if certain measures for cooling the thyristor are not taken its temperature will quickly exceed 150–160°C and voltage will cause a breakdown of the crystal structure.

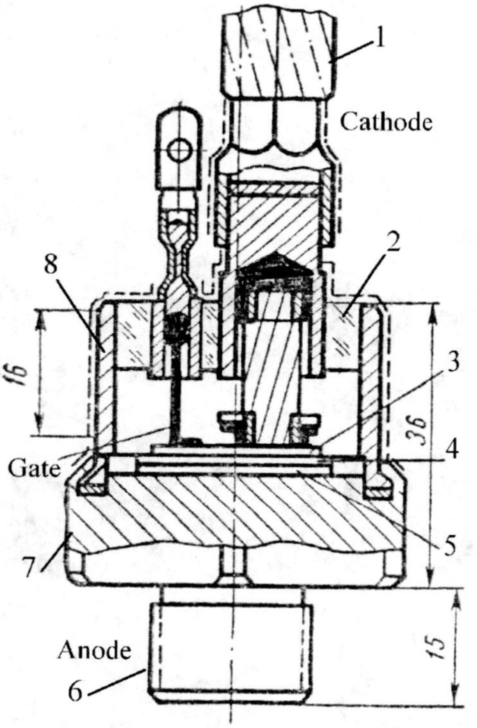

FIGURE 1.43b Construction of a power stud type thyristor.
1 – multiple-strand flexible copper braid with a point at the end; 2 – glass or ceramic insulator; 3 – layer n_2 of the semiconductor structure; 4 – silicon monocrystal (layer n_2); 5 – layer p_1 of the semiconductor structure; 6 – anode outlet made in the form of a screw-bolt; 7 – copper heel piece; 8 – steel cylindrical case.

That is why all high power thyristors are always equipped with heat sinks. These are big ribbed constructions from aluminum alloy for air cooling or more compact – for water cooling (Fig. 1.44).

Another problem with heating of thyristors was destruction of the joint points of the silicon crystal with a copper heelpiece and a cathode outlet, which were made with the help of standard tin solder. In the first power thyristors, already after several tens of thousands of switched OFF and ON cycles (when the thyristor was heated up to 100-120° C and then to be cooled up to 20–30°C) there was cracking of solder caused by the difference of linear expansion coefficients of the various materials.

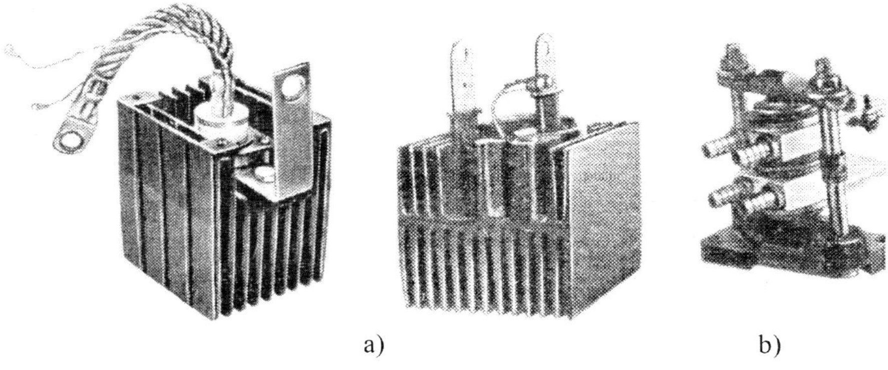

FIGURE 1.44 Air (a) and water (b) heat sinks for thyristors.

Later on, they managed to cope with this disadvantage by introducing special temperature compensators and using pressure contacts instead of soldered ones (Fig. 1.45).

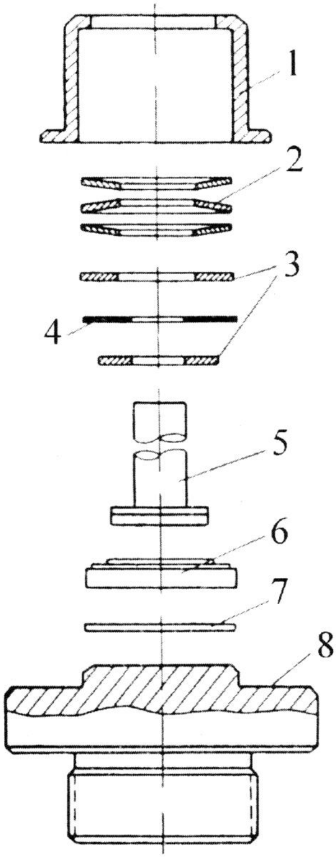

FIGURE 1.45 Construction of a stud-like thyristor with pressure contacts and a temperature compensator
1 – pressure cup; 2 – disk spring; 3 – metal disk; 4 – mica disk; 5 – contact stamp with a temperature compensating plate; 6 – semiconductor crystal structure on a temperature compensating plate; 7 – silver contact spacer; 8 – copper piece heel of the case.

Later it turned out that it is more convenient to produce and exploit tablet construction (called *capsule* types) in the form of a disk, as in construction of pressure contacts (Fig. 1.46). When such constructions began to be produced, high-power pin-like thyristors were almost entirely forced out of production. The stud-like construction remained only for low power and power thyristors (for currents up to a few tens of Amperes).

Modern thyristors have a distinguishing diversity of forms and sizes (Fig. 1.47).

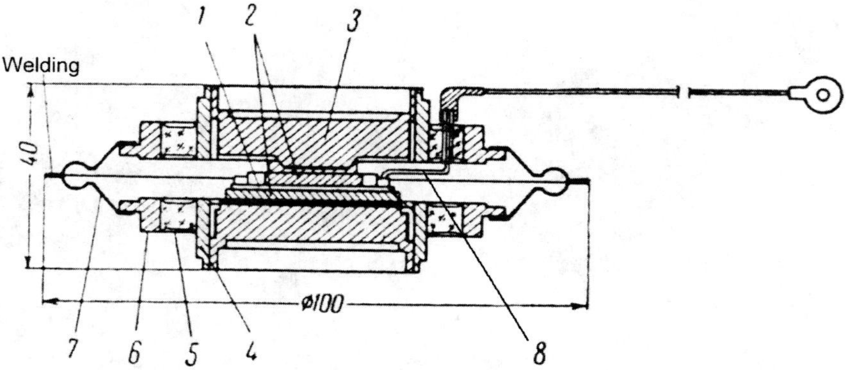

FIGURE 1.46 Earlier construction of a capsule type thyristor with pressure contacts
1 – semiconductor crystal structure; 2 – pressure tungsten disks; 3 – copper contact elements (anode and cathode); 4 and 6 – metal rings; 5 – glass insulator; 7 – spring goffered disk; 8 – gate.

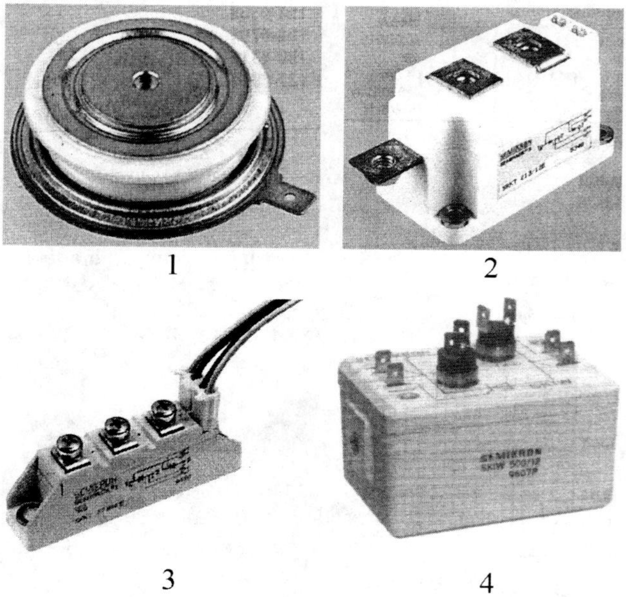

FIGURE 1.47a Modern high-power thyristors
1 – single thyristor; 2 and 3 – dual thyristor module with common anode or common cathode; 4 – high current (900 A) anti-parallel thyristors with isolated (via aluminum oxide – AlO_2) water flow.

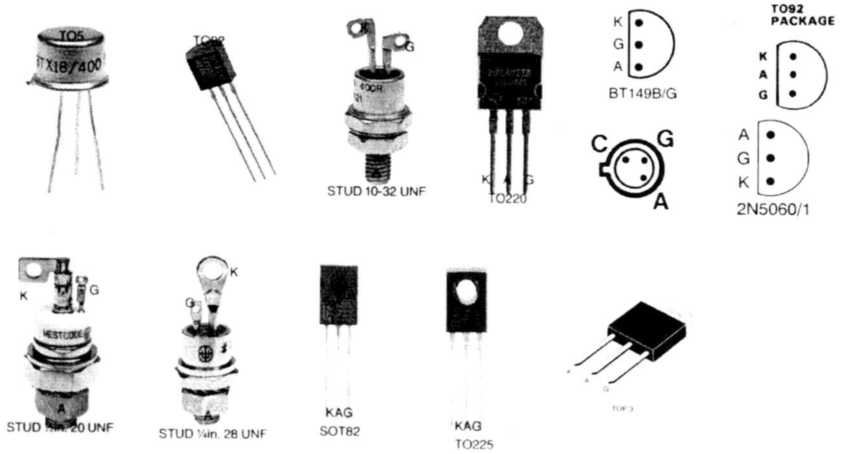

FIGURE 1.47b Modern low power and power thyristors.

1.7 CONTROL OF THYRISTORS ON DIRECT CURRENT

As has been mentioned above, the thyristor in the initial position is blocked in both current directions and for correct (non emergent) unblocking it is necessary to create certain conditions for current and voltage:

- Forward voltage not exceeding forward breakover voltage should be applied to the thyristor (Fig. 1.41);
- In the «Gate – Cathode» circuit there should be current of positive direction enough for thyristor unblocking both by value (0.05-0.2A for power thyristors) and by duration (tens and hundreds of microseconds).

Under such conditions the thyristor will be switched ON and current will flow through its main "anode-cathode" junction. The control junction (gate-cathode) will be shunted by forward current, and further operation of the thyristor will not depend on current in the gate circuit. The thyristor state after it has been enabled will be fully determined by the forward current value in the "anode-cathode" circuit, that is by load resistance. If this current exceeds the hold current (I_{HOLD} – Fig. 1.41), the thyristor will be conductive. If it is less than the hold current, the thyristor will be immediately switched OFF.

In the scheme shown in Fig. 1.48a, the thyristor *VS1* is switched ON at the moment when the resistance *R1* decreases to a value sufficient for the gate current in the circuit, corresponding to the unblocking control current of the given thyristor. When the thyristor is enabled, the resistance *R1* is shunted by low resistance of the main opened junction (anode-cathode) and doesn't affect the state of the thyristor.

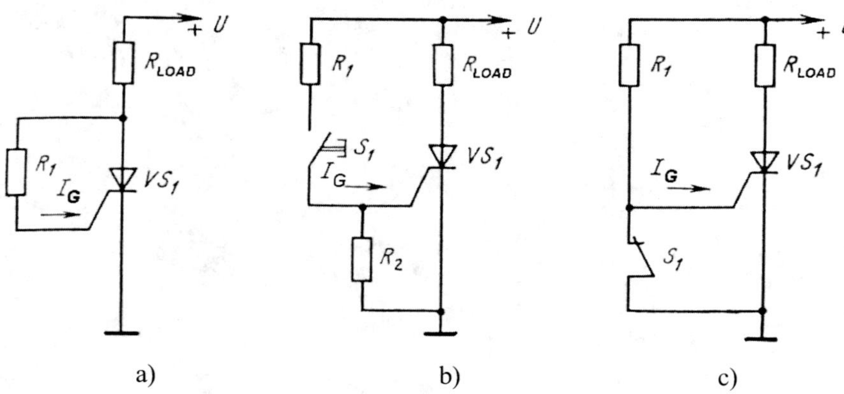

FIGURE 1.48 Connection and control of thyristors on DC.

In the circuit shown in Fig. 1.48b, current in the gate circuit arises only at the moment of closing of the control contact (*S1)*. The resistor *R2* is almost always used in such circuits to prevent the pulse noise from getting to the gate circuit, and spontaneous thyristor opening.

In the circuit shown in Fig. 1.48c, the control junction of the thyristor (gate-cathode) is constantly shunted by the contact (*S1)*. When the contact opens, current of the resistor *R1* changes its direction, passing to the circuit of the gate and opening the thyristor.

And how can an opened thyristor be switched OFF? It is not that easy to do on direct current (Fig. 1.49).

Methods applied in practice usually come to break circuits of anode current (a); shunting of the thyristor with an additional contact (b) or a transistor (c); reduction of anode current to a value less than the hold current (d); use of the charged capacitor C, which is connected parallel to the thyristor, at the moment when the thyristor must be switched OFF and runs down on it, creating current of the opposite polarity blocking the thyristor (e). All of these methods of forced closing of thyristors are called "forced commutation" (in contrast to "natural" commutation on AC). The method of closing of thyristors with the help of capacitors, like similar methods used for switching OFF of thyratrones, was the most popular. In the circuit in Fig. 1.49e the resistance of the resistor *R1* is much less than the load resistance R_L, which is why at the first moment after thyristor switching ON its node, the current passes not through the load but through the resistor *R1* charging the capacitor C. After that, the current ceases passing through the capacitor, which is why the anode current of the thyristor passes to the parallel branch with the load R_L. When the contact *S1* is closed (an additional thyristor *VS2* can be used instead of it, see Fig. 1.50), the voltage of the charged capacitor is applied to the thyristor with the opposite polarity ("plus" to the cathode, "minus" to the anode), causing blocking of the thyristor.

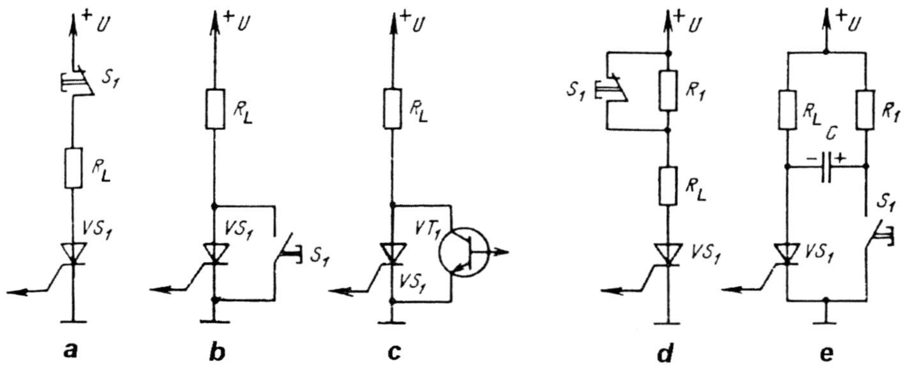

FIGURE 1.49 Principles of switching OFF of thyristors on direct current.

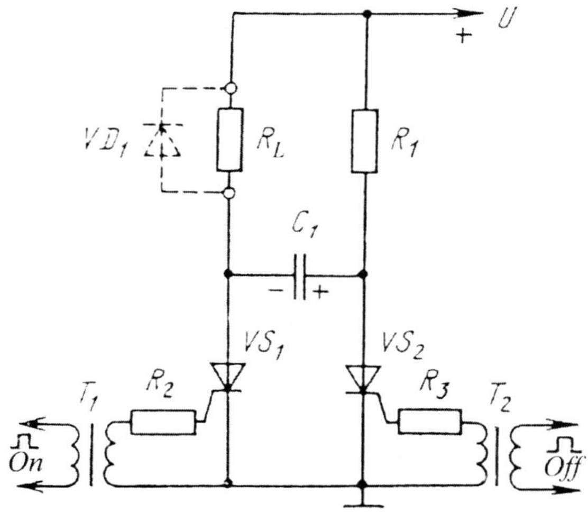

FIGURE 1.50 Pulse control circuit providing "forced commutation" of the main thyristor (*VS1*) on direct current.

Pulse circuits of thyristor control, with a transformer in the gate circuit, were very popular due to the fact that such small transformers allowed application of control pulses to the gate circuit of the power thyristor, which is under the full potential of the power source (which can be hundreds and even thousands of Volts), directly from low voltage microelectronic control devices, and also to control a group of thyristors connected in series and designed for work under high voltages (Fig. 1.51).

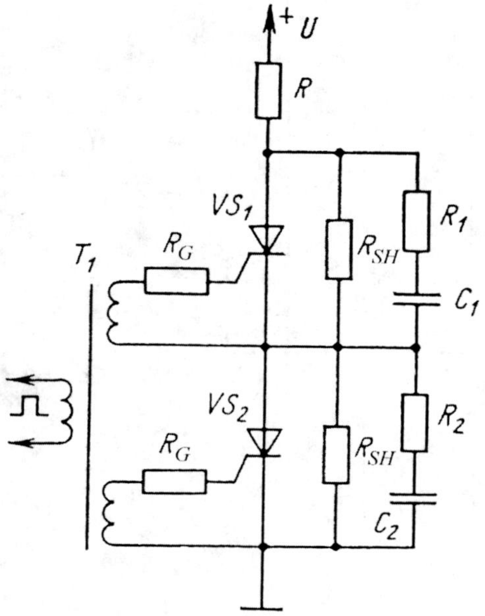

FIGURE 1.51 Series connection of thyristors with pulse control. R_{SH}– shunting resistors equalizing voltage distribution between thyristors connected in series; R_1C_1 and R_2C_2–circuits protecting thyristors from spikes during switching processes.

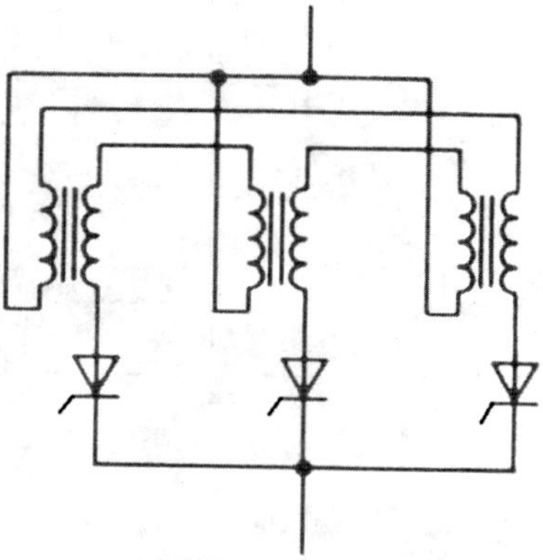

FIGURE 1.52 Parallel connection of thyristors with balancing reactors.

Sometimes it is necessary to connect thyristors in parallel in order to increase switched current. Like in the case of the series connection, one has to balance the work conditions of all the thyristors connected to the group because of the natural parameters of dispersion of the thyristors, but instead of balancing of voltage it is essential to balance currents passing through thyristors connected parallel, which is much more difficult. In such cases, more ponderous and expensive inductive reactors must be applied (Fig. 1.52).

1.8 CONTROL OF THYRISTORS ON ALTERNATING CURRENT

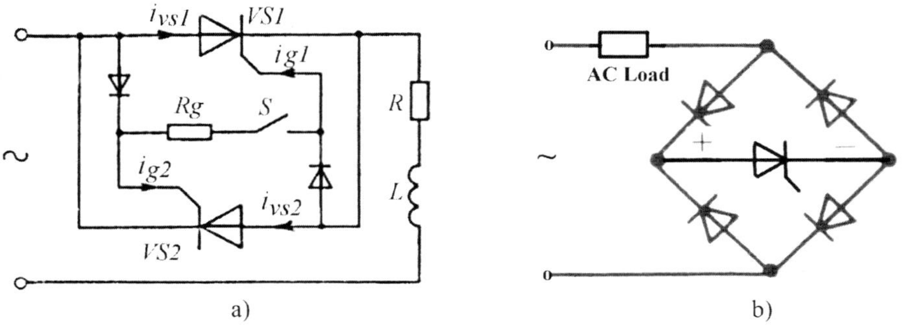

FIGURE 1.53 Thyristor AC switches.

In AC circuits the thyristor can be used without a forced cut-off, because every half-period the sinusoidal current passes through the zero value and at that moment conditions for thyristor cut-off are created, however for switching of both half-waves of current, two inverse-parallel connected thyristors (Fig. 1.53a), or a thyristor connected to the diagonal of the rectifier bridge (Fig. 1.53b), are required.

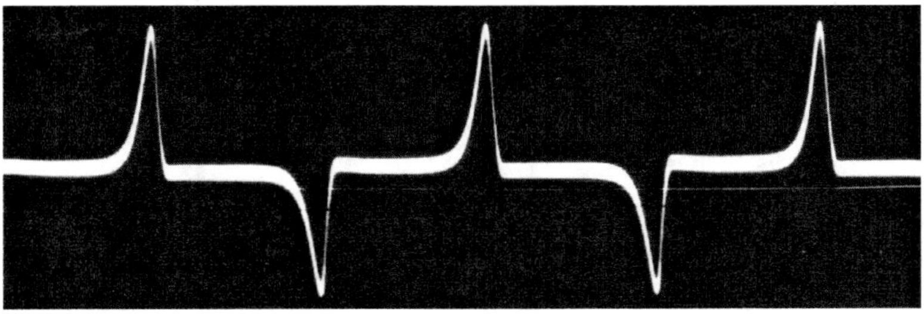

FIGURE 1.54 Oscillogram of pulses of the gate current automatically formed in the gate circuit of a thyristor AC switch.

In circuits of AC switches controlled by an additional contact (Fig. 1.53a – and this can be a reed switch) *in the closed position* of the contact in the gate circuit of thyristors, quite short control pulses are *automatically formed* from the anode voltage (Fig.

1.54). In order to switch three-phase loads, a three-phase switch constructed on the same principle is used (Fig. 1.55).

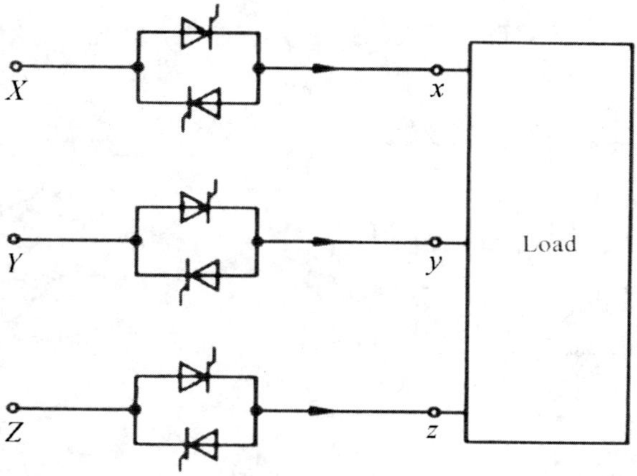

FIGURE 1.55 Three-phase thyristor switch based on inverse-parallel connected thyristors.

1.9 DIAC, TRIAC, QUADRAC ...

As in the case with transistors, there are several types of thyristors differing by their properties and characteristics. First of all, this is a so-called symmetrical thyristor – "triac" (the last two letters seem to stand for "alternating current").

FIGURE 1.56 Volt-Ampere characteristic (VAC) of a symmetrical thyristor "triac" (thyristor for alternating current); FB – forward branch; RB – reverse branch; I_G – gate current

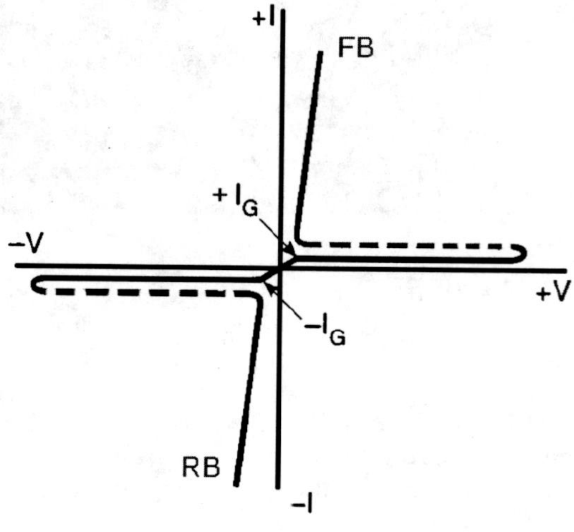

The symmetrical thyristor as it follows from its name has a symmetrical VAC (Fig. 1.56), that is when there is a control signal, it applies current in both directions and can replace two standard thyristors connected inverse-parallel (Fig. 1.57).

It is obvious that a triac has a more complex structure than a standard thyristor. It is no longer a four-layer device, like a thyristor. It is five-layer device, with a thyristor only as a part of a more complex structure.

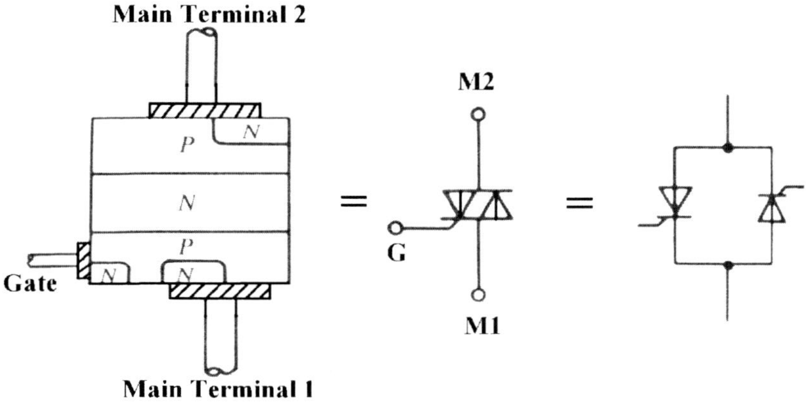

FIGURE 1.57 Triac structure, symbolic notation and thyristor equivalent.

In theory, the triac can be enabled at any combination of voltage polarities on main electrodes and on the gate. That's why it is quite senseless to indicate the main electrodes as an "anode" and a "cathode" and they are marked simply as M1 and M2. But it is correct for so-called "four-quadrant" (or 4Q) triac only. Such triac can be triggered in all four quadrants (Fig. 1.58). Three-quadrant triacs (3Q triac) allow triggering in quadrants I, II, and III only. 3Q triacs are more efficient in applications that have non-resistive loads, such as motor control applications, transformer loads, ignition circuits, etc. For these types of applications 4Q triacs must include additional protection components to minimize the effects of false triggering (uncontrolled triac conduction). These include RC snubbers across the main terminals of the triac and an inductor in series with the triac. 3Q triacs have eliminated or reduced the need for protection components, making system design for non-resistive loads more reliable, cheaper, and smaller. At more prevalent and more reliable 3Q triac VAC does not look as nice as in Fig. 1.56, and the symmetrical thyristor is not in fact all that symmetrical: the turn-on gate current with certain (reverse) voltage polarity on the main electrodes appears to be 3–5 times as much as with other (forward) polarity. Of course one can construct a control system capable of generating more powerful control pulses compensating for this difference in sensitivity. As it can be seen, rejection of the names of the main electrodes "anode" and "cathode" is not quite defensible, since despite triac "symmetry" it is essential to designate from which electrode the control signal will be applied to the gate.

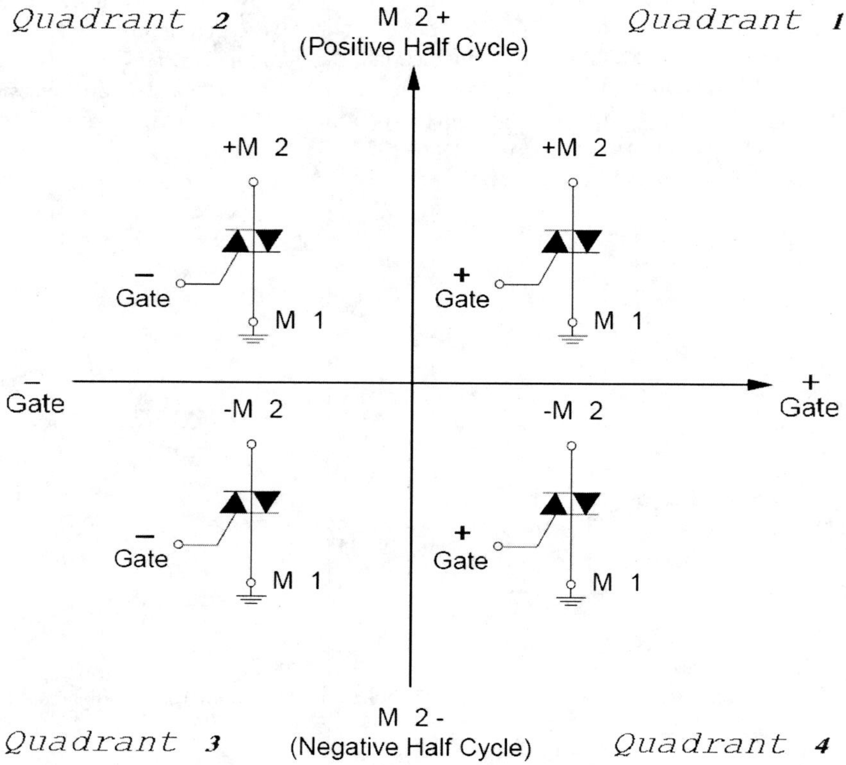

FIGURE 1.58 Triac quadrant definitions.

Like a standard thyristor, a triac can be controlled in different ways in real constructions of switching devices (Fig. 1.59).

It should be taken into account that *physically* the triac does not comprise two thyristors connected in parallel, as it is shown in Fig.1.57. It only *functions* as two inverse-parallel connected thyristors *on alternating current.* (Note: A*lternating Current;* a triac is not designed for work on DC and unlike a pair of inverse-parallel connected thyristors, does not operate very stably on DC).

Besides a triac there are some other variants of thyristors, like a "dinistor" (or "dynistor" in other transcription), for instance (Fig. 1.60a), which in fact is a standard thyristor without a gate outlet, being enabled when the voltage applied to it (between the anode and the cathode) is increased to the level of forward breakover voltage. Such devices are produced in Russia, but are not well known in the West. More popular are devices controlled by voltage without a gate based on a triacs, not thyristors. They are called a "diac" (Fig. 1.60b). Some firms produce semiconductor devices with a triac and a diac combined in its structure (Fig. 1.60c). Such devices are called a "quadric."

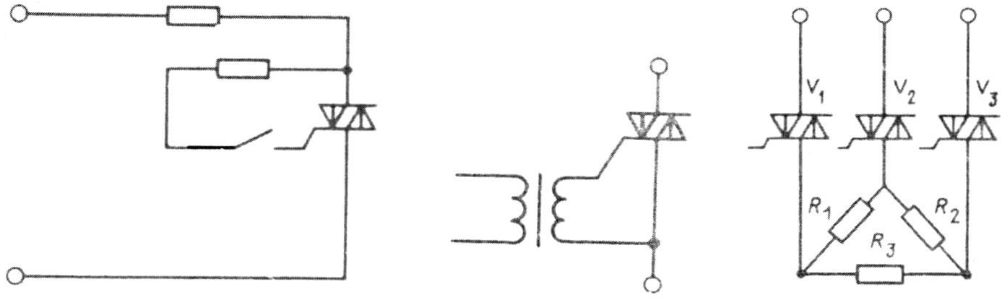

FIGURE 1.59 Some methods of triac control and its connection to a three-phase circuit.

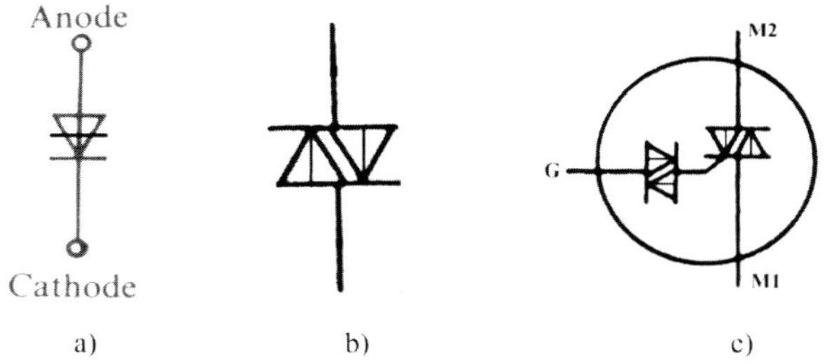

FIGURE 1.60 Different types of thyristors: dinistor (a), diac (b), quadrac (c).

The "sidac," Fig. 1.61 (also referred as "sydac" or bidirectional thyristor break-over diode) is one other semiconductor device of the thyristor family which is technically specified as bilateral voltage triggered switch. Its operation is similar to the diac, but in general, sidac have higher breakover voltages and current handling capacities than diac

FIGURE 1.61 Sidac's symbol on circuit diagrams.

2

Reed Switches

2.1 WHAT IS IT?

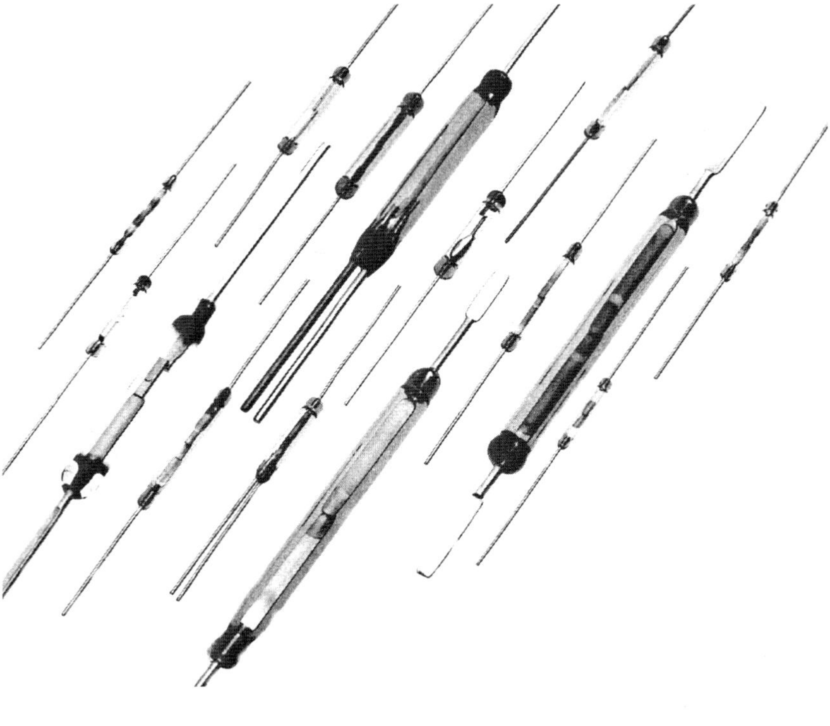

FIGURE 2.1 Different types of modern sealed reed switches.

Many engineers have come across original contact elements contained in a glass shell (Fig. 2.1). However not everyone knows that reed relays differ from ordinary ones not because of the hermetic shell (sealed relays are not necessarily reed ones), but because of the fact that in a reed relay a thin plate made of magnetic material functions as contacts, magnetic system, and springs at one time. One end of this plate is fixed while the other end is covered with some electroconductive material and can move freely under the effect of an external magnetic field. The free ends of these two plates directed towards each other, are overlapped for from 0.5 – 2mm and form a basis for a new type of switching device – a "hermetic magnetically controlled contact" (in Russian) or "reed switch" (in English). Such a contact is called a "magnetically controlled contact" because it closes under the influence of an external magnetic field, unlike contacts of ordinary relays which are switched with the help of mechanical force applied directly to them. The original idea of such a function mix, which was in fact the invention of the reed switch, was proposed in 1922 by a professor from Leningrad Electrotechnical University V. Kovalenkov, who lectured there on "Magnetic circuits" from 1920 until 1930. Kovalenkov received a USSR inventor's certificate registered under No. 466 (Fig. 2.2).

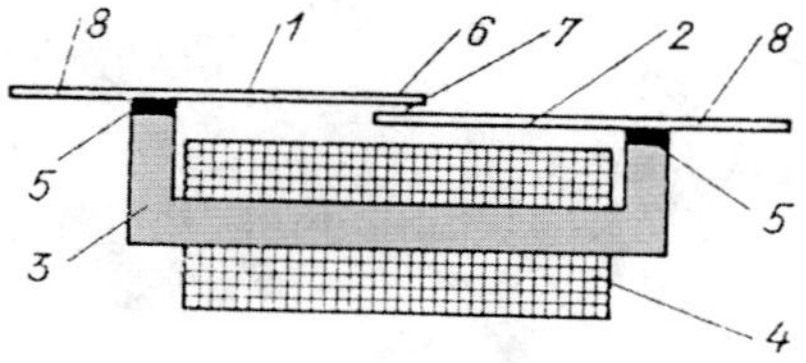

FIGURE 2.2 Kovalenkov's relay.
1 and 2 – contact elements (springs) made of magnetic material; 3 – external magnetic core (the core of the relay); 4 – control winding (external magnetic filed source); 5 – dielectric spacers; 6 – ends of contact elements; 7 – working gap in magnetic system and between contacts; 8 – contact outlets for connection of external circuit.

In 1936 the American company Bell Telephone Laboratories launched research work on reed switches. Already in 1938 an experimental model of a reed switch was used to switch the central coaxial cable conductor in a high-frequency telecommunication system, and already by 1940 the first production lot of these devices, called "Reed Switches," was released (Fig. 2.3).

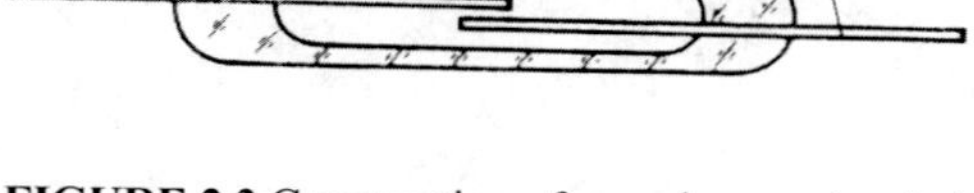

FIGURE 2.3 Construction of a modern reed switch
1 – contact elements (springs) from Permalloy; 2 – glass hermetic shell.

Reed switch relays (that is a reed switch supplied with a coil setting up a magnetic field – Fig. 2.4.), compared with electromagnetic armature relays which are similar in size, have higher operation speeds and durability, a higher stability of transient resistance, and a higher capability to withstand impacts of destabilizing factors (mechanical, climatic, specific), in spite of their relatively low switching power.

At the end of the 1950s some western countries launched construction of quasi-electronic exchanges with a speech channel (which occupied over 50% of the entire equipment of an exchange) based on reed switches and control circuits – on semiconductors.

In 1963 the Bell Company created the first quasi-electronic exchange of ESS-1 type designed for an intercity exchange. In a speech channel of such an exchange more than 690 thousand reed switches were used.

In the ensuing years the Western Electric Company arranged a lot production of telephone exchanges based on reed switches with a capacity from 10 up to 65 thousand numbers. By 1977 about 1000 electronic exchanges of this type had been put into operation in the USA.

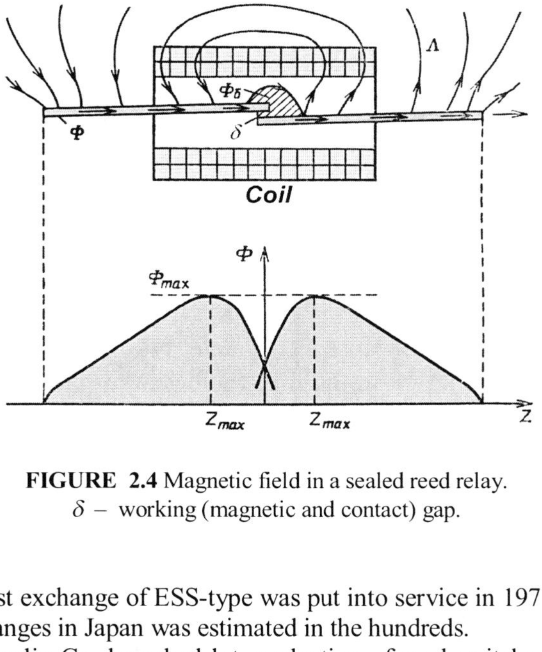

FIGURE 2.4 Magnetic field in a sealed reed relay.
δ – working (magnetic and contact) gap.

In Japan the first exchange of ESS-type was put into service in 1971. By 1977 the number of such exchanges in Japan was estimated in the hundreds.

In 1956 the Hamlin Co. launched lot production of reed switches and soon became the major producer and provider of reed switches for many relay firms. Within a few years this company built plants producing reed switches and relays based on them in France, Hong Kong, Taiwan, and South Korea. Under its licensed plants in Great Britain and Germany, it also started to produce reed switches in those countries. By 1977 Hamlin produced about 25 million reed switches, which was more than a half of all its production in the USA. Reed switches produced by this firm were widely used in space-qualified hardware, including man's first flight to the Moon (the Apollo program). The cost of each reed switch thoroughly selected and checked for this purpose reached $200 a piece.

In the former Soviet Union lot production of reed switches was launched in 1966 by the Ryazan Ceramic-Metal Plant (RCMP). Plants of the former Ministry of Telecommunication Industry (its Ninth Central Directorate in particular) were also involved in production of weak-current relays based on reed switches. At the end of the 1980s there were 60 types of reed relays produced in the USSR. The total amount of such relays reached 60-70 million a year. Economic crises in Russia led to a steep decline in production both of reed switches and of reed relays. In 2001 plants producing relays (those which were still working in Russia) ordered only about 0.4 million reed switches for relay production. Depending on the size of a reed switch, the working gap between contact-elements may vary within 0.05 – 0.8 mm (and more for high-voltage types) and the overlap of ends of contact-elements – within 0.2 – 2 mm. Due to the small gaps between contacts and a small total weight of movable parts, reed switches can be considered to be the most high speed type of electromagnetic switching equipment with a delay of 0.5 - 2 ms capable of switching electric circuits with frequencies of up to 200 Hz. Hermetic Switch, Inc (USA) produces the smallest reed switches in the world (Fig. 2.5).

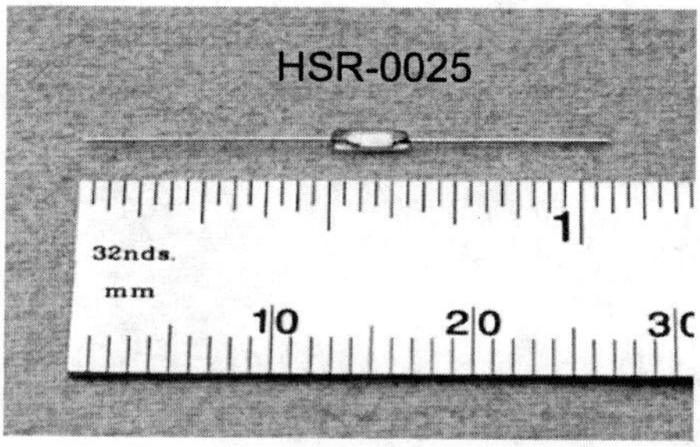

FIGURE 2.5 Smallest reed switch in the world, HSR-0025 type (Hermetic Switch, Inc).

The tiny oval shaped glass balloon of the HSR-0025 reed switch measures a mere 4.06 mm long, 1.22 mm wide, and 0.89 mm high. Maximum switching rating of HSR-0025: 30V, 0.01A, 0.25 W. Sensitivity ranges from 2 to 15 ampere-turns.

Bigger reed switches can switch higher current, as the contacting area of the contact-elements, their section, contact pressure, and thermal conductivity increase.

Most reed switches have round-shape shells (balloons), because they are cut from a tube (usually a glass one), the ends of which are sealed after installation of contact-points. Glass for tubes should be fusible with softening temperatures and coefficients of linear expansion similar to the ones for the material of contact-elements.

Contact-elements of reed switches are made from ferromagnetic materials with coefficients of linear expansion similar to glass. Most often it is Permalloy, an iron-nickel alloy (usually 25% nickel in alloy). Sometimes Kovar, a more high-temperature alloy is used. It allows application of more refractory glass for tubes (560-600°C) and as a result, more heat-resistant reed switches are obtained. To provide a better joint with the glass, contact-points are sometimes covered with materials providing better joints with glass than Permalloy. Sometimes contact-elements have more of a complex cover consisting of sections with different properties. Contact-elements may also contain two parts, one of which joins well with the tube glass and has the required flexibility, and the other one having the necessary magnetic properties. The contacting surfaces of contact-elements of average power reed switches are usually covered with rhodium or ruthenium; low-power reed switches designed for switching of dry circuits are covered with gold, and high power and high-voltage reed switches with tungsten or molybdenum. Covering is usually made by galvanization with further heat treatment to provide diffusion of atoms of a cover to the material. It can also be carried out by vacuum evaporation or other modern methods. Contact-elements of high-frequency reed-switches are fully covered with copper or silver to avoid loss of or attenuation of high-frequency signals, and after that the contacting surfaces are also covered with gold.

The tube of a medium or low power reed switch is usually filled with dry air or a mixture of 97% nitrogen and 3% hydrogen. A 50% helium-nitrogen mixture, carbonic acid, and other mixtures of carbon dioxide and carbonic acid can also be used.

Carefully selected gas environments effectively protect contact-elements from oxidation and provide better quenching of spark as low powers are being switched. Reed switches designed for switching of voltages from 600 up to 1000 V have a higher gas pressure in the tube, which may reach several atmospheres. High-voltage reed switches (more than 1000 V) are usually vacuumized.

The fact that there are no rubbing elements, full protection of contact-elements from environmental impact, and the possibility to create a favorable environment in the contact area, provides switching and mechanical wear resistance of reed switches estimated in the millions and even billions.

Reed switches which are in mass production and which are widely used in practice can be classified by the following characteristics:

1. *By size*:
- normal or standard reed switches with a tube about 50 mm in length and about 5 mm in diameter;
- subminiature reed switches with a tube 25 – 35 mm in length and about 4mm in diameter;
- miniature reed switches with a tube 13 – 20 mm in length and 2 – 3 mm in diameter;
- micro-miniature reed switches with a tube 4 – 9 mm in length and 1.5 – 2 mm in diameter

2. *By type of magnetic system*:
- neutral
- polarized

3. *By type of switching of electric circuit*:
- closing or normally open – A type;
- opening or normally closed – B type;
- changeover – C type

4. *By switched voltage level*:
- low voltage (up to 1000V)
- high-voltage (more than 1000 V)

5. *By switched power*:
- low-power (up to 60W)
- power (100 – 1000W)
- high-power (more than 1000 W)

6. By types of electric contacts:
- dry (the tube is filed with dry air, gas mixture or vacuumized);
- wetted (in the tube there is mercury wetting the surface of contact-elements)

7. By construction of contact-elements:
- console type (symmetrical or asymmetrical) with equal hardness of the movable unit (Fig.2.3. – main type of reed switches);
- with a stiff movable unit;
- ball type;
- powder type;
- membrane type, etc.

2.2 POLARIZED AND MEMORY REED SWITCHES

Polarized reed switches are those reed switches that are sensitive to the polarity of the control signal applied to the control coil, in other words to the vector direction of the magnetic field F (Fig.2.6). Such sensitivity is caused by an additional static magnetic field produced by a permanent magnet placed nearby (or by an additional polarized winding, which is rarer). The external magnetic field of the control signal may have the same direction as the magnetic field of the permanent magnet. In this case as their fluxes are summed up, causing operation of the reed switch, the sensitivity of the reed switch to the control signal increases considerably. If vectors of the magnetic fluxes have opposite directions, the resultant magnetic flux is so small that the reed switch cannot be energized.

One of the most important applications of such polarized reed switches is obtaining an opening (normally closed) contact out of a standard normally open one. In this case the magnet is selected in such a way that its magnetic field is enough to energize and hold a standard normally open reed switch in such a state. If the direction of the control magnetic field of the coil is opposite to the direction of the magnetic field of the permanent field, the total value of the magnetizing force affecting contact-elements will be less than their elastic force, and they will open, affected by these forces.

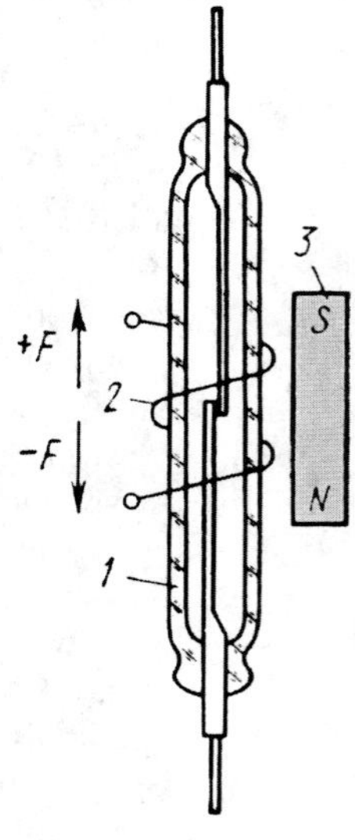

FIGURE 2.6a Polarized reed switch.
1 – neutral reed switch; 2 – control coil: 4 – polarized permanent magnet.

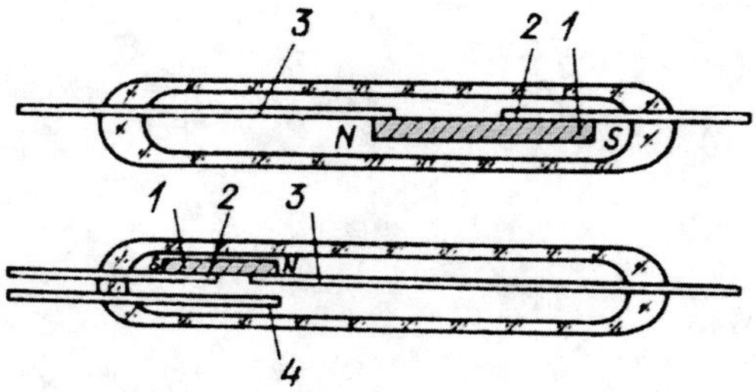

FIGURE 2.6b Polarized reed switches with internal arrangement of magnets. 1 – permanent magnet with electroconductive covering; 2 – stationary contact-element to which a magnet is welded; 3 – movable contact-element; 4 – second stationary contact-element.

As far as construction is concerned a magnet can be placed not only along the tube, but also outside it, as it is shown in Fig. 2.6.

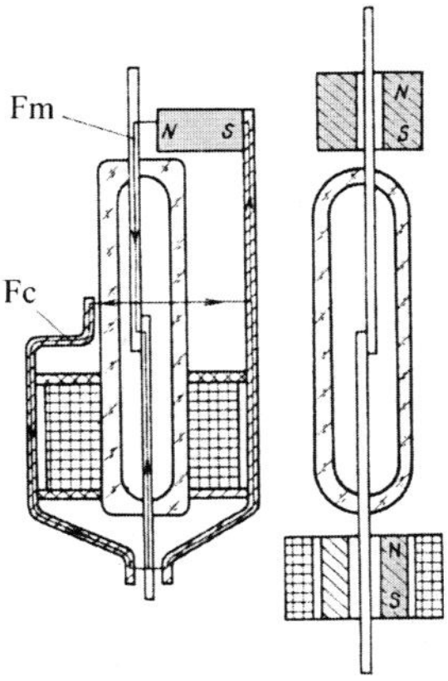

FIGURE 2.6c Polarized reed switches with external arrangement of magnets.
Fm – magnetic flux of the permanent magnet;
Fc – control magnetic flux.

There are a number of different constructions with original combinations of control coils and permanent magnets, some of which are shown in Fig. 2.6.

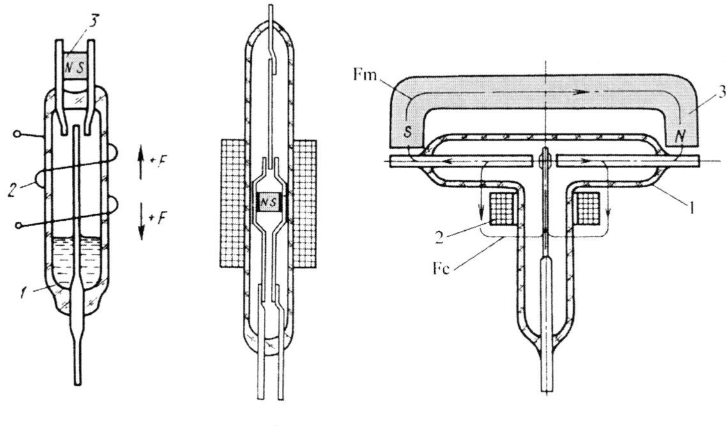

a **b** **c**

FIGURE 2.7 Three-position polarized reed switches:
a – mercury reed switch with an external magnet; b – dry reed switch with an internal magnet;
c – high-frequency reed switch.
1 – glass tube; 2 – control coil; 3 – permanent magnet with external insulating covering which can also be made of ferrite.

With the help of permanent magnets it is possible to produce a three-position reed switch with a neutral mid-position, which would switch this or that way under effect of the magnetic field of the control coil of this or that polarity (Fig. 2.7).

Using several control coils placed in different parts of a reed switch, instead of one, it is possible to produce reed switches capable of carrying out standard logical operations AND, OR, NOT, EXCLUSION, NOR (OR-NOT), NAND (AND-NOT), XOR (EXCLUSIVE-OR), etc. (Fig. 2.8).

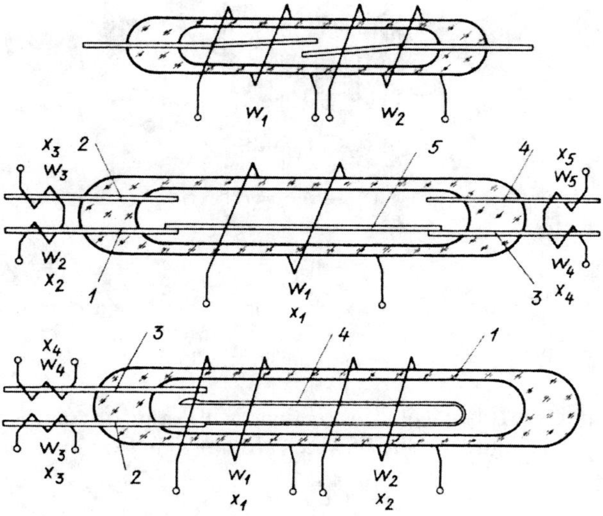

FIGURE 2.8 Multi-wound reed switches designed to carry out standard logical operations.

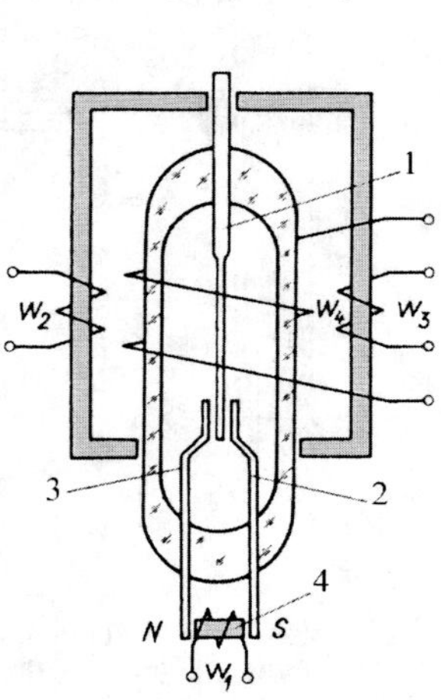

FIGURE 2.9 Combined switching logical device on a reed switch.
1, 2, 3 – contact-elements; 4 – permanent magnet.

If such multi-wound reed switches are combined with permanent magnets (Fig.2.9), it is possible to obtain quite complex functional elements with adjustable operation thresholds, and remote switching of certain options. The number of such combinations is practically endless. This allows designers to implement almost fantastic projects.

Taking into account that the reset ratio of reed switches is less than 1 (that is, for operation a stronger magnetizing force is needed than for release) one may try to choose a magnet of such a strength which is sufficient for operation of a reed switch, and at the same time capable of holding closed the contact-elements which have already been closed by the control coil field. In this case the reed switch is switched ON by a short current pulse in the control coil and remains in this state even after the control impulse stops affecting it (that is, it "memorizes" its state). The reed switch can be switched OFF by applying a control pulse of the opposite polarity to the coil.

Such a switching device, though in fact capable of operating, is not used in practice. There are several reasons for this. Firstly, such a device requires very accurate adjusting because the slightest excess of magnetizing force of the permanent magnet will cause spontaneous closing of a reed switch. If the magnetizing force is not strong enough, the reed switch will not remain closed after the control impulse stops affecting it. Taking into account great technologic differences between parameters of reed switches, magnets, and control coils, each device will require individual adjusting, which is impossible for mass production. That's why even a device adjusted beforehand to a certain temperature may malfunction at other temperatures.

In 1960 A. Feiner and his colleagues from Bell Laboratories published in the "Bell System Technical Journal" an article entitled "Ferreed – A New Switching Device," in which they suggested that the way to overcome these difficulties was by creating memory reed switches. The idea was that the permanent magnet should become a magnet only when it and the reed switch are affected by the control pulse of the coil. Other details were technical. They chose magnetic material with a medium coercive force, which could be magnetized during the time of affecting of the control pulse, and remain magnetized for a long time, until the pulse of the magnetic field of the opposite polarity affects it (such material is called *"remanent"*). This device, consisting of a reed switch and a ferrite element, was called *"ferreed"* (by the first letters of the words "ferrite" and "reed switch"). Later, for advertising purposes, some firms began to name devices operating on the same principle in a different way: *"remreed," "memoreed,"* etc.

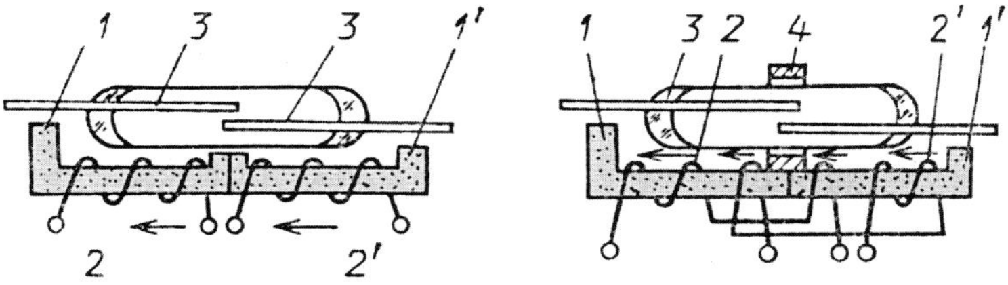

FIGURE 2.10 Ferreeds with two control windings
1 – core from remanent material; 2 and 2' – control windings; 3 – contact-elements; 4 – additional magnetic shunt.

It turns out that ferrite can be remagnetized for 10 – 50 μs while closing of contact-elements requires a time of 500 – 800 μs. This allows us to use very short pulses for ferreed control (in practice pulses with a reserve of up to 100 – 200 μs for closing are used). This means that contact-elements are not only held after magnetization of the ferrite, but also used for the contacts closing process where the remanent magnetic flux of the ferrite after the short control impulse stops affecting it.

It is obvious that ferreed with one control winding will be critical to the amplitude of the control pulse. When the amplitude of the switching current pulse (for switching OFF) is not high enough, the core in the control coil is not fully magnetized and the contacts will remain in a closed position.

If the control signal is quite strong the core can be reversely remagnetized and obtain the opposite polarity. In that case the contact-elements will remain in a closed position also. To avoid this two control windings are applied (Fig. 2.10).

The magneto-motive force of each winding is not enough to magnetize the core to the degree necessary for closing of the contact-elements. Only when switching-on current pulses of opposite polarity are applied to both windings is the total magnetizing force enough to magnetize the core, closing the contact-elements.

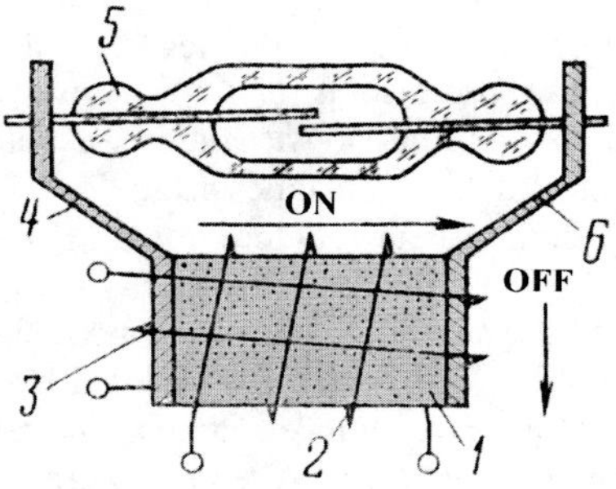

FIGURE 2.11 Ferreed with orthogonal control.
1 – core from remanent material; 2 – switching-on winding; 3 – switching-off winding; 5 – reed switch; 6 – magnetic core.

To open the contact-elements, switching-on current pulses of the same polarity are simultaneously applied to both windings. The polarity of magnetization of the halves of the ferrite core will be opposite, and both contact-elements are likely to be magnetized, therefore a repulsive force causing the contacts to open will arise between them. An additional shunt (4) made from soft magnetic material enhances configuration of the magnetic field in the overlap of the contact-points, and reliability of operation of the device.

In a ferreed with a so-called orthogonal control (Fig. 2.11), in order to change the state of the magnetization the vector turns by an angle of 90°, instead of 180° as in the previous case. The first such solution was patented by A. Feiner, from Bell Laboratories (USA patent No. 2992306). In his construction the magnetic flux of the winding (2) for switching ON passes through a magnetic gap between the contact-elements, and the magnetic flux of the winding (3) for switching OFF does not pass through the gap between the contact-elements, providing reliable switching OFF of the reed switch.

As in multi-wound reed switches, on ferreeds it is quite easy to implement single- or multi-circuit automation logic elements (Fig. 2.12). For example, in multi-circuit relays with a cross-shaped core (Fig. 2.12), there are 16 possible combinations of closed and opened reed switches, depending on what windings are switched ON. In some constructions one can enable or disable memorization options with the help of additional control signals (Fig. 2.13). In the construction described above a memory element of the external type is used.

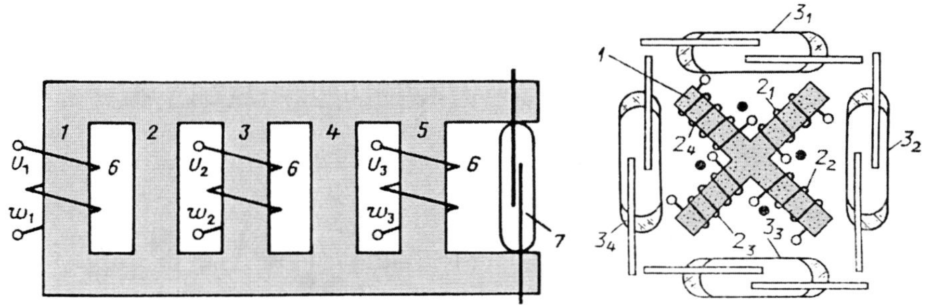

FIGURE 2.12 Automation logic elements on ferreeds.

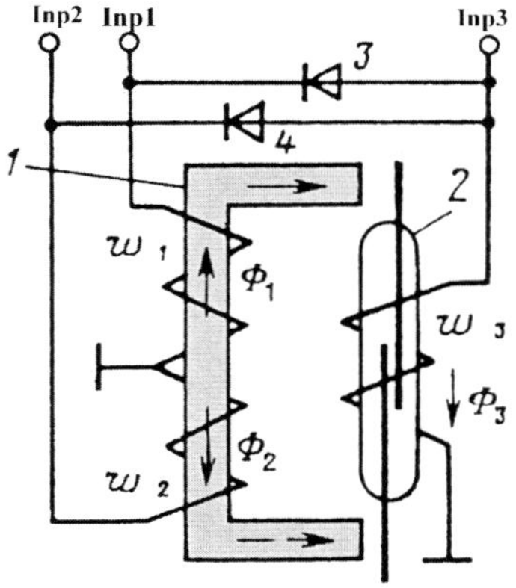

FIGURE 2.13 Device in which memorization option can be disabled.

Since the 1970s one can observe rapid development of ferreeds with internal memory, which were produced by Hamlin, FR Electronics, and Fujitsu. Their external design was practically identical to that of dry reed switches, but their contact-elements were made of special alloys, providing sealing of the reed switch after it was affected by the pulsed magnetic field. Thus no external elements are needed for such ferreeds.

Originally, contact-elements of such ferreeds consisted of two parts: an elastic one and a hard magnetic one (from remanent material), but there were also excess joints with increased magnetic resistance. Later on, hard magnetic alloys were invented and used, so that the contact-elements could be made more flexible and elastic enough. Such an alloy consists of 49% cobalt, 3% vanadium, and 48% iron, or of 30% cobalt, 15% chromium, 0.03% carbon, and the rest iron. There are also bimetallic contact-elements (USA patent No.3828828), the internal rod of which consists of an alloy of 81.7% iron, 14.5% nickel, 2.4% aluminum, 1% titanium, and 0.4% manganese, with the shell of the section made of an alloy containing 42% iron, 49% cobalt, and 9% vanadium.

2.3 POWER REED SWITCHES

The American branch office of the Yaskawa Company advertises its R14U and R15U type power reed switches, produced under the brand name "Bestact" (Fig. 2.14), which belong to the same class of reed switches with partially detached magnetic and contact systems.

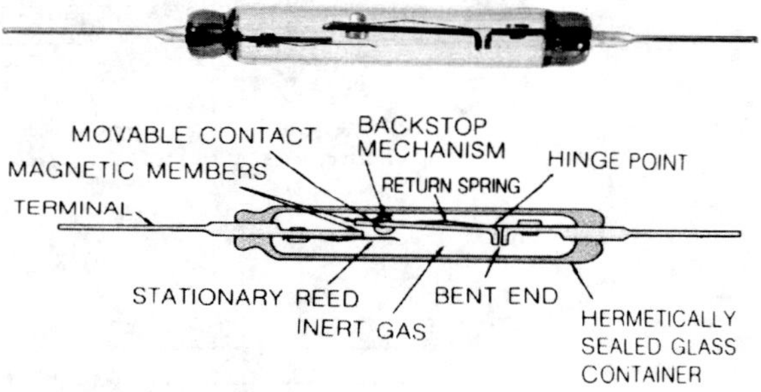

FIGURE 2.14a Power reed switch Bestact™ produced by Yaskawa Electric America Dimensions of glass tube: dia. 6 mm, length 37 mm.

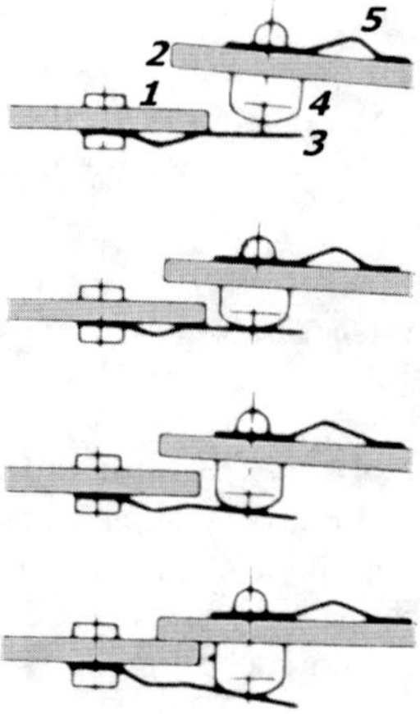

FIGURE 2.14b The closing/opening process of power reed switch Bestact™.
1 and 2 – main silver coated contacts; 3 and 4 – auxiliary contacts from tungsten; 5 – spring.

This reed switch has current carrying capacity up to 30A. It can break similar current (as emergency one) 25 times in an AC circuit with a power factor of 0.7. Switched power in the AC circuit is 360 VA (inductive load), maximum switched current is 5 A, maximum switched voltage is 240 V.

The electric strength of the gap between contacts is 800 V AC, mechanical life 100,000 operations for R15U and 50,000 for R14U. Operating and release time is 3 ms.

On the basis of such reed switches the Yaskawa Company produces a great number of different types of switching devices: relays, starters, push-buttons, etc.

Inquisitive readers can find many interesting descriptions of power reed switches in the author's previous book, *Electric Relays: Principles and Applications*.

3

High-Voltage Reed Relays

3.1 HV REED RELAYS FOR LOW CURRENT DC CIRCUITS

High-Voltage Reed Relays (HVRR) are a new type of high-voltage device designed by the author for automation systems (overload protection, fault indicating, interlocking of HV equipment) as well as for transfer of control signals from ground potential to HV potential (reverse connection).

FIGURE 3.1 High voltage relays RG-series for industrial and military applications.

The RG-series (Relays of Gurevich) consists of the following devices: RG-15, RG-25, RG-50, and RG-75, which are designed to operate under voltages of 15, 25, 50, and 75 kV DC, respectively (refer to Fig. 3.1).

The operation of these devices is based on the separation between the electric and magnetic electromagnetic field components. Each device is based on a magnetic field source (coil), connected in a high potential current circuit, a reed switch, and a layer of high voltage insulation, transparent for the magnetic component of the field and completely insulating the reed switch from the electric field component (Fig. 3.2 – 3.5). The current trip levels can be adjusted up to 50% (for each subtype).

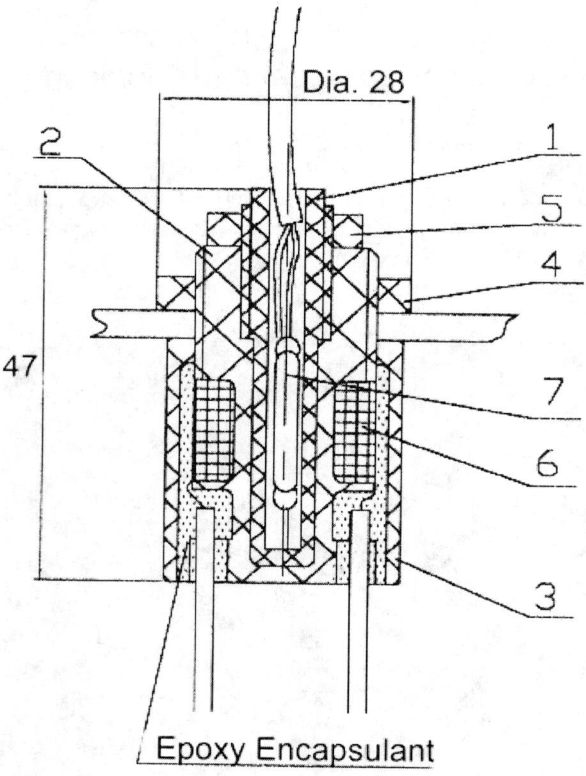

FIGURE 3.2 RG-15 series design for compact HV power supplies. 1 – moving insert; 2 – main insulator; 3 – external bushing; 4, 5 – nuts; 6 – coil; 7 – reed switch.

Material for all construction elements is “Ultem-1000.” Leads – HV Teflon cable (etched) 178-8195 type.

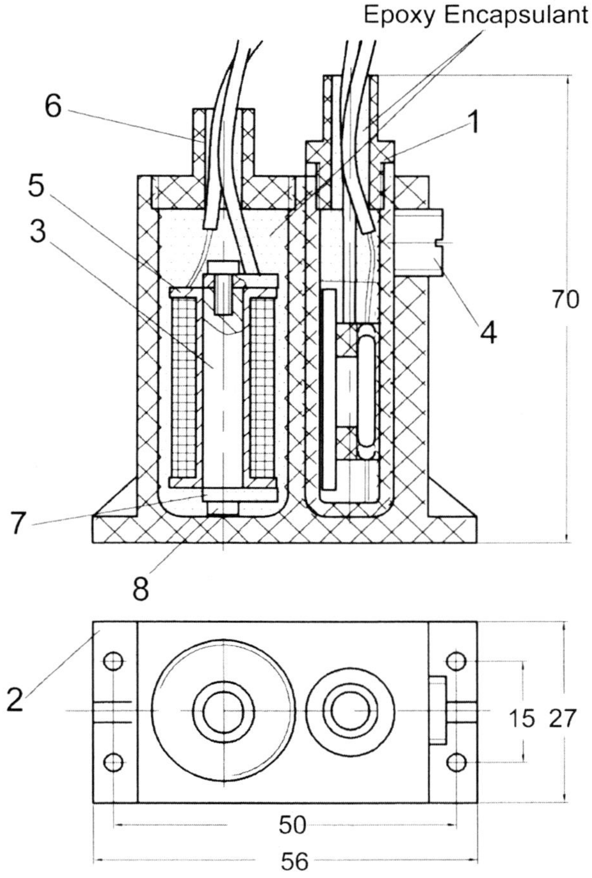

FIGURE 3.3 RG-25 series design for power lasers, industrial microwave ovens, medium power radar.
1, 6 – bushing; 2 – main insulator; 3 – ferromagnetic core; 4 – plastic screw; 5 – coil; 7 – pole.

The RG-75 (and RG-50) relay (Fig. 3.5) is comprised of the main insulator 1 formed as a dielectric glass, whose cylindrical part is extended beyond flange 2.

The flat external surface of bottom 3 of this glass smoothly mates with the extended cylindrical part 4 having threaded internal 5 and external 6 surfaces. The relay also includes control coil 7 with a Π-shaped ferro-magnetic core 8 located inside the main insulator and reed switch 9 located in an element for reed switch rotation through 90°, 10. This element 10 is formed as an additional thin-walled dielectric glass with walls grading into the bottom and mating with the inner surface of cylindrical part 4. These mated surfaces are coated with conducting material 11.

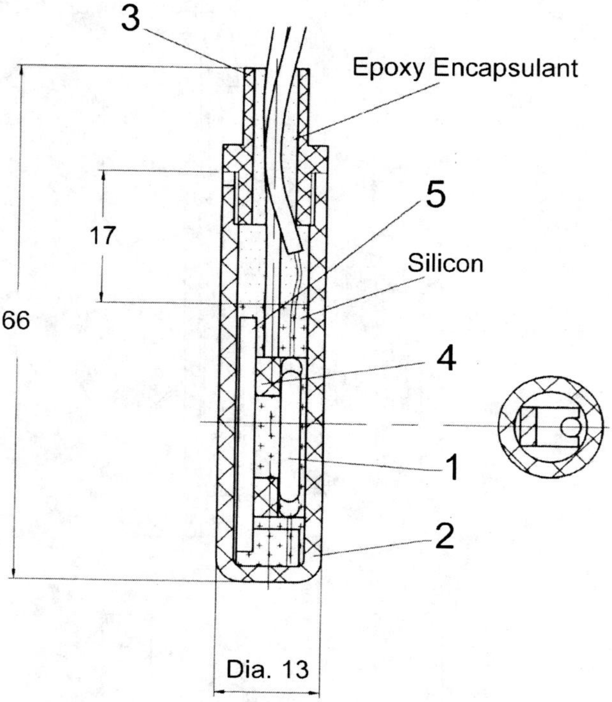

FIGURE 3.4 Revolving Assembly Part of RG-25.
1 – reed switch; 2 – insulator; 3 – bushing;
4 – support; 5 – ferromagnetic plate.

Reed switch outputs 9 are conveyed through additional insulator 12 formed as a tube extending beyond reed rotation element body 10. The lower end of this tube is graded into oval plate 13 covering the reed formed with the conducting external coating. Control coil 7 outputs are also conveyed through tube-shaped insulator 14 extending beyond the main insulator. The reed switch position fixation element is formed as disk 15 with a threaded side surface and a central hole with insulator 12 conveyed through it. External attachment of the device is effected with dielectric nut 16. Lower layer 17 of epoxy compound filling the main insulator to the control winding performs conduction by the addition of copper powder (60 –70% of the volume). The rest of the filling compound 18 has been made dielectric. Element space 10 is filled with the same dielectric epoxy compound.

The shape of the main insulator and the reed switch rotation element are chosen so that their mating surfaces, which contact with the conducting coating, do not form sharp edges emerging on the main insulator surface and, at the same time, provide for safe shunting of the air layer between them and removing the thin conducting sharp-edged layer from the design.

Significantly reducing the field intensity generated by the sharp outputs of the reed switch is achieved by adding one more tube-shaped insulator extending beyond the main insulator used to convey the reed switch outputs and executing the inner end of this tube as a plate with conducting coatings covering the reed switch.

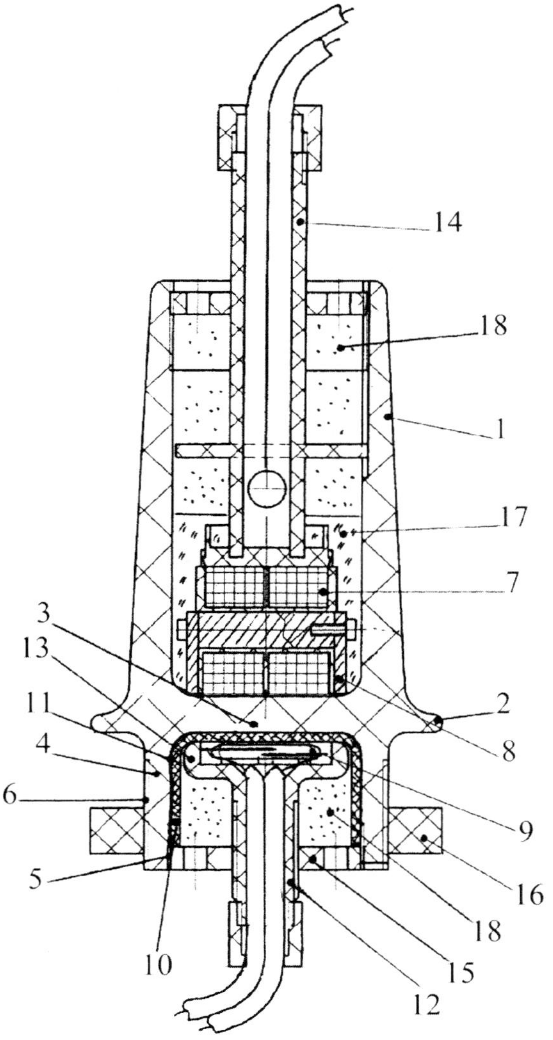

FIGURE 3.5 RG-75 and RG-50 series relay design.

Applying the lower layer of epoxy compound, which fills the main insulator conducting space (holding the control coil with a ferromagnetic core), thus reduces the intensity of the field generated by the winding outputs and neutralizes the action of the air bubbles remaining between the coil windings.

Implementing the reed fixation element as a simple threaded disk, which is threaded into the respective part of the main insulator, forces the reed rotation element. Use is made of an additional dielectric nut threaded on the appropriate part of the main insulator as an element of the relay external attachment assembly, and the main insulator flange is used as a stop for this attachment assembly.

Device operation is based on the action of the magnetic field of the control coil (penetrating through bottom 3 of high voltage insulator 1) to reed switch 9. When the reed switch threshold magnetic flux value is attained, it becomes engaged and appropriately switches the external circuits of the installation.

The reed switch engagement threshold value is adjusted by changing its position relative to the magnetic field source. This change is effected by rotation of element 10 with reed switch 9 by an angle of 90° relative to the poles of Π-shaped ferromagnetic core 8. The position of element 10 with the reed is fixed by forcing element 10 as disk 15 is screwed in.

Each device from this series (Table 3.1) functions as four separate devices:

- *current level meter in an HV circuit*
- *trip level adjustment unit*
- *galvanic isolation assembly between the HV and LV circuits*
- *fast response output relay in LV circuit*

Table 3.1 Main parameters of the RG devices.

RG Relays	**RG-15** (Subtypes **A – G)**	**RG-25** (Subtypes **A – G)**	**RG-50** (Subtypes **A – D)**	**RG-75** (Subtypes **A – D)**
Nominal voltage, kV	15	25	50	75
Test DC voltage 1 min, kV	20	35	70	90
Response currents range for subtypes, A	**A**: 0.01…0.02 **B**: 0.02…0.03 **C**: 0.03…0.05 **D**: 0.05…0.10 **E**: 0.10…0.20 **F**: 0.20…0.50 **G**: 0.50…1.00	**A**: 0.025…0.05 **B**: 0.05…0.10 **C**: 0.10…0.25 **D**: 0.25…0.50 **E**: 0.50….1.00 **F**: 1.00…2.50 **G**: 2.50…5.00	**A**: 0.25…0.5 **B**: 0.5… 1.0 **C**: 1.0… 3.00 **D**: 2.0… 5.0	**A**: 0.05…0.03 **B**: 0.15… 0.5 **C**: 0.25… 2.5 **D**: 1.0…. 5.0

Limit current in control circuit with 1 s duration	Ten-fold value of the maximal response current for each version			
Control signal power, W	0.2… 0.4	0.2…0.5	0.5	0.9
Maximal switching voltage in the output circuit, V: DC AC	600 400			
Maximal switching output circuit current, A	0.5			
Maximal switching output circuit power, W	25			
Maximal response frequency, Hz	100			
Maximal response time, ms	0.5…0.8			
Control circuit parameters for different versions: R, Ohm, L, mGH	15…1300 23…400	1…600 0.12…900	0.8…50 0.26…70	0.8…50 0.26…70
Maximal dimensions, mm	∅26×47	56×27×70	∅75×150	∅75×190
Weight, g	45	130	370	620

The devices withstand the action of external environmental factors according to MIL-STD-202 requirements:

- Operation temperatures range –55 to +85°C
- Cyclical temperature change in range –55 to +85°C
- Air humidity 87% at a temperature of 40°C
- Low air pressure with high voltage applied – 87 mmHg
- Vibrations resistance 10g at an oscillation amplitude frequency in range 10-500 Hz
- Repeated shock: 55 g, duration 2 ms, 10,000 impacts
- Single mechanical impacts: 30 g/s, ½ sinusoid with duration 11 msec.

The tests were carried out in the Environmental Engineering Center of RAFAEL (Haifa, Israel).

The overload protection system based on RG devices was tested at Optomic Lasers (Israel) as a part of powerful industrial CO_2 type laser with the output beam power equal to 1600W.

The total time of the protection system operation was 240 hours. During this time reoccurring emergency cases were simulated. The developed protection system proved to be reliable and effective.

In addition, a protection system was built and passed over to Elta Electronics Industries (Israel) for testing as a part of a powerful ship radar system.

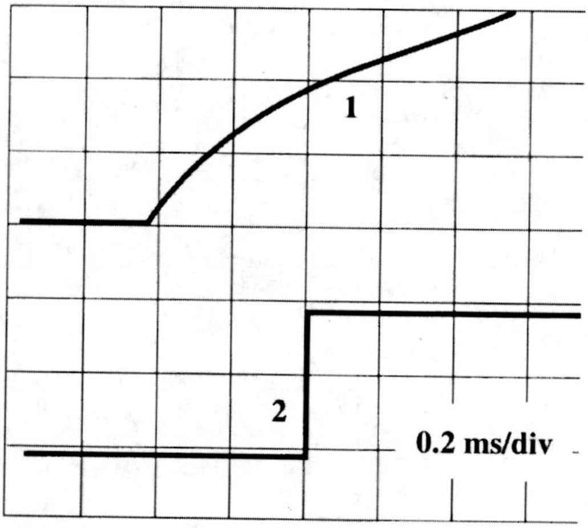

FIGURE 3.6 Time/Current Characteristics of RG-relay.
1 – current in a control coil;
2 – current in an output circuit.

In current overload protection systems, the RG Relays are usually connected as series circuits between the rectifier bridge and filter capacitor in the HV power supply, when the acting current does not exceed 10 A (pulsating current amplitude up to 30 A).

However, when the current is above 10A, they are connected by means of the shunt. The LV RG output is usually connected to the trip circuit of the LV solid-state contactor on the LV side of the power supply. The RG Relay is triggered when the current in the HV circuit exceeds the trip level (Fig. 3.6).

The RG response time (Fig. 3.7) depends, to a large degree, on the overload-to-nominal current ratio (I/I_{NOM}).

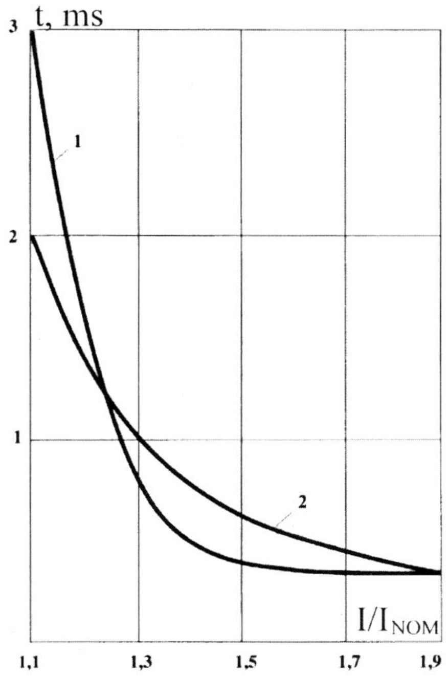

FIGURE 3.7 Dependence of operating time of the RG from current in the control coil.
I – operating current, I_{NOM} – nominal operating current
1 – RG for I_{NOM} = 0.25 A;
2 – RG for I_{NOM} = 3A.

3.2 HV REED RELAYS FOR HIGH CURRENT APPLICATIONS

The RG-24-bus device, Fig. 3.8, is designed to be used in overload protection units for 3 – 24 kV AC power networks, powerful electric motors, etc. The device output is a 100 Hz signal with 100 – 250 V DC or a standard "on-off" type relay protection signal.

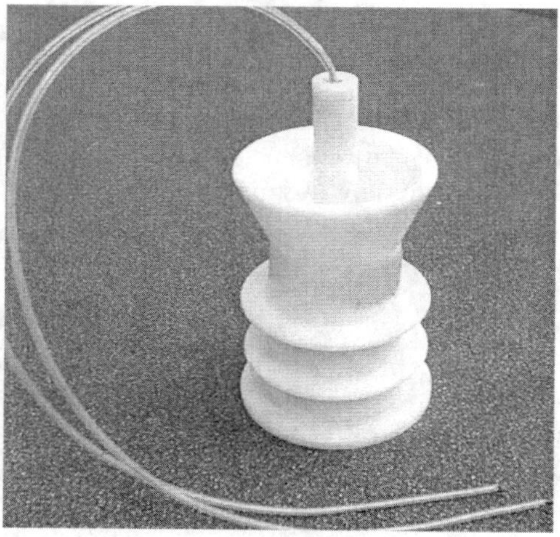

FIGURE 3.8 RG-24-bus device.

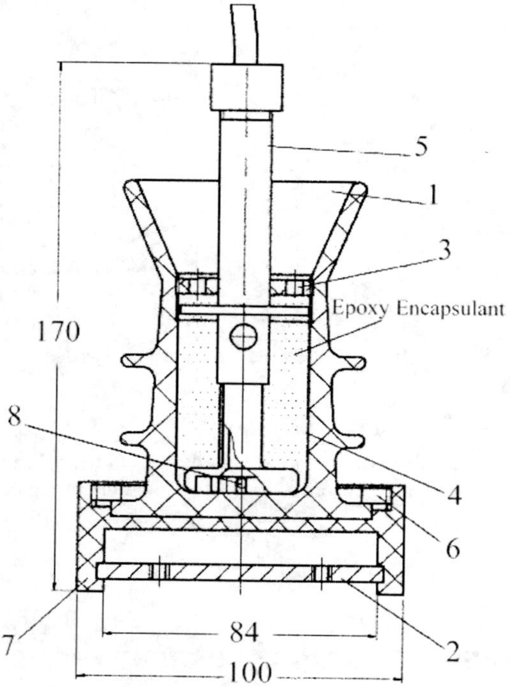

FIGURE 3.9 Construction of RG-24-bus device.
1 – main insulator; 2 – fixative plate; 3 – inside nut; 4 – semi-conductive cover; 5 – bushing; 6 – fixative nut; 7 – fastener; 8 – reed switch.

The device design (Figs. 3.9, 3.10) envisages its installation directly on a high voltage current-carrying bus or cable, as well as allowing for the possibility of wide range variations of the operation threshold (5 – 5000 A). Operate time – 1 ms.

FIGURE 3.10 Installation of RG-24-bus device on a high voltage bus bar.

The main advantage of these devices, as compared to those available on the market, is their possible direct installation on HV buses and output connection to low voltage control circuits. Medium voltage compact switchboard and switchgear cubicle systems (including SF_6 filled) can be significantly improved by using these devices.

Built-in fault detectors and other automatic systems can now be produced as factory-standard equipment and at affordable prices and are obtained without any alteration whatsoever of the HV equipment design.

A high voltage analog output current transducer can be made based on an RG-24-bus device (by using a coil instead of a reed switch) and is intended for direct measurement of current in HV bus bar or wiring. The output signal (voltage or current) is linear and directly proportional to bus bar current (Fig. 3.11).

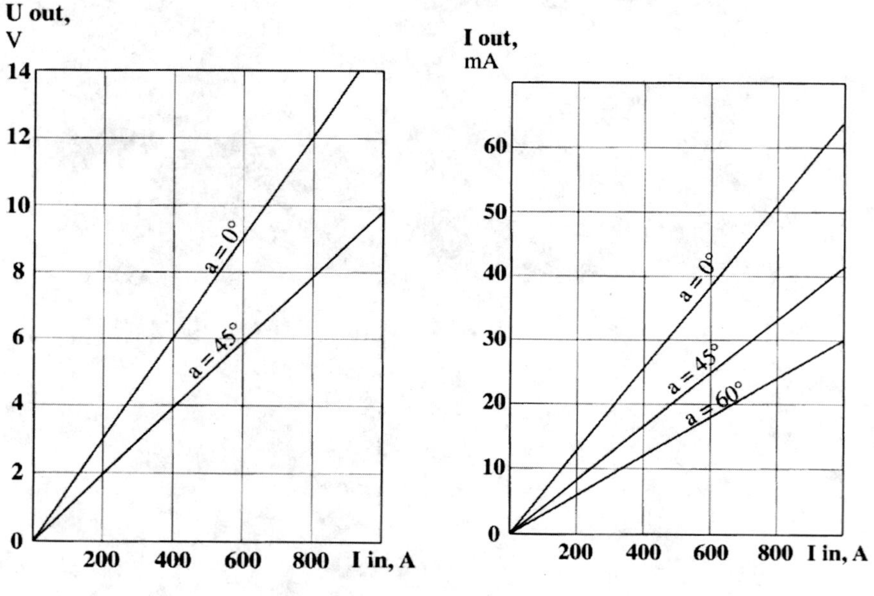

FIGURE 3.11 Typical characteristics of the analog output current transducer on an RG-24-bus device base.

The linear function slope is proportional to the transducer's relative angle position (angle "a") on the bus bar.

The transducer output may be connected to any electronic measurement device.

3.3 RELAY RESPONDING TO THE CURRENT CHANGING RATE

In circuits with load varying in a wide range a need in protection devices distinguishing between emergency and load increase regime is generated. For this purpose high voltage RG-relay with additional winding 3 and additional RC-circuit (4 and 5) can be used (Fig. 3.12).

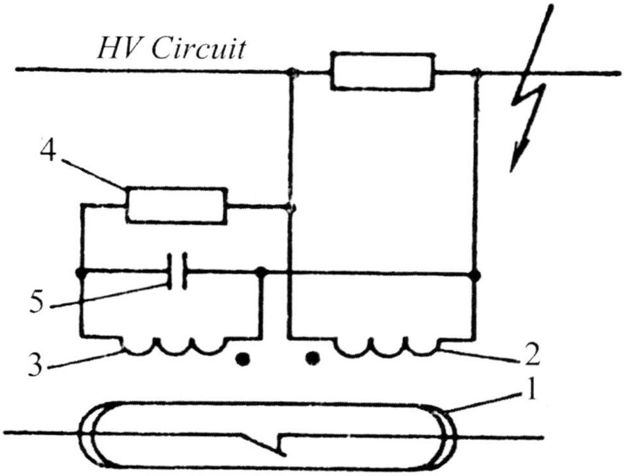

FIGURE 3.12 Circuit diagram for RG-relay responding to the current change rate.

In this device the magnetic fluxes generated in windings 2 and 3 are opposing and compensate for each other, moreover, all the changes in magnetic flux generated by winding 3 are delayed in time relative to the magnetic flux of main winding 2. When the current increase in the device input circuit is relatively slow, magnetic flux of additional winding 3 has the opportunity to compensate for the magnetic flux of main winding 2, thus reed switch 1 is not operated even if considerable current increase occurs.

Upon fast current jumps in the input circuit the magnetic flux in the main winding remains uncompensated and the reed switch has the time for a short-term switching before the magnetic fluxes in both windings become balanced. In order to provide for steady balance the reed switch can be connected to a thyristor or latching relay circuit.

3.4 DIFFERENTIAL HV REED RELAY

The reed switch used as a final control output element in this device (Fig. 3.13), is switched when a difference in the magnetic fluxes generated by the two differential windings occurs.

When internal insulator 5 is turned together with asymmetrically mounted reed switch 3, the relative position of the reed switch to the windings is changed: it is drawn to one of them and withdrawn from the other. As a result the reed switch sensitivity to the signals of either one of the inputs becomes different.

The device can be used also for summation of signals from two inputs.

3.5 REED-BASED DEVICES FOR CURRENT MEASUREMENT IN HIGH POTENTIAL CIRCUITS

In this device (Fig. 3.14) the reed switch is switched under the sum of magnetic fluxes generated by the current measured in high potential circuit (winding 2) and the magnetic flux in winding 3, which is smoothly increased when source 6 is activated.

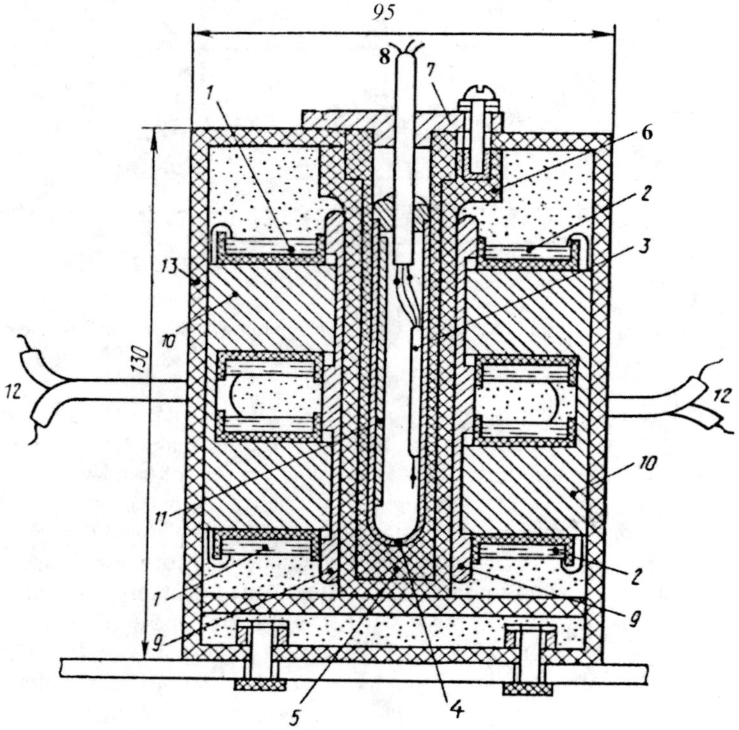

FIGURE 3.13 Differential HVRR.
1, 2 – differential windings; 3 – reed switch; 4 – aluminum capsule; 5 – internal insulator; 6 – immovable insulator; 7 – limb with fixing unit; 8 – reed switch leads; 9 – aluminum shields; 10 – ferromagnetic cores; 11 – ferromagnetic shield; 12 – differential inputs (HV cables); 13 – insulated body frame, filling with epoxy compound.

When the sum of these magnetic fluxes becomes equal to the magnetomotive force of the reed switch operation, it is switched and connects capacitor C to voltmeter 5.

Given the reed switch operation magnetomotive force, which is stable in time, the voltage measured with voltmeter 5 provides a quite unique current value in the high potential circuit. In order to revert the device to its initial position power supply 6 must be disconnected.

For continuous current monitoring in the high potential circuit a second type of device containing a low frequency saw-tooth voltage generator (6) as a power supply is used (Fig. 3.15).

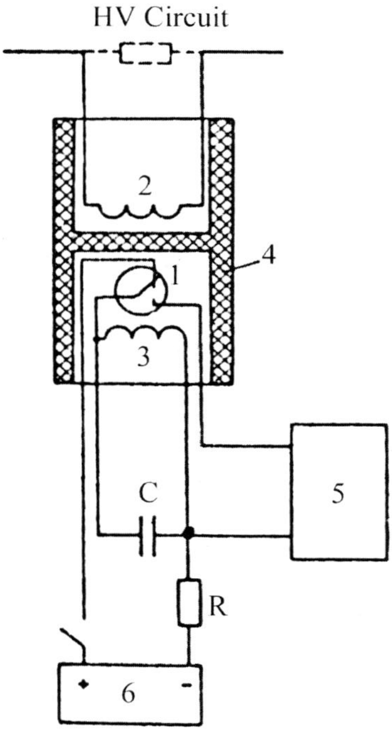

FIGURE 3.14 RG-relay based device for current measurement in high potential circuits.

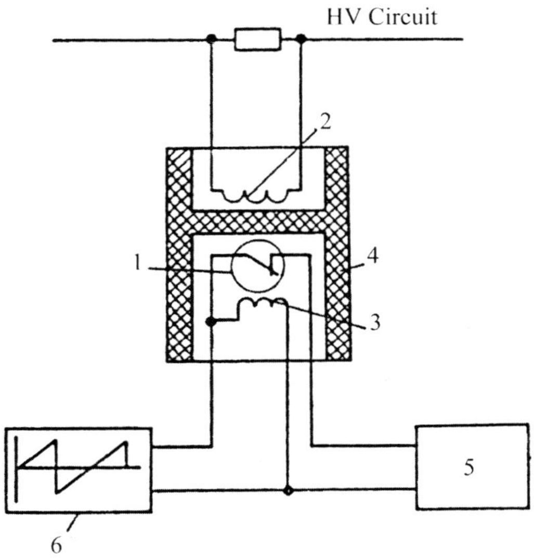

FIGURE 3.15 RG-based device for continuous current monitoring in high potential circuit.

3.6 SPARK-ARRESTING CIRCUITS FOR REED RELAYS

Without going into detailed classification of the principles of spark-arresting circuits designed for low voltage, low-powered DC contacts (according to switching type, composition of bypass circuit, the use of auxiliary switching contacts, etc.), two groups can be pointed out: active and passive spark-arresting (arc – arresting). The first group contains devices having active electronic elements such as transistors and thyristors. The other group offers the contacts shunting (bypassing) with spark-arresting circuit containing a capacitor and a resistor (linear, non-linear, or a combination of both).

The simplest way of passive arc arresting on contacts switching a DC circuit with inductance is to bypass them with an active linear resistance R, Fig. 3.16a.

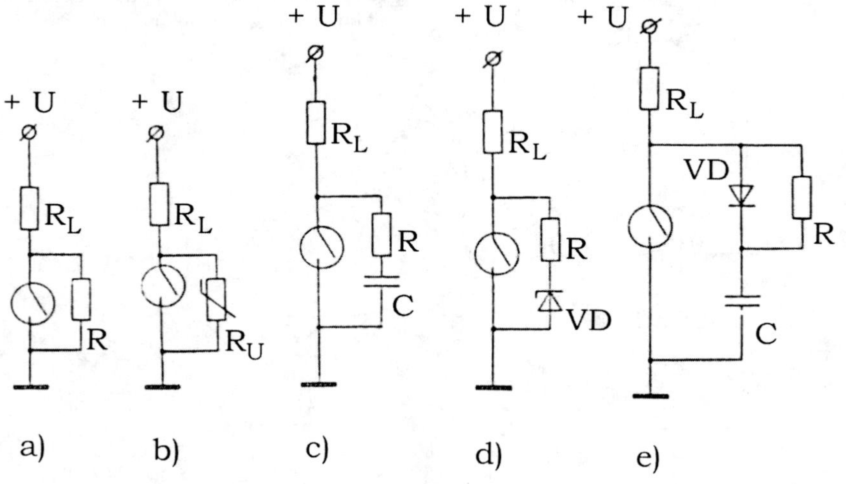

FIGURE 3.16 Embodiments of passive arc-arresting circuits for the DC circuit contacts.

In order to limit over-voltage at switching the R-value must meet the requirement:

$$R \leq \frac{U_{BR} R_L}{U}, \tag{3.1}$$

where U_{BR} – minimal value of breakdown voltage between contacts.

If control winding of electro-magnetic devices (relays, contactors) is used as the discussed contacts load (e.g. power amplifier), the following condition must be preserved for safe release of this device:

$$R \geq \frac{K_R R_L (1 - K_S)}{K_S}, \tag{3.2}$$

where K_R – return coefficient of electromagnetic device used as a load;
K_S – device release safety factor.

Substitution of standard coefficient values $K_R = 3$ and $K_S = 2$ to (3.2) will yield $R \approx 5\ R_L$. With lower resistance value the mentioned electro-magnetic device is not switched off upon contacts breaking, moreover the arc arresting becomes inefficient as the resistance is increased. In this case it is very advantageous to use a non-linear resistor – varistor (Fig. 3.16b), or Zener doide included in this circuit and acting in a similar way (Fig. 3.16d), featuring high resistance in the initial state which is automatically reduced with the increase of the voltage applied to them.

Often the scheme with resistor-capacitor circuit shown in Fig. 3.16c provides for adequate arc-arresting only for low values of R:

$$R < \frac{R_L U_{BR}}{U}$$

However, in this case the switching contacts will be subjected to overload caused by the capacitance discharge current at the instant of contacting and are due to be welded.

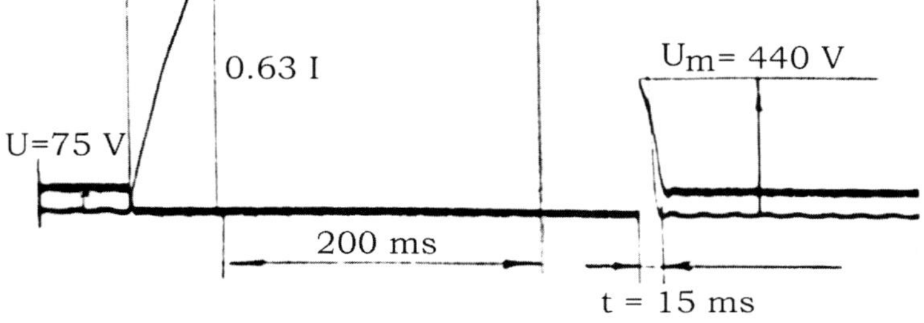

FIGURE 3.17 Switching process oscillogram in control coil circuit of powerful electromagnetic coil with arc-arresting circuit shown in Fig. 3.16b. Nominal supply voltage U = 75 V. Nominal current I = 2.5 A. Time constant of the switched circuit T = 54 sec.

The circuit illustrated in Fig. 3.16e is the most advantageous in the discussed group of circuits. At contact breaking capacitor *C* is rapidly discharged via very low resistance of direct diode junction *VD*, bypassing the contacts and preventing generation of electric arc over them. In contacting process the charged capacitor is discharged via resistance *R* that limits the discharge current to the value safe for the contacts.

The use of film-type capacitors with capacity of 1– 2 μF and working voltage of 630 V and resistor of about 500 Ohm allows for the use of reed switches for controlling of intermediate relay windings, contactors with nominal voltage up to 220 V DC.

Experimental study of this circuit with high-inductivity big electromagnetic coils used as a load showed that quick and efficient arresting of the arc process across the contacts at the instant of contact breaking results in considerable overloads across the load (in this case 440 V at power supply voltage of 75 V), Fig. 3.17. Therefore in this case one varistor is to be connected in parallel to the load and the other one – in parallel to the capacitor (Fig. 3.18).

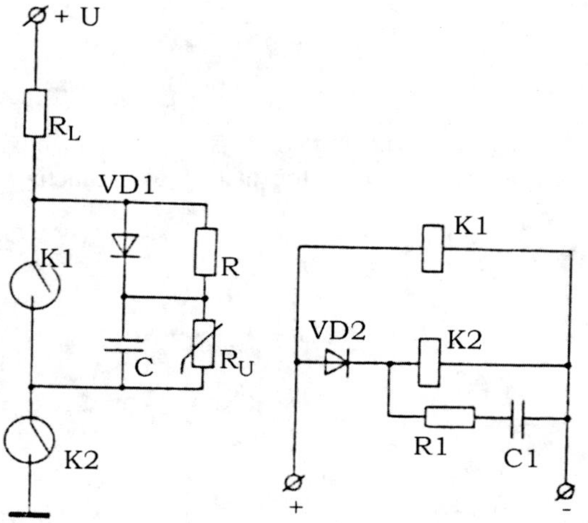

FIGURE 3.18 Arc-arresting device circuit for switching of high-power inductance loads.

If for some reason (say for safety reasons) the varistor cannot be left permanently connected in parallel to the contacts, an auxiliary relay K2 can be connected in the device, engaged with a certain delay relative to switching contacts K1. This delay is provided with the standard R1C1 circuit.

If necessary, additional relay R2 with an RC-circuit must be connected in the active arc-arresting device for safe galvanic decoupling of load circuit (Fig, 3.19).

In the device illustrated in Fig. 2.22a at contact K1 breaking the VT transistor collector-emitter junction is bypassed along with capacitor C_B charging circuit which provides for current flow in the transistor base, hence it becomes opened until saturation is reached, shunting the contact during switching. After the capacitor is charged the transistor is cut off (since the base current flow ceases) and contact K2 can be cut off without current.

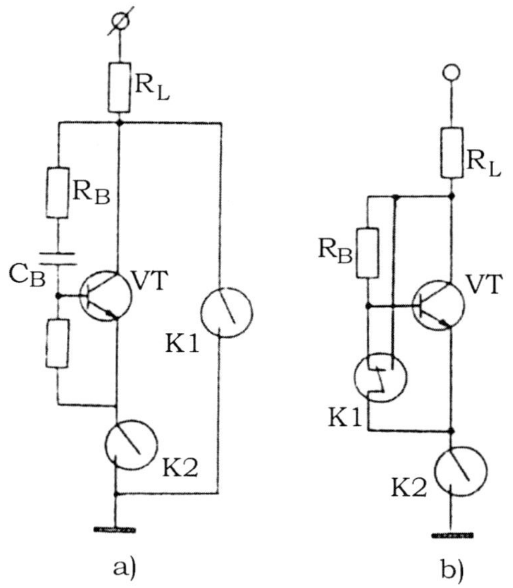

FIGURE 3.19 Active type arc-arresting device circuits.

Circuit illustrated in Fig. 3.19b does not have a capacitor. When a control signal is generated, contact K2 is closed and then contact K1 is switched, allowing for current flow in the load. During this time period the transistor is cut off because of collector-emitter junction shunting. As signal to control coil K1 is removed the contact is returned to its initial state in which the transistor is also cut off, then contact K2 is disconnected with very low current (limited with high resistance resistor R_B) de-energizing the load. In this device the transistor is activated only during contact K1 switching, providing for arcless commutation.

In both the examined circuits the transistors are activated in single short pulse regimes and at low commutation frequency can be operated without radiators, however, when working voltages are increased to 250 V, transistors with allowed collector-emitter voltage of 700-900 V shall be used; moreover at switching of high inductivity load it must be bypassed by a varistor.

4

Elementary Function Modules

In this chapter the elementary functional modules of automation devices are described. They are implemented on discrete electronic components: high-voltage transistors and thyristors, and powerful reed switches. Today it is somewhat difficult to find the descriptions of these devices since for the most part they are implemented by integrated microchips. However, as it turns out, interesting functional modules may be constructed without resorting to microchips, power supplies for them, optocouplers at their input and output, amongst other difficulties. In fact, it is possible to simplify automation devices significantly and to increase their reliability.

The modules described in this chapter certainly have their own meaning; however, as we think, their most important purpose is to present the reader their principles of construction, such that the reader may use them to design a desired device having necessary characteristics.

4.1 SWITCHING DEVICES

Switching devices (relays, contactors, switches) are the most widespread units found in automation devices. Combining them with highly sensitive miniature reed switches able to response to the coil magnetic field with low current or to the sensor with permanent magnet with powerful semiconductor elements, one may achieve very effective results. One gets simple and reliable devices ensuring load switching without arcing. When powerful load switching with current of dozens of amperes by means of simple n-p-n transistors of middle power is needed, the transistors are connected in parallel, as shown in Fig. 4.1.

Resistors, R_E, are necessary for balancing the current distribution in the parallel-connected transistors having a natural dispersion of parameters. The value of these resistors is selected so that the load current maximum value, voltage drop at each resistor, is not less than 0.2 V.

It is easy to calculate that with a load current of, for example, 100 A, each of these resistor's capacity will have power of about 4 W. In order to achieve a reliable device and to decrease the resistor's temperature one has to select their power to be 3 – 4 times more than the calculated dissipated one.

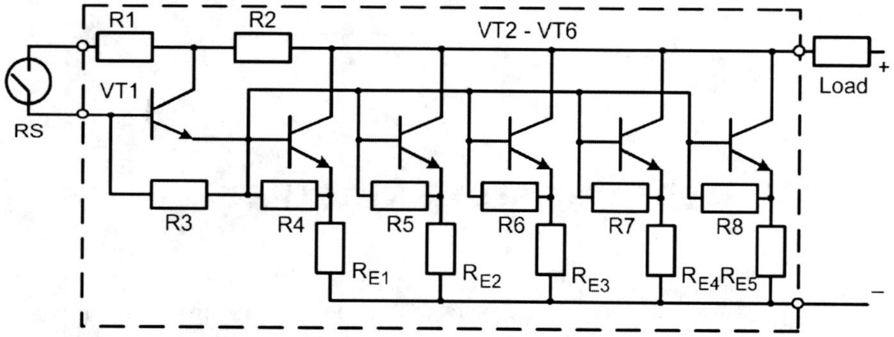

FIGURE 4.1 High power hybrid reed-transistor switching device on bipolar *n-p-n* transistors.

In addition, one has to take into account that the resistance of these resistors is very low (in our example only 0.01 Ω) and more likely one will have to make such resistors on his own of high-ohm wire or tape.

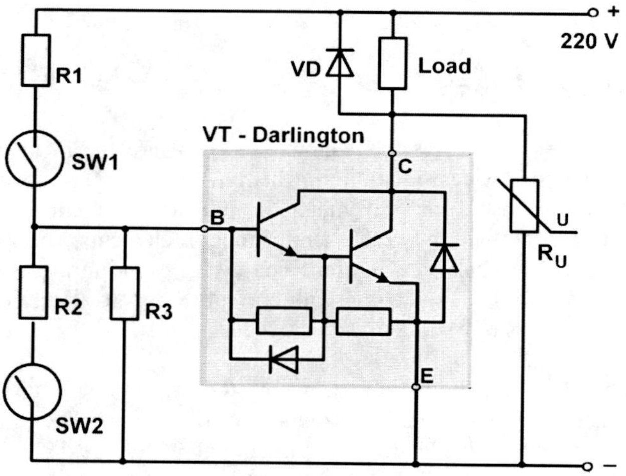

FIGURE 4.2 Hybrid switching device with Darlington transistor.

Powerful bipolar transistors suitable for functioning in industrial electrical equipment with a voltage of 220 V, i.e., sustaining the voltage of 1000 V, have very low gain (usually not more than 5 – 8) making it impossible to use low-current reed switches for powerful loads control. Darlington transistors have increased amplification coefficient. Their structure consists of two transistors and other elements, Fig. 4.2. Unfortunately, on the electronic components market one may find only several types of high-voltage Darlington transistors suitable for functioning in industrial equipment at voltage of 220 V. First of all, they are transistors of BU808DF1 (BU808DFH) type produced by STMicroelectronics and NTE2558 (NEC Electronics). They have an inductive load making it necessary to shunt them with diodes as shown in the circuit diagram. In this case one should select a varistor, R_U, with reduced clamping voltage level (not more than 500 – 600 V).

A hybrid analog of a powerful arcless changeover contact suitable for functioning at voltages of 125 – 250 V is shown in Fig. 4.3. In the device high-voltage Darlington transistors are used.

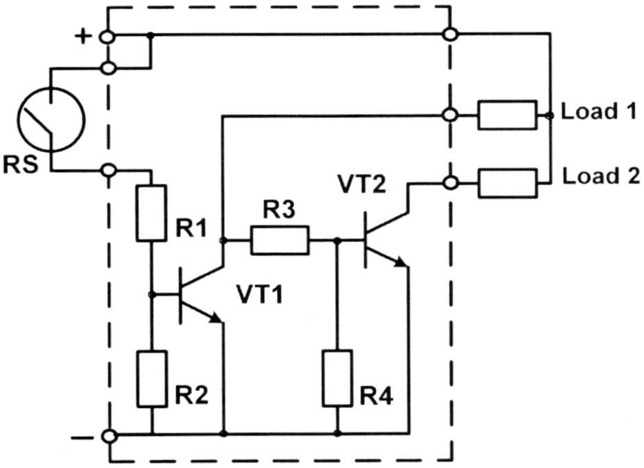

FIGURE 4.3 Hybrid analogue of the changeover switch

When the reed switch is opened, the VT1 transistor is turned off, first consumer (Load 1) is de-energized and VT2 is turned on by low leakage current flowing through the load number 1. Load number 2 is turned on.

When reed switch is closed, the VT1 transistor is turned on and it turns on load number 1 simultaneously with short-circuiting the VT2 transistor base circuit. The VT2 is turned off and it turns off load number 2.

When the load is inductive, they have to be shunted with opposite connected diodes. Transistors have to be overvoltage-protected by varistors.

IGBT transistors have much better switching parameters than simple bipolar ones. Also, they do not require high control current. In general, IGBT is the transistor, which is a hybrid of a bipolar transistor and field-effect transistors with insulated gates, and is theoretically controlled by electric field (like field-effect transistor), but not by the control current.

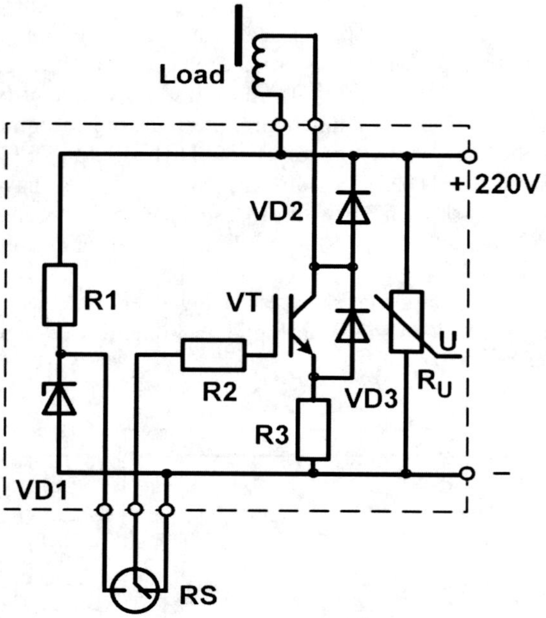

FIGURE 4.4 Switching changeover device on IGBT.

However, practice shows that a fairly insulated gate circuit has significant capacity. So to turn on the transistor one needs a current pulse to charge it. To turn the transistor off one has to discharge this capacitance, otherwise the transistor remains in the open state even in the absence of the gate trigger voltage. In order to discharge the inner capacitance quickly it is necessary to provide the gate with an external voltage that is more negative than the voltage on the emitter. Usually, to turn the IGBT transistor on, one uses on the gate a positive voltage +15 V relatively to emitter, and to turn it off, a negative voltage -8 V. Devices forming such heteropolar pulses are called drivers of IGBT transistors and are produced by many companies as small separate units. These drivers are meant to ensure the transistor turning on and off with high voltage in different invertors and generators. In most cases one does not need to use these drivers as switching amplifiers controlled by auxiliary relay contact or reed switch.

In circuit diagram Fig. 4.4 the supply of the heteropolar voltages (in regards to the emitter) to the gate is ensured by a changeover reed switch. The turned off gate is constantly short-circuited to negative of the power supply through a low-ohm R2 resistor. When the reed switch actuates, a positive voltage of 15 V limited by the Zener diode, VD1, is applied to the gate. With this, through the limiting R2 resistor the charging current impulse is limited to the level acceptable for reed switch flows. The limitation of the charging current naturally increases the period of transistor turning on and heat loss; however, in this concrete application, unlike with the high frequency inverter, it is not important as we are not faced with a nanosecond range and switching frequency of hundreds of kHz. Upon the transistor turning on, through resistor R3, the load current with resistance of several ohms will flow and create a voltage drop of not more than 5 – 8 V, i.e., increasing the emitter potential by 5 – 8 V in relation to the minus of power supply. Due to this resistor when reed switch returns to the initial position, on the gate turn off a negative voltage appears which turns off the transistor. R3 resistor has to be selected with respect to the rating and power appropriate to the concrete load.

For example, for a load with 5 A current, the resistance of the resistor has to be about 1 Ohm and the power of the resistor has to be approximately 3 – 5 times more than the calculated dissipated power, i.e., in the given case about 25W. Often the VD3 protective diode is already installed in the body of the IGBT transistor. In this case its application is not needed. VD2 diode preventing reverse voltage surge on the inductive load by transistor turning off is meant for voltages not less than 1000 V and currents of not less than 1A. Finally, the whole device has to be properly protected against over-voltage by varistor R_U, with adequate clamping voltage.

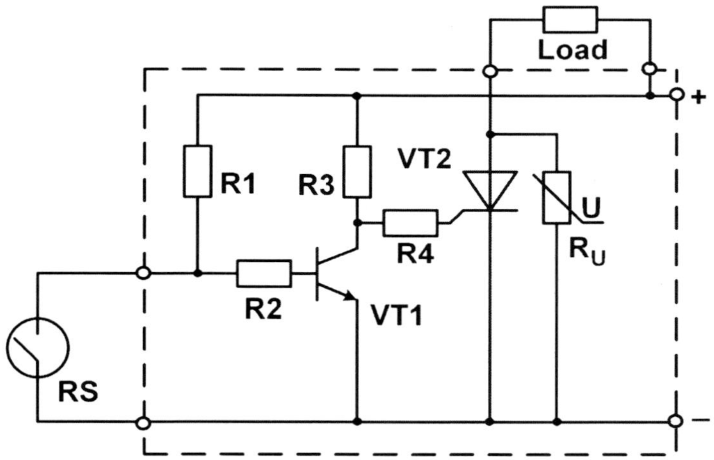

FIGURE 4.5 Hybrid reed-transistor-thyristor switching device.

Thyristors have even higher switching capacity and this makes them attractive for switching devices. In the circuit diagram Fig. 4.5 with a nominal voltage of 220 – 250 V the thyristor should be high-voltage and the transistor low-voltage as when it is turned off it is shunted with the low resistance of the thyristor's gate circuit (resistor R4 is as low as 50 – 100 ohm).

The transistor is turned on when the reed switch opens and it shunts the thyristor gate circuit with its low resistance, preventing the thyristor from activating. The transistor is turned off (and de-shunts the gate circuit) when the reed switch closes; as a result, the control voltage appears on the thyristor gate, turns it on and energizes the load. The R3 resistor has to ensure necessary current in the thyristor gate circuit (50 – 100 mA) and it depends on the level of supply voltage. A small disadvantage of this scheme is the permanent flow of low current through the transistor (about 50 – 100 mA).

In classic DC switching devices using two thyristors, one of them turns the load on and the other turns the first thyristor (and the load, as a result) off forcibly by means of discharging a previously charged capacitor as described in Chapter 1 (see Fig. 1.50). A practical circuit diagram implementing this operating algorithm is shown in Fig. 4.6.

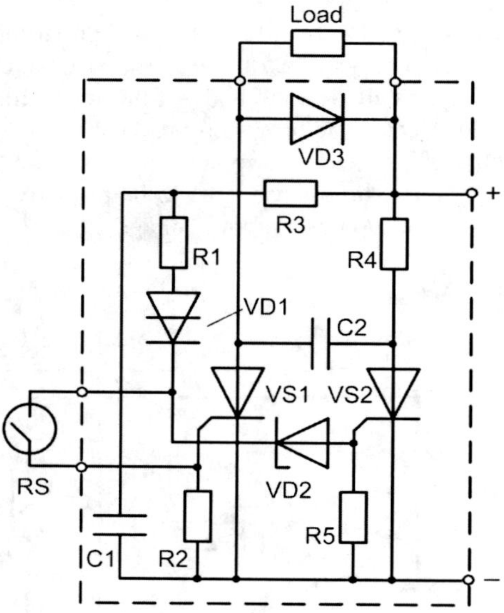

FIGURE 4.6 DC switching device on thyristors.

The main power thyristor VS1 (Fig. 4.6) is controlled by a reed switch and the auxiliary thyristor VS2 by the dynistor VD1. The supply voltage feed, C1, is charged until the breakdown voltage of dynistor VD1 resulting in thyristor VS2 turning on. Under the charging rate through resistor R3 and the discharge rate down the circuit: R1, VD1 dynistor, Zener diode VD2, and VS2 thyristor gate, capacitor C1 is charged up to a voltage equal to nominal voltage of the Zener diode VD2. At that, VS2 thyristor and VD1 dynistor are kept turned on. The device is in the waiting mode. The switching capacitor C2 is temporally discharged.

With the reed switch RS being closed, a new circuit of capacitor discharge appears through the VS1 thyristor gate. Resistance of this new circuit of C1 capacitor discharge is much lower than it was until the reed switch was closed because of the exclusion of the series-connected Zener diode; this is why the reed switch is closed. The C1 capacitor discharge current impulse is formed through the VS1 thyristor gate.

The thyristor is turned on thereby turning the load on and charging the capacitor C2 through resistor R4. At this point, thyristor VS2 is closed as during capacitor C2 charges its anode circuit is shunted with the power thyristor VS1 and the gate circuit is de-energized due to the closed Zener diode VD2.

The reed switch RS being turned off, the thyristor VS1 gate circuit is disconnected and the thyristor VS2 gate recovers. VS2 is activated according to the above-described algorithm, but as capacitor C2 is fully charged and the thyristor VS2 has been activated, it discharges voltage of reverse polarity on thyristor VS1 and this leads to its turning off and returning the device to its initial state. It is necessary to choose resistances of resistors for concrete supply voltage and concrete elements in order to ensure the operating algorithm as described.

Another variant of a simple but very reliable switching thyristor amplifier for reed switch or simple auxiliary relay contact is shown in Fig. 4.7.

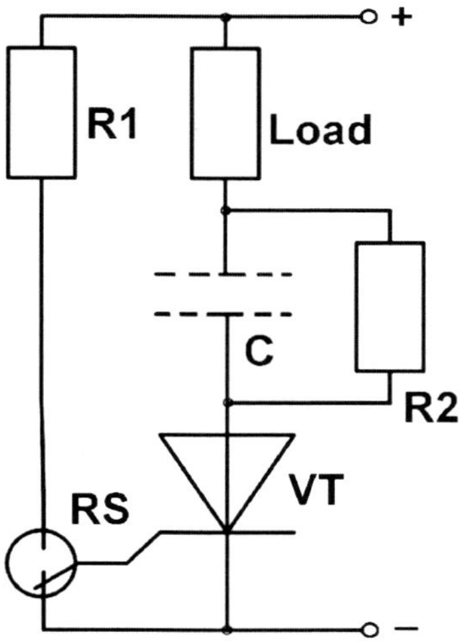

FIGURE 4.7 Very simple but reliable switching device.

Enhanced reliability and the immunity from interference of this device are provided due to the fact that the thyristor gate in the waiting mode is short-circuited to the cathode. One may profitably use the scheme when the reed switch is situated near the thyristor, in the same case. The scheme cannot be used if the reed switch (or auxiliary relay contact) is connected to the thyristor by long wires (several meters) due to the risk of spontaneous thyristor activation by the interferences in the long wires connected to the thyristor gate or its non-turning off when these wires have resistances that are too high.

If one adds to the circuit a capacitor C of large capacitance and an extra R2 resistor, the switching device turns into an impulse one: the reed switch being closed, the thyristor is activated only for the period of capacitor C charge, after this it turns off. After capacitance discharging through the discharging resistor R2, the circuit returns into its initial state (in same reed switch condition). In the impulse mode, the device allows controlling of low-power loads, for example, activation of an auxiliary relay or contactor. The impulse width is directly proportional to the C2 capacitance and inversely proportional to load capacity.

AC switching prevents problems connected with the necessity to turn the thyristors off by force (see Chapter 1).

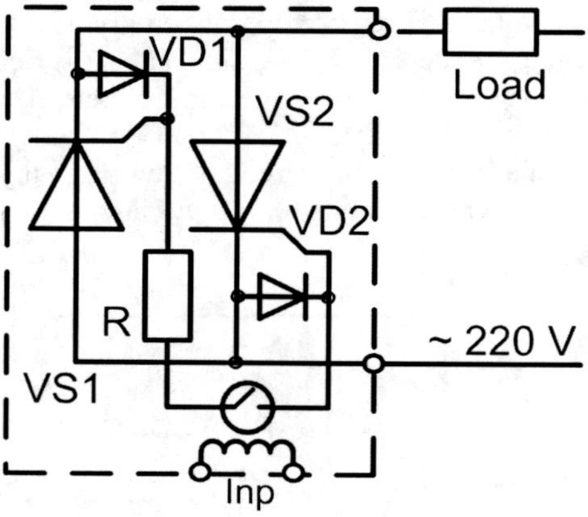

FIGURE 4.8 Simplest AC switching device.

This is why thyristors are widely used in AC switches. In the simplest case the switch consists of two powerful inverse-parallel connected thyristors, with gates connected to one another through the limiting resistor and controlling contact of reed switch or low-power auxiliary relay (Fig. 4.8).

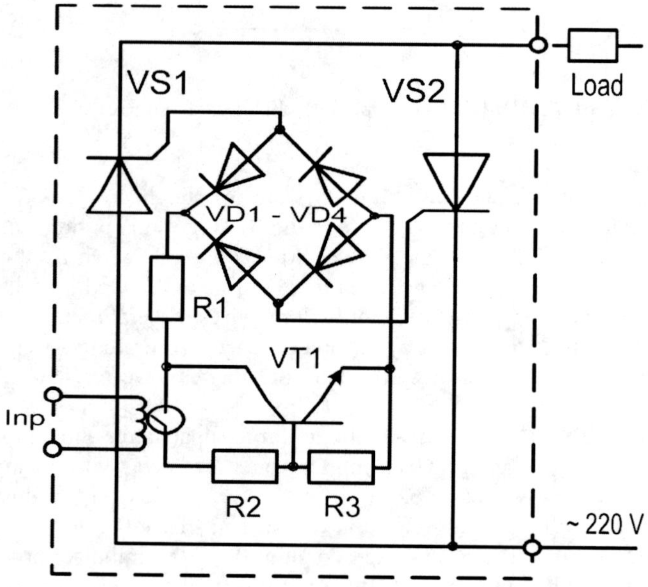

FIGURE 4.9 High power AC switching device with low power control element.

Using powerful thyristors with a nominal network voltage of 120-230 V to control them one may need a rather high power of the controlling contact which exceeds switching capability of miniature reed switches. In this case, between the reed switch and thyristors' gates, an extra amplifier is inserted as high-voltage transistor (Fig. 4.9).

A small diode bridge, a simple bipolar transistor, included in this device functions in the AC circuit. In the device one may use the high-voltage bipolar transistor with switched current of 0.5 – 0.8 A and also a low-power vacuum reed switch.

Having united three single-phase switches, we get a three-phase switching device – three-phase contactor, Fig. 4.10.

The reed switch in this device should be selected for functioning at 400 VAC, i.e., it should be high-voltage. The current flowing through it is equal to the sum of gate currents of the six thyristors (certainly in the form of phase-shifted pulses), this is why one requires much from reed switches also concerning its switching capacity.

Instead of a reed switch, it is better to use a simple electromagnetic relay with more powerful contacts in such low-voltage applications. If a reed switch has to be used and powerful thyristors with high gate currents, the reed switch should be supplied with the amplifier on a high-voltage transistor analogous to the way it is made in the device shown in Fig. 4.9.

Fig. 4.11 shows the circuit diagram of a switching device with two independent outputs for controlling of two powerful independent consumers.

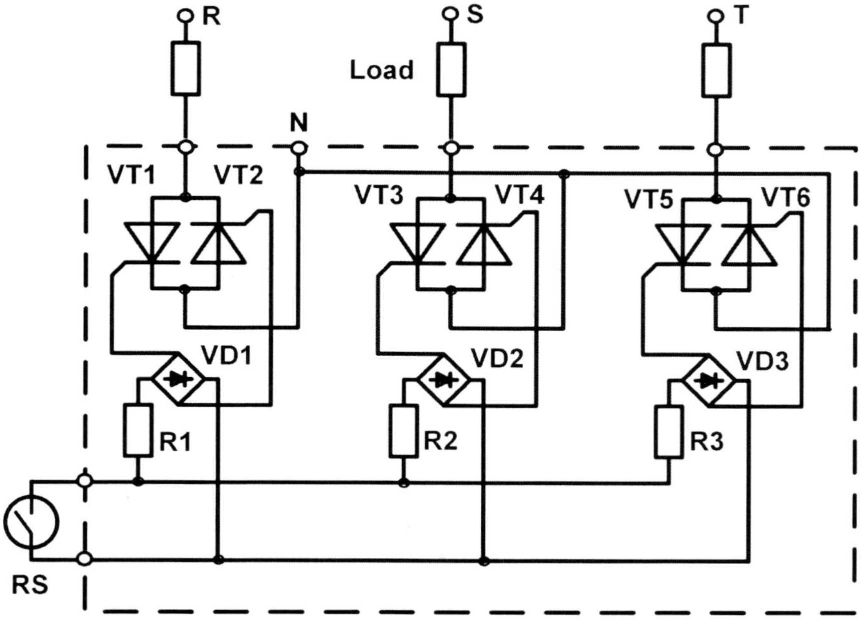

FIGURE 4.10 Three-phase hybrid switching device.

The device allows one input to control two fully independent loads, which can even be meant for different voltages.

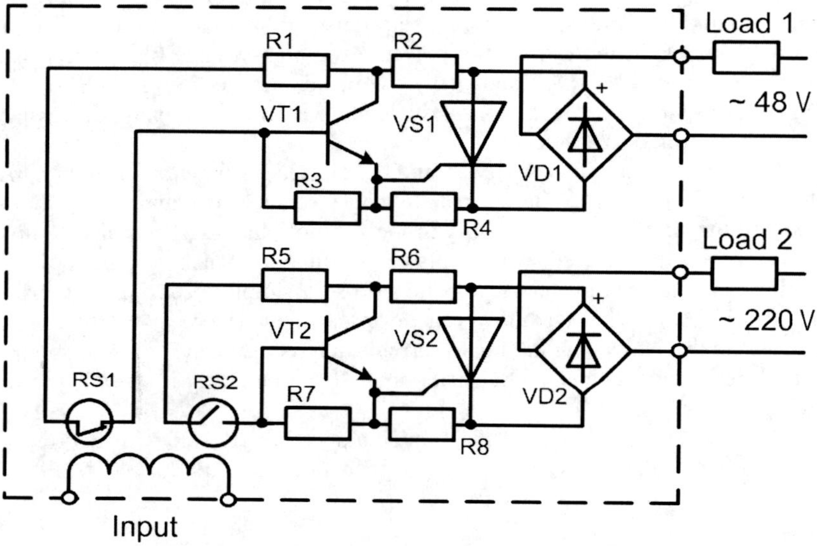

FIGURE 4.11 Power changeover with two independent outputs.

In this device the normally closed RS1 and normally opened RS2 reed switches are insulated from each other, but located in one shared control coil. One may also use two separate control coils for each reed switch by way of connecting them in parallel to each other. All semiconductor elements are high-voltage. Thyristors and diode bridges should be designed for full load current. Transistors may be low-current (designed for current of 0.5 – 0.8 A), but high-voltages are obligatory, i.e., meant for full line voltage (with appropriate reserves). Diode bridges and thyristors should be meant for load full current (again, with appropriate reserves and with respect to starting current). It is necessary to install varistors in parallel to the output terminals.

The circuit diagram shown in Fig. 4.12 has one interesting peculiarity. Having only one diode bridge, VD2, the device makes up a simple AC contactor. The thyristor is turned on and current in the load flows until the starting contact closes. As to the thyristor not direct but pulsating current is applied, in every half-period it is turned off (when the instantaneous value of current is less than the one of holding current), and at the beginning of a new half-period it is turned on (the starting contact being closed). The situation changes with the presence of a rectifier bridge, VD1. The low current flowing through this diode bridge, which is phase-shifted in regard to the main current flowing through the thyristor, keeps the thyristor turned on in the periods when main current falls to zero. If this extra current is higher than the thyristor holding current (i.e., higher than 100 – 300 mA), the thyristor remains turned on even after the disconnecting of the starting contact.

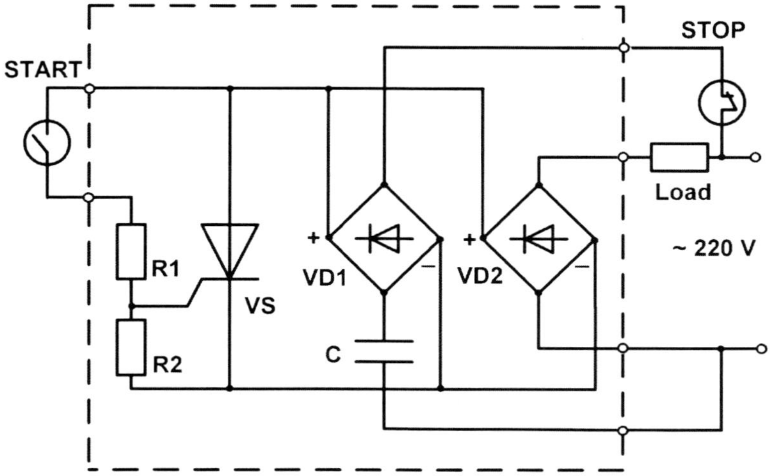

FIGURE 4.12 Pulse controlled latching AC contactor.

Thus, short-time (several milliseconds) activation of starting contact is enough for the constant turning on of the load. In order to turn the load off it will be necessary to open contact STOP for a period of not less than 10 ms. At this period the main current reducing to the value lower than the value of holding current of the thyristor and load will disconnect successfully.

Thus, the contactor allows turning the load on and off using short impulse signals with a width up to 10 ms.

The diode bridge VD2 is meant for load full current and the diode bridge VD1 is low-current (up to 1A).

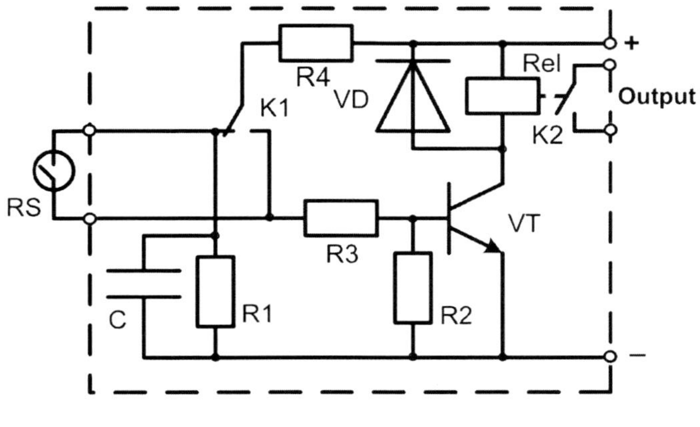

FIGURE 4.13a Two-position device based on bipolar transistor and controlled by single normally open contact.

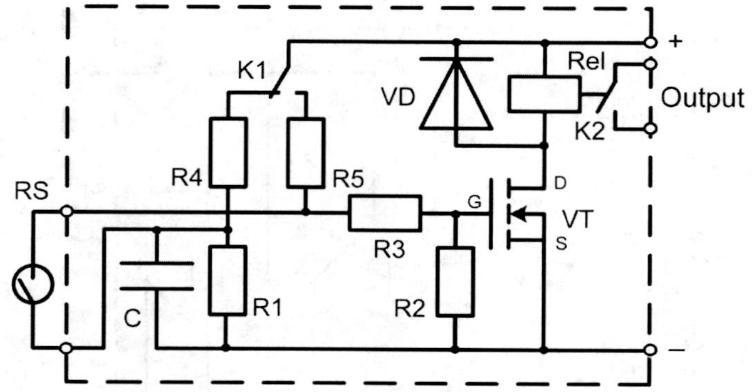

FIGURE 4.13b Two-position device based on FET and controlled by single normally open contact.

The distinctive feature of devices shown in Fig. 4.13 is the possibility of turning load on and off with the help of one and the same normally opened controlling contact, for example, push button. The devices differ in the transistor types but they have similar operating principles.

Capacitor C is charged to the voltage a little lower than that of the power supply voltage when the power supply is turned on. Transistor VT is turned off, output relay is also turned off. The base circuit of the transistor gets power through the current-limiting resistor R3 at the closing of controlling contact RS for a short time, so the transistor is turned on and it turns the output relay on. When energized, the relay locks the controlling contact by means of its contact K1.

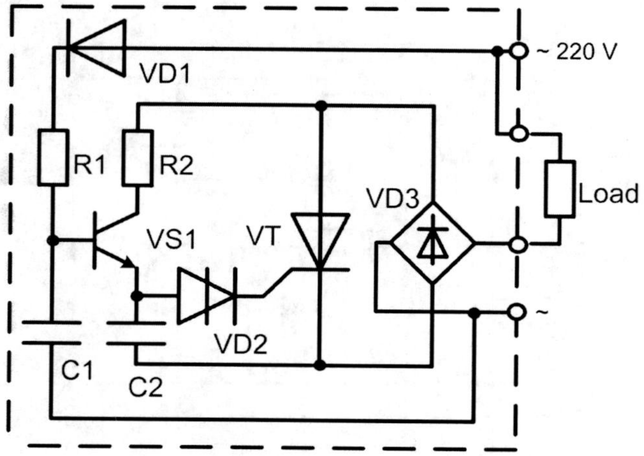

FIGURE 4.14a Simple soft-start contactor, based on power thyristor.

Owing to this, the circuit remains activated after the releasing of the RS contact. It is necessary to close the same contact momentarily for load disconnecting.

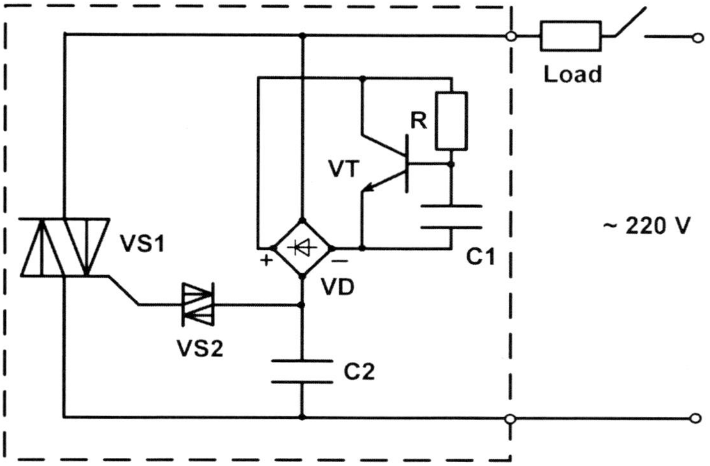

FIGURE 4.14b Simple soft-start contactor, based on power transistor.

At this moment, the discharged capacitor C appears to be connected to resistor R3 in the transistor base circuit and shunting it with its low resistance. The transistor is turned off, de-energizing the relay and returning contact K1 to its initial position. The closing of the controlling contact should be brief. It should be released before capacitor C will be charged through resistor R4. When needed, there may be several such contacts in parallel connections and these contacts may be located in different places. With any of these contacts, no matter which of them, it is possible to turn the load on and with any – to turn it off.

In some cases one does not need a jump-start, rather a soft-start load turning on is sufficient. As a rule, such turning on is used for loads with high starting current: asynchronous short-circuited motors, strong incandescent lamps, etc. Such starting current limitations allow reducing the switching and protective equipment and wiring overload, on one hand, and the increase of service life of power consumers (for example, this increase is significant for strong incandescent lamps), on the other hand. The soft-start contactor (Fig. 4.14a) is, in fact, a voltage regulator where transistor VS1 is used instead of a variable resistor. The resistance of the transistor is smoothly decreased during capacitor C1 charging when the device is energized. The resistance of the transistor being decreased as the capacitor C2 is charging and the dynistor (diac) VD2 is turning on changes in a sinusoidal curve of the applied voltage in the same way as it happens in the voltage regulator. Thus, during the device turning on, the angle of thyristor (or triac, Fig. 4.14b) activation is changed. The device functions well for resistive loads and loads containing small inductive components. As in other cases, the circuit should be protected against over-voltages in the power network by means of varistors.

A de-multiplexing unit, Fig. 4.15, may be used to energize several strong loads simultaneously when the power of the switching elements is lower than the total power of all loads.

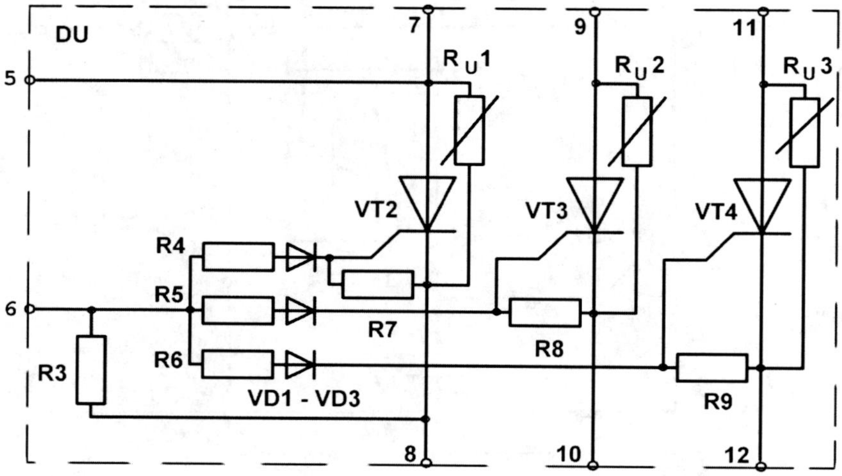

FIGURE 4.15 Demultiplexing unit.

All loads must be connected to outlets 7, 9, 11, and common plus of power supply if one controls the device by means of a starting contact (powerful reed switch, auxiliary relay contact, transistor) connected between outlets 5 and 6.

High-voltage vacuum reed switches with nominal voltages of 10 – 15 kV may withstand high voltage (50 – 70 kV) if one connects them in series and places them in an insulating case with the internal diaphragms filled with a non-rigid epoxy resin. However, a construction such as this does not allow absolutely the simultaneous closing of all reed switches, this is why there is risk that the whole voltage in some intervals of time may be applied to a small part of the series-connected reed switches and this will lead to their breakdown or damage.

In addition, the current switching capability of the high-voltage reed switches is very low (a matter of milliamperes) and often it is not enough for their practical application.

The circuit diagram for control of series-connected reed switches (shown in Fig. 4.16) realizes a rapid break of the high-voltage power supply in the low voltage stage when Start or Stop commands occur. In this case, the reed switches closing and opening take place without current.

The powerful high-voltage loads with currents of a few amperes can be controlled by means of this device.

Relay Rel1 is a two-position, double-stable with two windings (set and reset). Rel2 is a simple relay with one contact. VS is low-voltage thyristor interrupter of load current.

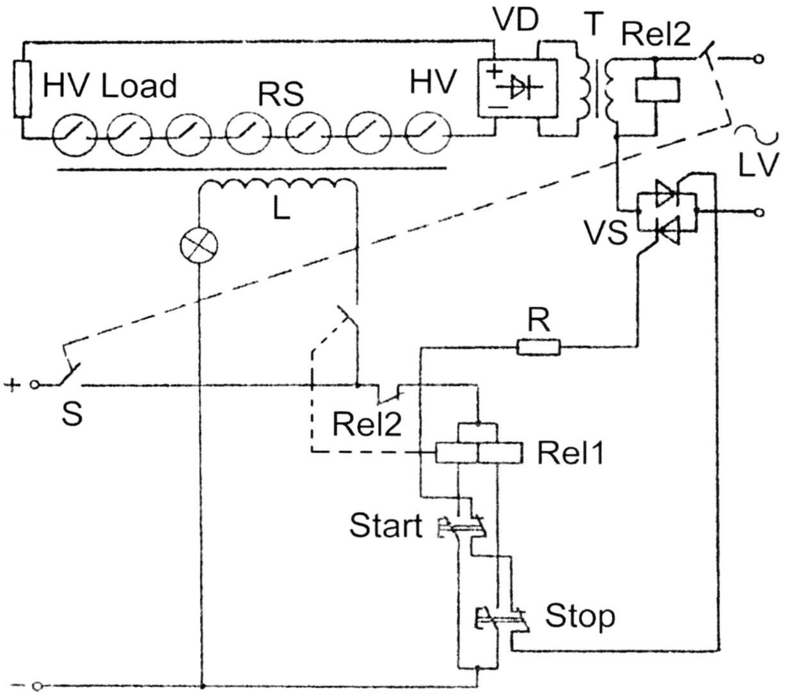

FIGURE 4.16 Control unit for HV reed switch contactor.

4.2 GENERATORS, MULTIVIBRATORS, PULSE-PAIRS

Generators, multivibrators, and pulse-pairs are widely used elements of industrial automation devices.

We will start our examination of this class of devices with the so-called monostable multivibrator, Fig. 4.17. In technical literature devices of this kind are called differently: Monostable Multivibrator, Monostable Flip-Flop, Monovibrator, One-Shot Timer.

The operating principle of this circuit is as follows: it generates a single pulse with a duration that is proportional to both the capacitor capacitance and the resistance of the relay winding when the power supply is turned on.

The relay resistance should correspond with the voltage of the power supply, and the capacitor rated voltage should be 1.5 – 2 times more than supply voltage. The device may be used directly in a 230 V network.

The Zener diode, VD, is needed to ensure more accurate relay actuation during a smooth change of voltage on the charging capacitor. When it is needed to obtain pulses of long duration (several tens of seconds) one may supplement the circuit with a VT transistor having a high DC current gate (Darlington).

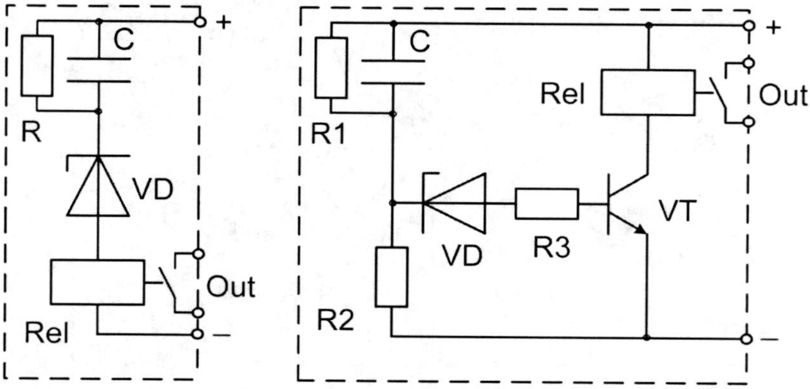

FIGURE 4.17 Monostable multivibrators.

Increasing of capacitor charging time is reached in this circuit due to the fact that resistance of the resistors R2 and R3 may be much more than that of the relay winding (the determinant of capacitor time charging in the first circuit diagram).

It is preferable to shunt the relay winding, usually with an opposite connected diode.

The circuit as described is very simple in comparison with microelectronic analogues.

Another variant of the monostable multivibrators is shown in Fig. 4.18.

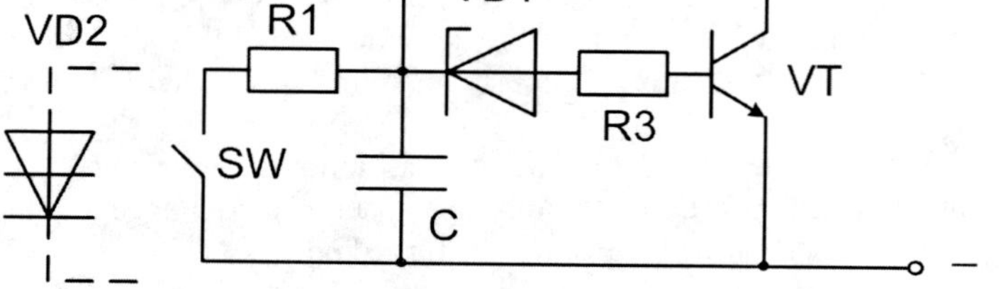

FIGURE 4.18 Monovibrator with disconnection function.

The purpose of this device is to turn an external circuit off for some time in accordance with an external command and then, some time later, automatically turn it back on.

Capacitor C is charged up to the Zener diode VD1 rated (breakdown) voltage when the device is energized. After this, transistor VT is turned on and the output relay actuates. The device is in the waiting state.

Capacitor C discharges quickly through a small current-limiting resistor, R1, when the control contact SW is closed briefly. When this happens, the voltage on the capacitor falls quickly and the Zener diode VD1 turns off. The transistor passes into cutoff mode and the relay is turned off, thereby disconnecting the external circuit.

The capacitor begins charging slowly through high-ohm resistor R2 when the controlling contact SW is closed. The external circuit remains turned off during the capacitor charging period.

The Zener diode transitions into the conducting state and transistor VT is turned on, which, in turn, turns on the output relay. When the voltage on the capacitor reaches the rated voltage of the Zener diode, the circuit returns to its initial state.

The device may be used for any voltage of power networks, up to 250 V.

An external load may be inserted directly into the transistor circuit (instead of the relay coil), but this transistor VT should have high DC current gate (Darlington).

If one uses dynistor or diac VD2 instead of the controlling contact, the device turns into a low-frequency multivibrator, which periodically turns the load on and off with a sequence of pulse periods of several seconds or several tens of seconds. To get this device to function normally one has to select a rather high-resistance resistor R2 so that with the rated voltage of power supply current flowing through, it would be lower than the dynistor (diac) holding current.

It is preferably to shunt the output relay, usually with an opposite connected diode.

A simple low-power relaxation oscillator circuit diagram is shown in Fig. 4.19.

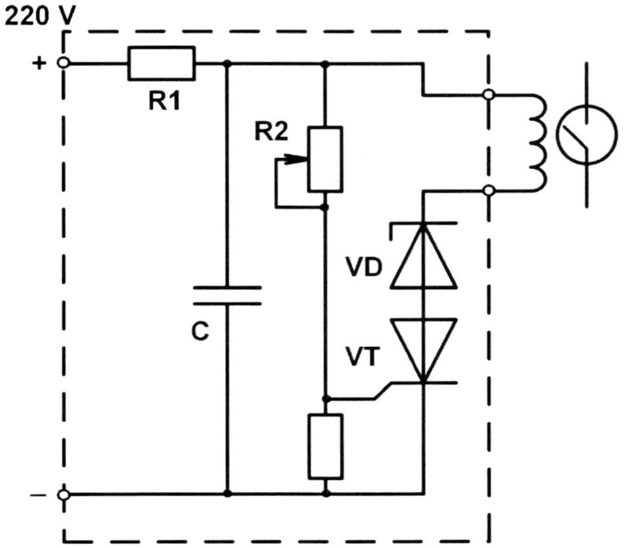

FIGURE 4.19 Simple low power oscillator.

Capacitor C begins to charge when the circuit is energized. During its charging, the current in the thyristor circuit gate increases. The thyristor is turned on at a determined value of the gate current, adjusted by potentiometer R2. Capacitor C discharges on the load (for example, a relay coil with a mercury reed switch) through the thyristor and Zener diode. At this point, the capacitor voltage begins to decrease quickly. When the voltage falls below the rated voltage of Zener diode, its resistance increases sharply and, as a result, the load current decreases almost to zero and the thyristor is turned off. The device returns into its initial state and the process repeats.

The Zener diode rated voltage should be about 20 per cent of the power supply voltage.

As a power supply one may use networks of 230V. The Zener diode should have a power rating of not less than 5 W. Thyristor VT is high voltage (1000 – 1200 V).

A very simple circuit diagram for a pulse generator that may be implemented on the basis of semiconductor element, called a Sidac, is shown in Fig. 4.20.

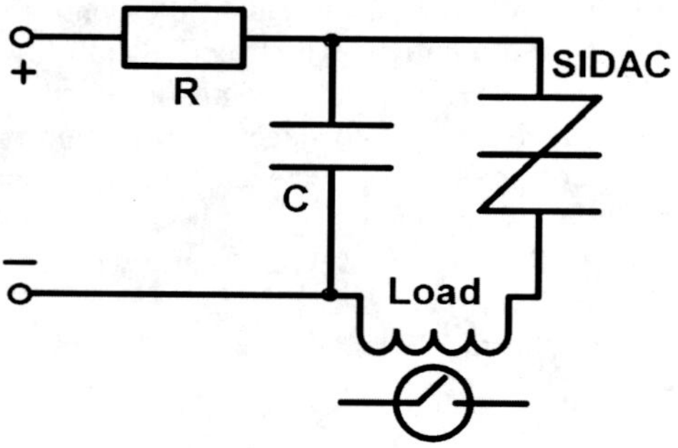

FIGURE 4.20 Very simple pulse generator with Sidac.

The Sidac is a silicon bilateral voltage triggered switch with greater power-handling capabilities than the standard diac. Upon application of a voltage exceeding the Sidac breakover voltage point, the Sidac switches on through a negative resistance region to a low on-state voltage. Conduction will continue until the current is interrupted or drops below the minimum holding current of the device.

Teccor company offers the complete voltage range (95-330) over three different packages:

- TO-92 (95-280 volts)
- Axial lead DO-15X (95-280 volts)
- Surface Mount DO-214AA (95-280 volts)
- TO-202AB (190-330 volts)

In the circuit diagram shown in Fig. 4.20 the capacitor charges faster and the frequency of the pulsation increases with the decreasing of R resistance.

The disadvantage of this scheme (in contrast to the generator on the Zener diode and thyristor) is that the resistance of the resistor R has to be rather high (depending on the voltage of the power supply) to ensure that the current through the resistor after the capacitor discharges will be lower than the holding current of the Sidac (it is several tens of milliamperes), otherwise the Sidac will not be turned off and the circuit will not return to its initial state.

Unlike the simple dynistor or diac, the Sidac has a normalized and rather accurate barrier voltage with its slow change.

In order to increase switched power one may supplement the circuit of generator with a powerful transistor (Fig. 4.21).

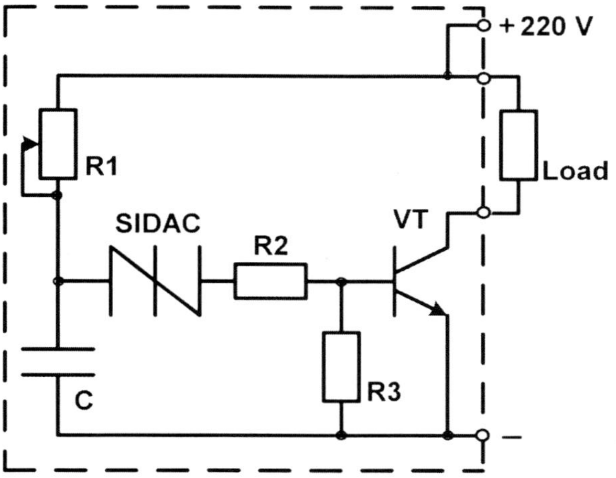

FIGURE 4.21 High power oscillator with sidac and transistor.

All elements of this device should be high-voltage. If one uses a Darlington transistor as the transistor VT, capacitor C may have a small capacitance. Moreover, even with respect to the necessity of resistor R1 having high resistance (the current through it should be lower than holding current of the Sidac) one may get a rather high oscillation frequency.

In the case of an inductive load, it should be shunted, usually with an opposite connected diode.

The device scheme given in Fig. 4.22 is called a pulse-coupled oscillator or pulse-pair. The purpose of the device is the periodic changeover switching of two loads in automatic mode.

This device functions as follows. Transistor VT1 is turned on and the first load is turned on when the circuit is energized. The charging of capacitor C1 begins through the timing resistor R1. The dynistor VD1 is turned on and capacitor C1 discharges in thyristor VS gate circuit when the capacitor breakdown voltage of the dynistor is reached.

The thyristor is turned on and energizes the coil of reed switch relay Rel through the non-charged capacitor C2. At this point, the relay contact changes position: turning transistor VT1 off and turning transistor VT2 on. Instead of the first load, second one is turned on.

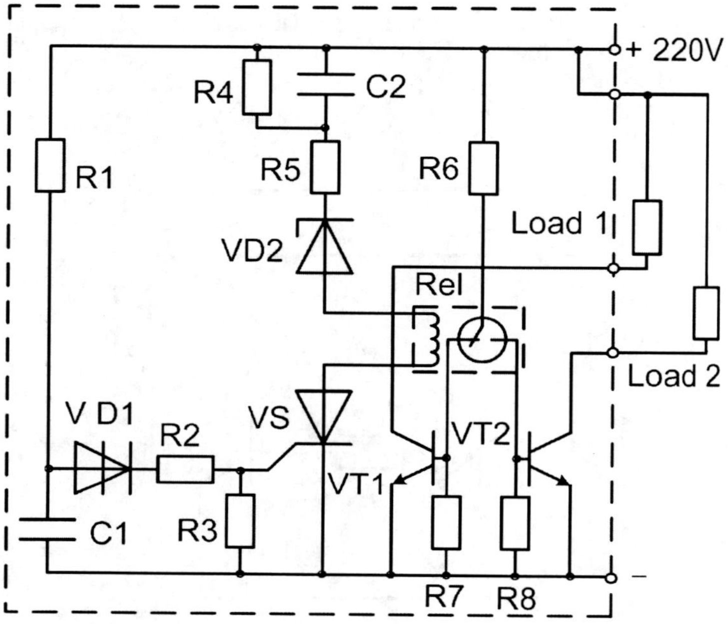

FIGURE 4.22 Pulse-coupled oscillator (pulse-pair).

The device remains in this state until the process of capacitor C2 charging goes on and until the voltage on the Zener diode is higher than its rated voltage. When the voltage on the Zener diode decreases, the current through the Zener diode decreases also up to some threshold level (higher than the relay release threshold), the Zener diode is turned off, interrupting the current in the relay coil and turning off thyristor VS regardless of presence or lack of current in the gate. At this point the relay contact changeovers (by means of transistors) loads, i.e., turns load 2 off and turns load 1 on. The device will be in this state until dynistor VD1 is turned off (if it has not been turned off before this) and the process of capacitor C1 charging begins.

The moment of the first changeover from load 1 to load 2 is determined by the parameters of the R1C1 circuit. The period of the device being in this state and the moment of switch reversal is determined by the capacitance of the capacitor C2 and Zener diode VD2 rated voltage. Here the dynistor VD1 state and the presence or absences of controlling signal on the thyristor VS gate do not matter.

Darlington transistors should be used to switch heavy loads. The thyristor is low-current but it is meant for full network voltage. It is preferable to protect the thyristor and transistors by means of varistors.

The semiconductor unit with the circuit diagram shown in Fig. 4.23 may turn any relay and even a powerful electromagnetic contactor into a pulsator controlling load.

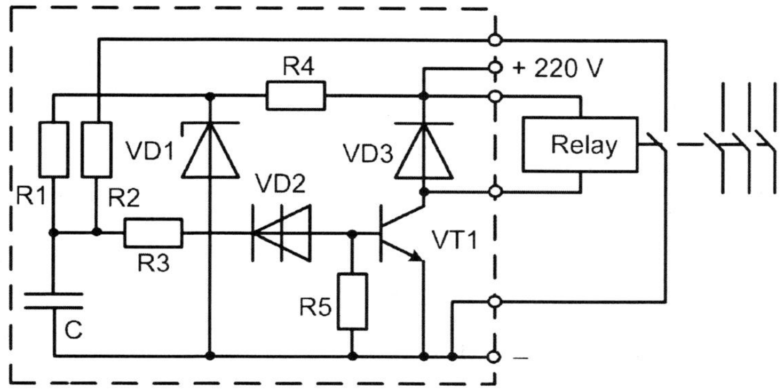

FIGURE 4.23 Control supplement to electromechanical contactor.

When the circuit is energized, capacitor C charging begins through the timing resistor R1. After the capacitor has charged up to the dynistor VD2 breakdown voltage, the dynistor is turned on and the capacitor discharges through the limiting resistor R3 and base circuit of transistor VT1. The transistor is turned on and, in turn, turns on the electromechanical relay or contactor, the contacts of which turn the load on. One of the contacts of this relay is used to control the circuit.

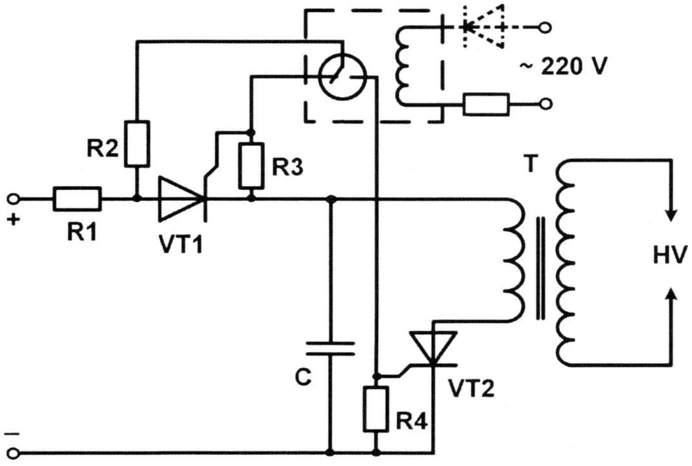

FIGURE 4.24 Simple high-voltage pulse generator.

This contact connects the discharging resistor R2 in parallel to capacitor C. Capacitor discharging begins. After the capacitor voltage decreases lower than a level at which current through the dynistor is lower than its holding current, the dynistor is turned off and the transistor is turned off thereby turning off the relay. The relay contacts are disconnected and the process begins again.

The relay switching frequency depends on the parameters of resistor R1 and capacitor C1. The relay duty pulse is determined by the R2 resistance value. The Zener diode VD1 and resistor R4 make up parametric voltage stabilizer providing capacitor C charge with reduced (in comparison to voltage in power network) and stable voltage. Resistor R3 has a low resistance (50 – 100 Ohm) and it is meant for limiting the transistor base current when capacitor C is being discharged. The dynistor VD2 barrier voltage should be lower than Zener diode VD1 rated voltage.

In concluding this section we will consider a simple high-voltage pulse generator (Fig. 4.24).

In initial condition when the circuit is energized the thyristor VT1 is turned on as its controlling circuit is structured such that the thyristor is turned on by the normally closed contact of reed switch relay. Through this turned on thyristor capacitor C charges quickly. The thyristor turns off after capacitor charging and the current in the thyristor circuit drops. The gate circuit of thyristor VT1 breaks and the gate circuit of thyristor VT2 connects when the reed switch relay picks up. As a result, thyristor VT2 is turned on and capacitor C discharges quickly through the low resistance primary winding of high-voltage transformer creating a high-voltage pulse on its secondary winding. After the capacitor discharges, thyristor VT2 turns off automatically as there is no current in its circuit (remember thyristor VT1 is turned off this time).

Then the process repeats again and again synchronously with the reed switch switching cycles. Mercury reed switches have sufficient switching capacity for a very long service life. One should definitely consider using mercury reed switches in this device.

Alternating current mains may be used as generator of control pulses. In this case the pulse repetition frequency will be 100 Hz. If it is necessary to reduce this frequency by half, one may insert a diode in series with the coil of control relay (in the diagram it is indicated by dashed lines).

If one needs lower switching frequency or it should be changeable, one should use simplest relaxation generator.

Instead of step-up, high-voltage transformer one may use any load requiring pulse power supply.

Instead of reed switch relay one may use its transistor analogue.

4.3 TIMERS

A circuit diagram of a simple timer suitable both for independent application and for using as part of various automation devices is given in Fig. 4. 25. When the circuit is energized, thyristor VT is turned off and that is why the whole voltage of the power supply is applied to the thyristor and charging circuit R1C. The capacitor C being charged, the voltage on it is increased. Having achieved the Sidac barrier voltage, capacitor C will be discharged through the Sidac and thyristor VT gate circuit and will turn it on. After thyristor activation and the load being turned on, the voltage on the thyristor falls sharply almost to zero and the capacitor stops charging. In order to return the circuit to its initial state it is necessary to disconnect the power supply for a short time, for example, with the help of a push button.

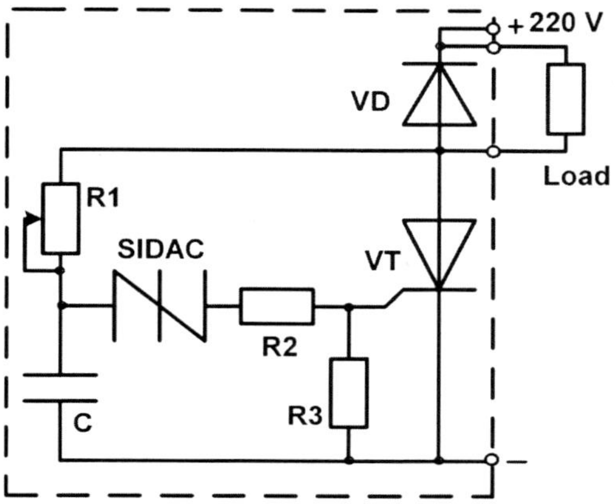

FIGURE 4.25 Simple timer for ON time delay.

Since after the first actuation the device self-blocks, one makes no demands to the R1 resistance, except demands to ensure necessary time delay.

The more R1C circuits, the more the time delays. To increase time delays one may shunt capacitor C with additional resistance, higher than R1 to slow down the process of capacitor charge.

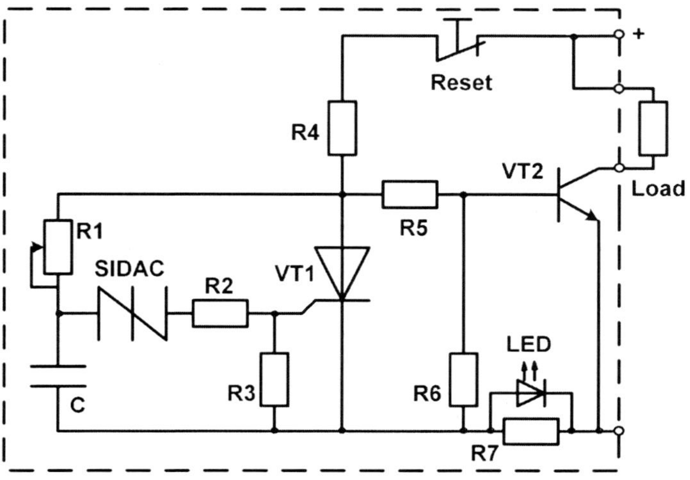

FIGURE 4.26 Simple timer for OFF time delay.

Another variant of a turn-off delay timer is given in Fig. 4.26.

When the circuit is energized, transistor VT2 is turned on, thyristor VT1 is turned off, and the load is being energized through the turned on transistor VT2.

Capacitor C charge begins through the timing resistor R1. After it is charged to the Sidac barrier voltage, thyristor VT1 is turned on and it sharply decreases the positive potential of transistor VT2 base leading it to turning off and de-energizing of the load.

The circuit is used when it is necessary to turn the load off automatically some time after its activation.

The transistor and thyristor in this device should be high-voltage, if 120-230 V is to be used as power supply. The transistor may be rated for low currents (for currents of 0.5 – 1 A), and the thyristor should be meant for switching the full load current and, in addition, have necessary reserve.

The circuit diagram for timer turning on the load for a definite period of time and then turning it off automatically is given in Fig. 4.27.

When the circuit is initially powered up, capacitor C is discharged, transistor VT1 is turned off, and VT2 is turned on. Therefore, thyristor VS is turned on and the load is energized at once. However, the load is turned on for only a short time. After capacitor C charges through resistor R2 to the Zener diode VD2 rated voltage, the Zener diode is turned on and current flows through it turning on transistor VT1. At this point, transistor VT2 is turned off and thyristor VS is turned off afterwards by the first transition over the zero of the sinusoidal load current. The device is in the waiting state and is ready to work.

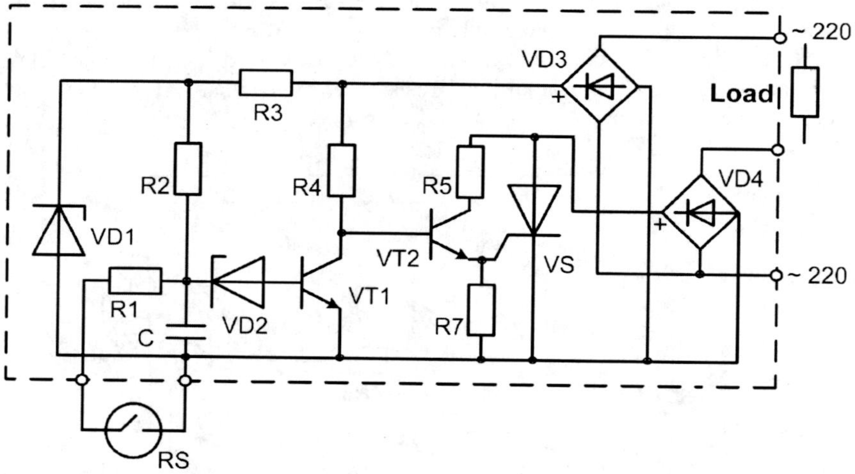

FIGURE 4.27 Self-reset timer.

By briefly closing the starting contact RS, capacitor C discharges quickly to zero; by this, transistor VT1 is turned off and transistor VT2 is turned on turning thyristor VS on which in its turn turns the load on. After returning of starting contact RS into its initial (opened) condition, capacitor C begins to charge through resistor R2 to the Zener diode VD2 rated voltage. Values of R2 resistance and C capacity determine time delay.

After capacitor C charges, transistor VT1is turned on and transistor VT2 together with thyristor VS are turned off, thereby turning the load off. Thus, after briefly closing and releasing the starting contact, the period of the energizing condition of the load will be determined by the time it takes capacitor C to charge.

Thyristor VS and diode bridge VD4 should be used for the full current of the load. Zener diode VD1 provides stable level of capacitor C charge voltage, i.e., stabilizes the time delay. Transistor VT1 should be high-voltage and have high gain (Darlington).

4.4 LOGIC ELEMENTS

In automation devices logic elements are widely used. They realize different logic functions. The INHIBIT function is one of often used functions of this type.

The logic element realizing this function (Fig. 4.28) has two inputs: main (triggering), initiating signal at the element output, and an inhibiting (blocking) signal at the main input that inhibits the signal occurrence at the device's output.

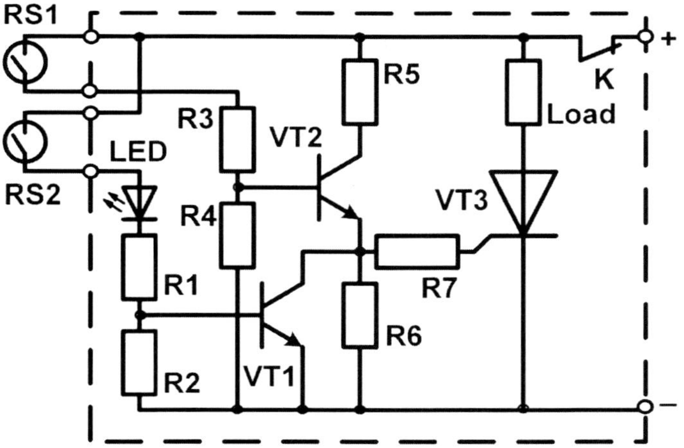

FIGURE 4.28 Simple high-voltage logic element with INHIBIT function.

Thus, signal occurrence at the output of this element is possible only when there is signal at the main input and there is no signal at the inhibiting input.

In our device RS1 is main triggering reed switch, and RS2 is inhibiting reed switch. Closing RS1 turns on transistor VT2 and thyristor VT3 if RS2 is opened.

Closing RS1 turns on transistor VT2, but it cannot turn on output thyristor VT3 until the reed switch of RS2 closes and transistor VT1 is turned on.

The LED indicates the inhibiting signal by which the input signal cannot change circuit condition. With a voltage of 220 V Darlington high-voltage transistors should be used. To protect electronic components against network over-voltage, a varistor should be included in the device's output terminals.

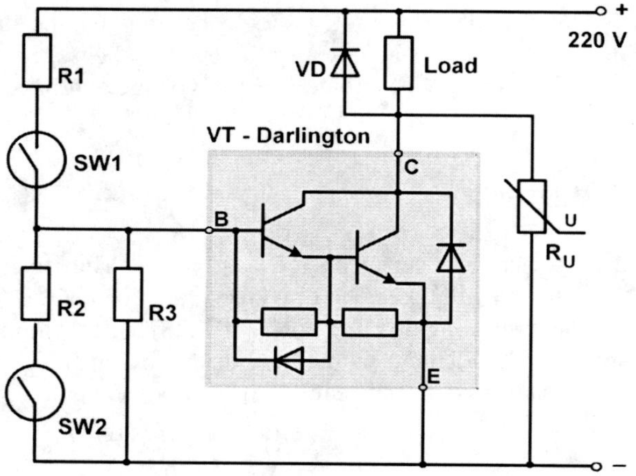

FIGURE 4.29 Logic element INHIBIT on single Darlington transistor.

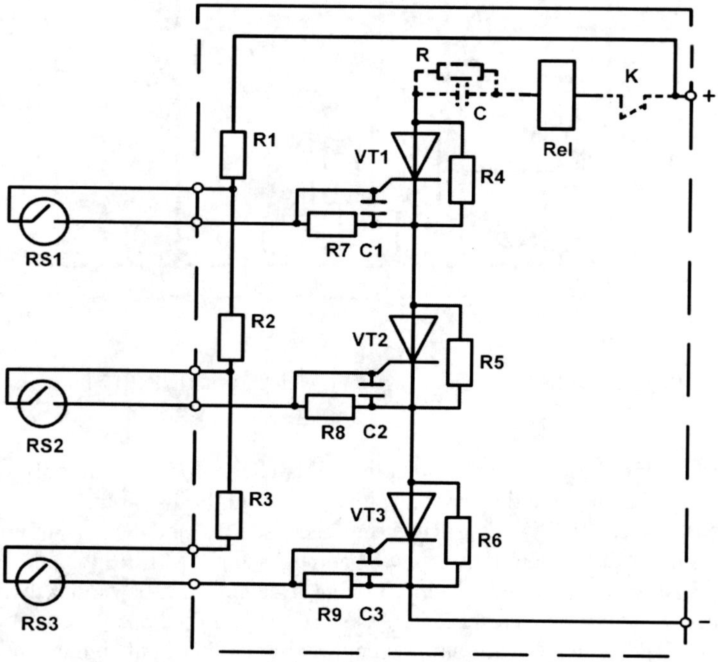

FIGURE 4.30 Power synchronous logic element AN.

A circuit diagram of one more variant of logic element with INHIBIT function is shown in Fig. 4.29. In this device input with reed switch SW1 is the main one (triggering) and the input with reed switch SW2 is the inhibiting one.

Unlike the previous circuit with thyristor, in this circuit the inhibiting signal may appear at any moment (even after main triggering signal has occurred) and thereby the functioning of the triggering signal will be blocked regardless of the circuit's previous condition.

Resistance of R2 should be much less than that of R3 and less than that of R1.

The logical AND function may be realized by means of the device shown in Fig. 4.30. In the logical "AND" circuit the output signal appears only when there are signals at all its inputs. This device may be used to energizer loads (output relay, contactor) if there is a synchronous presence of signals at all inputs. If input signals are received asynchronously, i.e., do not coincide, the output relay does not respond. After pick ups, the output relay remains energized and the device may be returned to its initial state by way of momentarily disconnecting the K contact (manually or by remote control). The device may be additionally provided with capacitor C and discharge resistor R if it is necessary to automatically return the device to its initial position after pick ups. After this, when all input signals are synchronous and the thyristors are turned on, capacitor C is charged fast and output relay K is actuated by the charging current of capacitor C. Relay K is released (if it has no latching) when the charging current drops in final charging stage.

On the expiration of several seconds capacitor C is discharged through the resistor R and the device is returned to its initial state. Resistors R4 – R6 are needed to equalize the voltage at the series-connected thyristors and they have resistances of several tens of kOhm. Resistors R1 – R3 are needed to limit current in the gate circuits of the thyristors to the level of about 150 – 200 milliamp and they depend on power supply voltage. R7C1, R8C2, and R9C3 are anti-jamming circuits. Wires to triggering elements should be shielded.

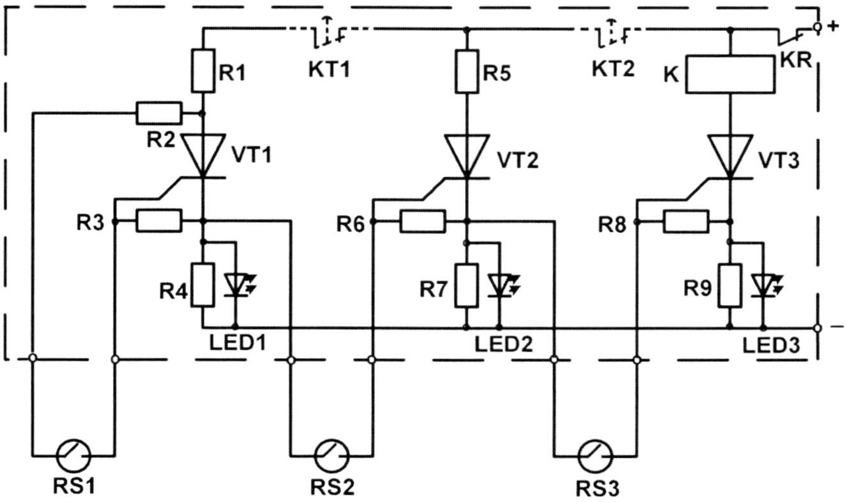

FIGURE 4.31 Non synchronous power logic element AND.

Another type of the logical AND is shown in Fig. 4.31. Unlike previous devices requiring the simultaneous and synchronous presence of signals at all inputs for output activation, for the device shown in Fig. 4.31 the output signal occurs after all inputs have been activated regardless of their synchronicity. This means that the device "remembers" that at its inputs there were signals and it activates an output even after the signals disappear at inputs. One could say that it is a "non-synchronous" logical AND.

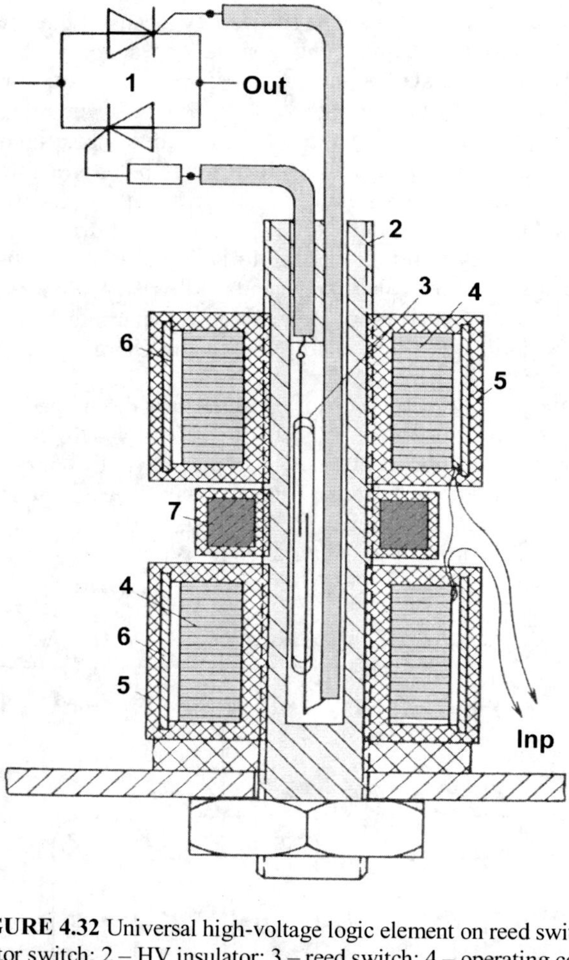

FIGURE 4.32 Universal high-voltage logic element on reed switch.
1 – output thyristor switch; 2 – HV insulator; 3 – reed switch; 4 – operating coils; 5 – external insulators; 6 – ferromagnetic shields; 7 – constant magnet.

In this device a single activation of every separate reed switch, except the first one, does not lead to a change of the device's state. After the first reed switch has been closed, even momentarily, the first thyristor is turned on and remains turned on (the first LED, LED1, is activated). It is equivalent to the removal of the first protection stage.

Now the positive power supply is applied to the second reed switch through the opened first thyristor. When the second reed switch, RS2, is closed, the second thyristor, VT2, turns on and remains in this state (now both LED1 and LED2 are activated). Then the third reed switch has received the positive power supply and the device is in waiting state until the moment of closure of the third reed switch. After this occurs, the output relay, K, activates. Contact KR serves for the device's return to the initial state. Additional contacts, KT1 and KT2, may be used in the device. They serve to return, separately, each activated protection stage to the initial state in the event that during a specific period of time, the reed switch of the next stage has not been activated. To realize this function instead of resistors R1 and R2 (or parallel, or in series with them) one can insert inputs of simple timers with output contacts: KT1 and KT2.

Lastly, we will describe the construction of a multipurpose logic element having high-voltage insulation among all input and output circuits; this is shown in Fig. 4.32.

This device contains two controlling coils (4), a permanent magnet (7), a reed switch (3), and a high-voltage insulator (2).

The coils may be connected in series or parallel, and also they may be controlled separately. The coils and magnet may move by means of the thread along the reed switch.

Many different combinations of reed switch statuses are possible depending on the mutual location of elements and the combination of electric signals at the input. For example, by means of this simple device implementation the following logic functions are possible: AND; AND-NOT; OR; OR-NOT; NOT; MEMORY; INHIBIT; EQUIVALENCE, etc. We leave it as an exercise for the reader to think of ways to employ the elements we have just presented in order to realize these logical functions.

4.5 OVERCURRENT AND OVERVOLTAGE PROTECTION MODULES

A circuit diagram for simple relay pick ups at the increase of input voltage higher than the determined level is shown in Fig. 4.33.

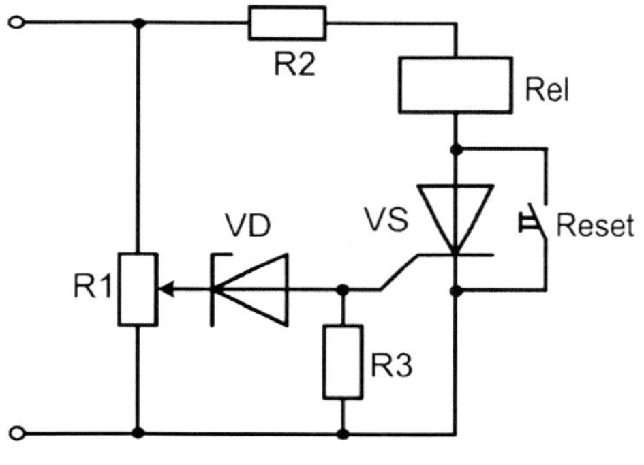

FIGURE 4.33 Simple overvoltage relay.

The operating level of the device pick ups is adjusted with potentiometer R1. By means of resistor R2 the rated voltage of electromagnetic relay may be much lower than the controlled voltage (but it cannot be higher than this voltage). Resistor R2 may be excluded if the rated voltage of the electromagnetic relay is equal to controlled voltage.

The device is turned on when a portion of the input voltage on the lower arm of the potentiometer exceeds the rated voltage of Zener diode VD.

The analogous operating principle is depicted in the circuit shown in Fig. 4.34. Its difference is in two throw-over loads (Rel1 and Rel2 2).

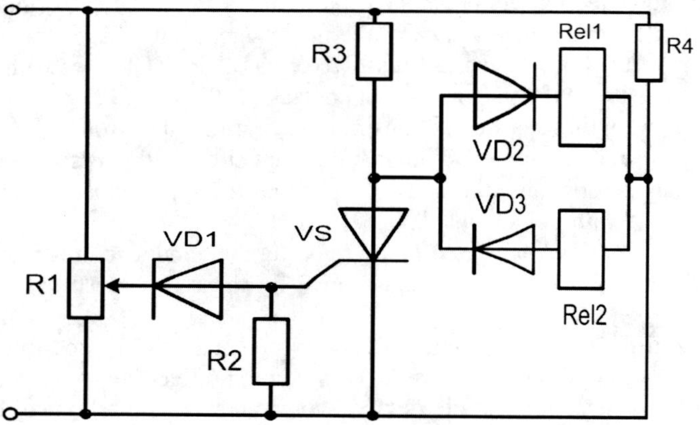

FIGURE 4.34 Overvoltage relay with two outputs.

When the voltage is lower than the Zener diode operating voltage, the thyristor is turned off and relay Rel1 is turned on through the VD2 diode and R3 resistor. When the voltage rises to the Zener rated value, the thyristor VS is triggered, the Rel1 relay is released, and the Rel2 relay is picked up.

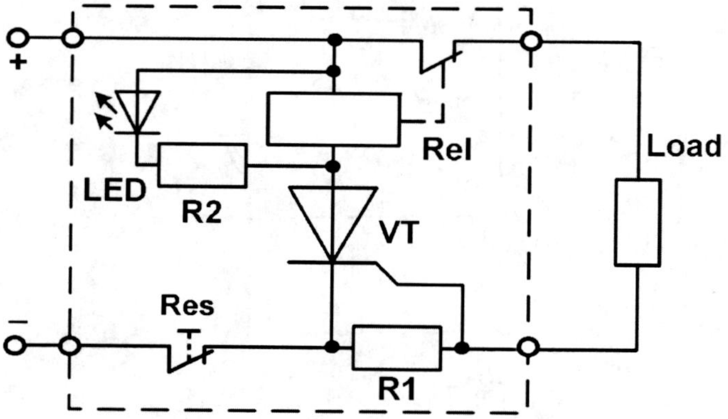

FIGURE 4.35 Overcurrent protection module.

The circuit diagram of a module for a consumer's over-current protection is shown in Fig. 4.35.

The device is inserted between the power supply and a consumer (load).

In this device the voltage drop on resistor R1 increases and the positive potential on the thyristor VT gate also increases in respect to its cathode when high current passes through the load.

The thyristor VT is turned on and activates relay Rel disconnecting the load from power supply as the load current reaches a specific threshold that is dependent on R1 resistance and thyristor VT triggering gate current. In order to increase device accuracy, one may connect a Zener diode into the gate circuit of the thyristor. One should keep in mind that with voltages of 125 – 250VDC, contacts of most miniature electromagnetic relays are able to break only resistive loads.

After actuation of this protection device, it is self-latched and it illuminates the LED signal. To return the device to its initial state, a "Reset" button has to be pressed.

Sometimes the operational speed of a simple electromagnetic relay (10 – 15 ms) is not enough to protect loads that are especially sensitive to overloads. Fig. 4.36 shows a circuit diagram of a high-speed protection device able to de-energize the consumer at current overload in the range of microseconds (time of thyristor turning on).

The purpose of this device is to cause an accelerated firing of fuse F by means of short circuit surge current generated by thyristor VT when actuated. In this case the time delay until the load is de-energized is determined not by the time the fuse burns, but by the time of the thyristor turning on, as after its activation the voltage on the load drops almost to zero. The thyristor for this device should be chosen to withstand peak non-repetitive current surges more than the short-circuit current surge that actually occurs in circuit and is determined by the supply mains capacitance and resistance of connecting wires.

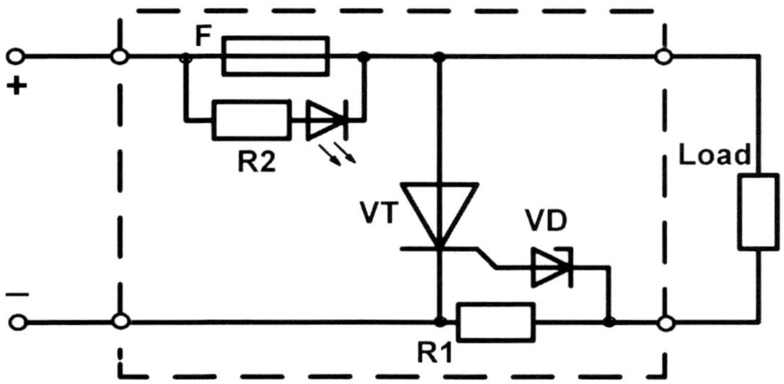

FIGURE 4.36 Super faster overload protection device.

The LED illuminates (indicating device actuation) and the fuse burns.

The device allows combining relatively slow protection against insignificant overloads caused by the fuse and high-speed protection in case of significant overload or short-circuiting.

The circuit diagram for full-fledged over-current protective relay with high release ratio is shown in Fig. 4.37.

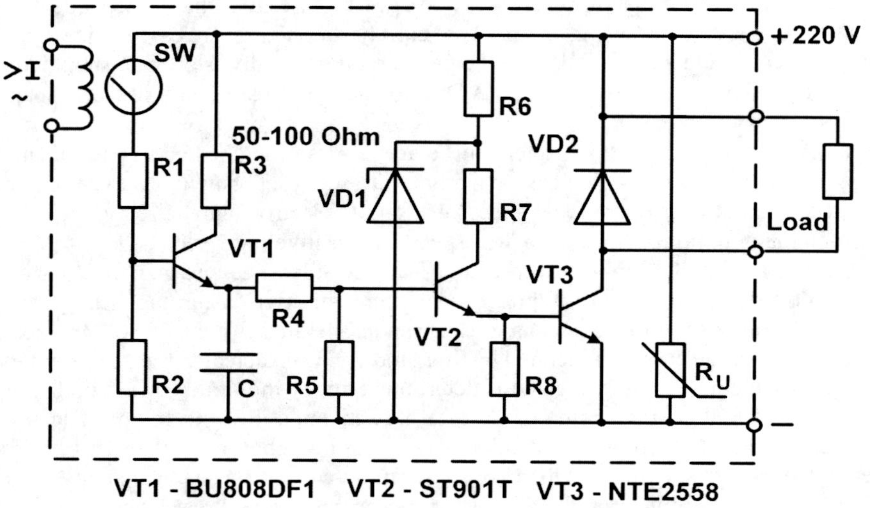

FIGURE 4.37 Overcurrent relay with high release ratio.

In the device the reed switch SW begins to vibrate (closing and opening) with a frequency of 100 Hz (for a network with a 50 Hz rated frequency) when current applied to the device input reaches reed switch actuation threshold.

Transistor VT1 fully repeats the reed switch switching cycles. When this transistor is turned on, a positive potential is supplied to the transistor VT2 base and at the same time capacitor C charges. Transistor VT1 is meant for ensuring the high charging current of capacitor. This transistor should have a relatively high gain for unloading the reed switch and also it should have a high emitter-collector voltage suitable for functioning with 125-250 VDC and relatively high collector current to provide a rapid capacitor charge. The Darlington transistor BU808DF1 type fits these requirements. With this transistor and limiting resistor R3 (50 – 100 Ohm) capacitor C charges very quickly even with the large capacitance, and this may ensure a relay operation speed of 5 – 15 ms.

Transistor VT2 acts as a filter connected between the input transistor VT1 operating in the pulsating mode and output transistor VT3 in which circuit pulsations are not acceptable. This transistor should have very high gain to ensure a minimum C capacitance for keeping this transistor in the conductive mode at the disappearance of the input voltage when the reed switch contact opens, actuated with frequency 100 Hz by the AC magnetic field with frequency of 50 Hz.

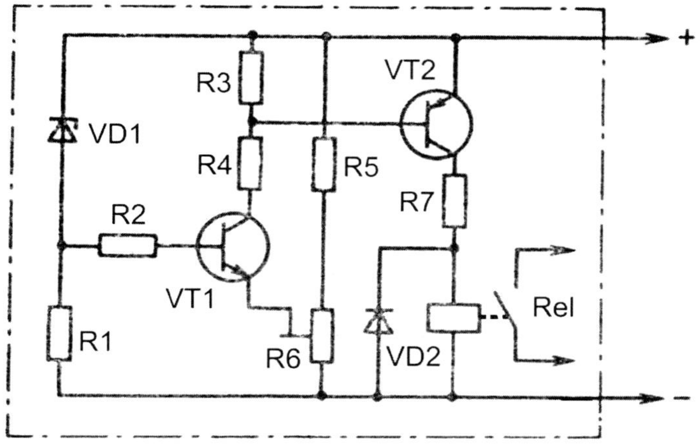

FIGURE 4.38 Small gisteresis relay.

As there are no transistors with gain of about 1000 with collector-emitter voltage of 800 – 1000 V, Darlington transistor ST901T with minimal gain 1500 and emitter-collector voltage of 350 (500) V was chosen to be such a transistor. To provide the required voltage reserves and device reliability, the collector voltage on this transistor is reduced to 50 – 100 V with the help of the Zener diode VD1. Using the Zener diode for reducing the operating voltage is acceptable only for this transistor, as it is the only one to be used in the mode of relatively low current.

The Darlington transistor is preferred for transistor VT3 based on its need to work with relatively powerful loads (several amperes) since the base current would not be high.

A powerful output transistor allows using not only auxiliary relays, but also powerful contactors, actuators of the trip coil of high-voltage circuit breakers, etc. as the load.

Devices of this type are usually not meant for extended functioning under current; this is why the heat sinks are not required for transistors.

High release ratio (0.97 – 0.99) is provided by the reed switch.

The high release ratio suitable for protective relays may be provided not only by using reed switches, but also by means of simple electromagnetic relay actuated according to a special circuit diagram (Fig. 4.38).

As is well known, electromagnetic relays usually have large hysteresis, i.e., high ratio of pick up voltage to release voltage. Thus, pick up voltage of simple electromagnetic relays with rated voltage of 24 V may be 15 – 18 V, and release voltage may be 5 – 8 V. In many cases one needs a relay to pick up and release by a much closer voltage range, i.e., it would have small hysteresis.

The device consists of parametric voltage stabilizer formed by the Zener diode VD1 and the resistor R1 and a DC amplifier on two transistors of n-p-n (VT1) and p-n-p (VT2) types. The output electromagnetic relay in this device picks up as the supply voltage increases relative to the nominal value and returns into its initial state when the supply voltage slightly decreases.

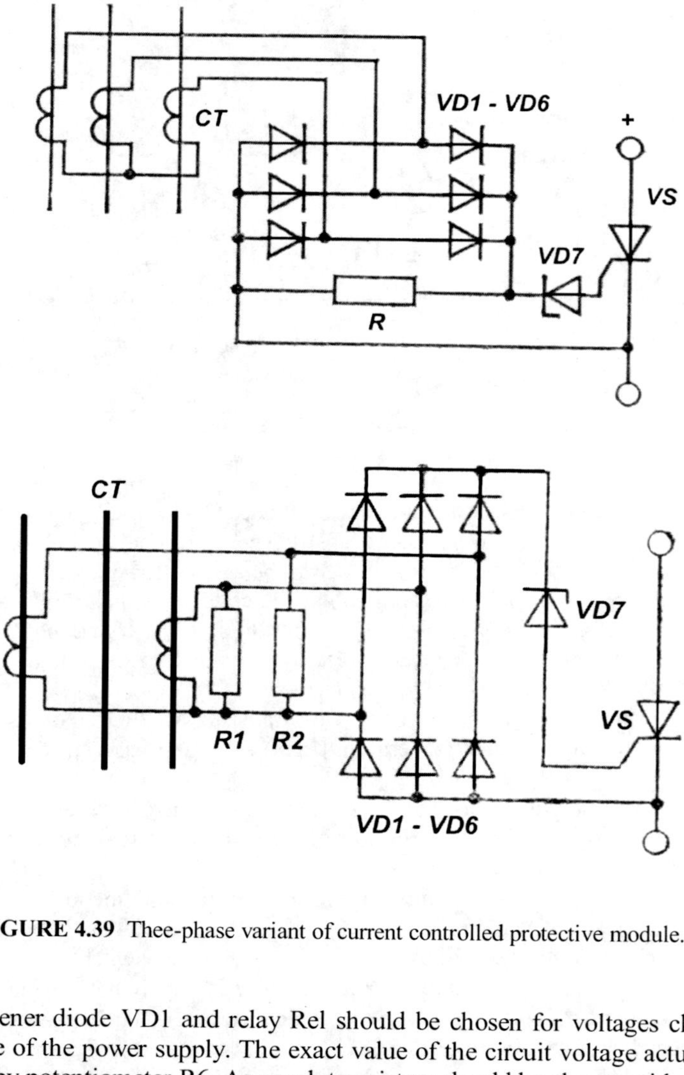

FIGURE 4.39 Thee-phase variant of current controlled protective module.

The Zener diode VD1 and relay Rel should be chosen for voltages close to the rated voltage of the power supply. The exact value of the circuit voltage actuation may be adjusted by potentiometer R6. As usual, transistors should be chosen with large voltage reserves.

One may use several such devices adapted to different voltage actuation thresholds for stepped load control.

The current relay for protection of three-phase circuits may contain separate measuring units for independent current control in each phase, and it may contain one measuring unit common for all three phases. This last case can be realized using one of the circuit diagrams shown in Fig. 4.39.

4.6 VOLTAGE STABILIZERS AND REGULATORS

The parametric stabilizer is the simplest stabilizer type. A Zener diode is its basis. A typical simple circuit diagram of a parametric stabilizer is shown in Fig. 4.40.

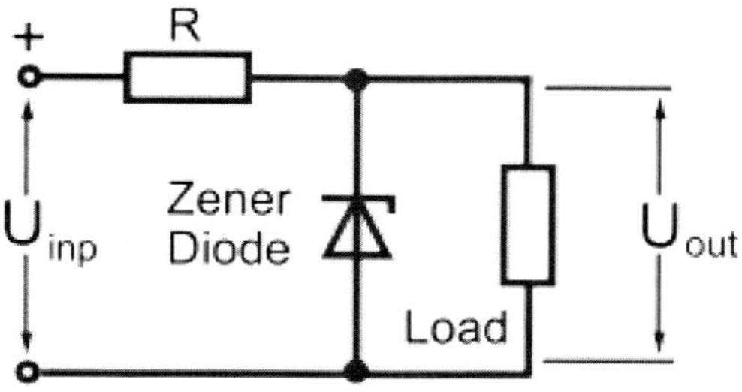

FIGURE 4.40 Simplest parametric stabilizer on Zener diode.

In the Zener diode one uses the effect of electric avalanche breakdown. At breakdown, in wide range of current change through the diode, the voltage changes on it very slightly.

Even when the input voltage increases slightly, the current through the Zener diode increases sharply and this results in a voltage drop increase on the limiting (ballast) resistor R. When this occurs, the voltage on the Zener diode itself, and therefore, on the load remains unchangeable.

If voltage U_{INP} changes from U_{MIN} to U_{MAX}, one may use the following equation to calculate R:

$$R = \frac{0.5(U_{INP-MAX} + U_{INP-MIN}) - U_Z}{I_{Z-AV} + I_{LOAD}},$$

where U_Z – rated voltage of Zener diode;

$I_{Z\text{-}AV}$ – average current through Zener diode.

Since the entire increase of the source voltage is dampened on resistor R due to current increase through the Zener diode, then as the most acceptable voltage at the circuit input one may apply any voltage with which the current through the Zener diode will not exceed the maximum allowable for it (otherwise stabilization would be impossible). This condition may be expressed in the following form:

$$\Delta U_{INP} \leq \Delta I_Z R,$$

where ΔU_{INP} – maximum change of input voltage;

ΔI_Z – maximum allowable current change through the Zener diode.

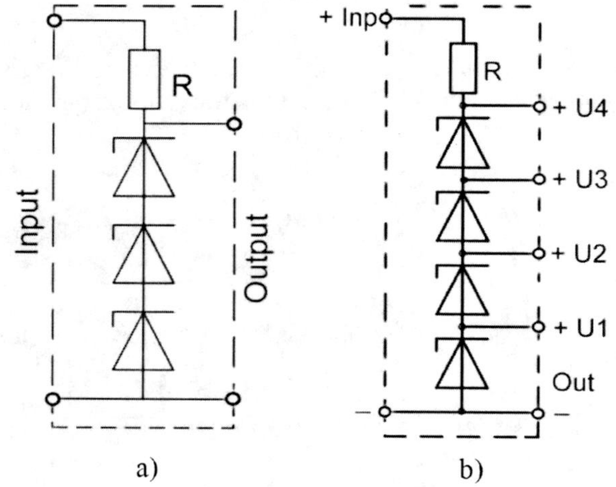

FIGURE 4.41 Parametric stabilizer on series connected Zener diodes.

Several Zener diodes can be connected in series if the operating voltage exceeds the rated voltage of each separate diode, as shown in Fig. 4.41a. The Zener voltage of such a circuit complies with the total rated voltages of serial connected diodes. Several different levels of stabilized voltage in a single circuit can be realized if in such a circuit one makes outlets from every Zener diode (Fig. 4.41b).

In practice a reverse problem also appears: obtaining stable voltages of less than rated voltage of the Zener diode. In this case one may use the circuit shown in Fig. 4.42.

FIGURE 4.42 Parametric stabilizer for voltage less than nominal voltages (U_1 and U_2) of the Zener diodes.

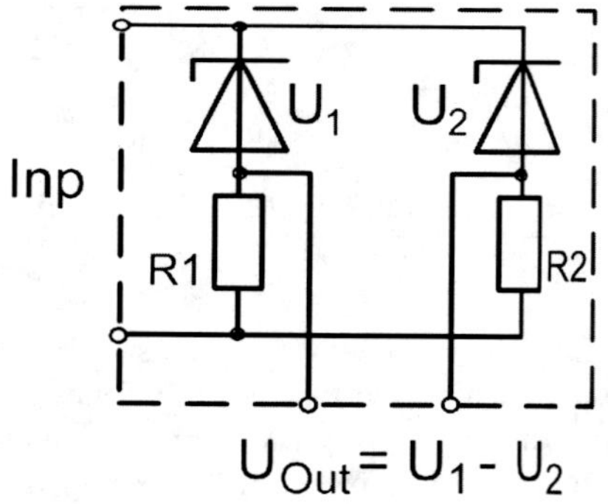

On the basis of the simplest parametric stabilizer circuit one may implement an adjustable stabilized power supply (Fig. 4.43). In this scheme the first Zener diode VD1 should be chosen with a rated voltage complying with the maximum limit of output voltage level, and the second Zener diode VD2 should be selected for the rated voltage complying with minimal limit of output voltage.

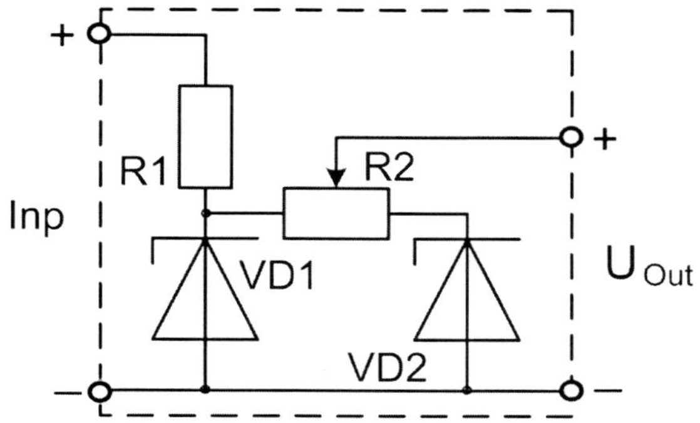

FIGURE 4.43 Adjustable parametric stabilizer.

To increase the power of a parametric voltage stabilizer, one connects it to the base circuit of a power bipolar transistor connected on the common collector scheme (Fig. 4.44).

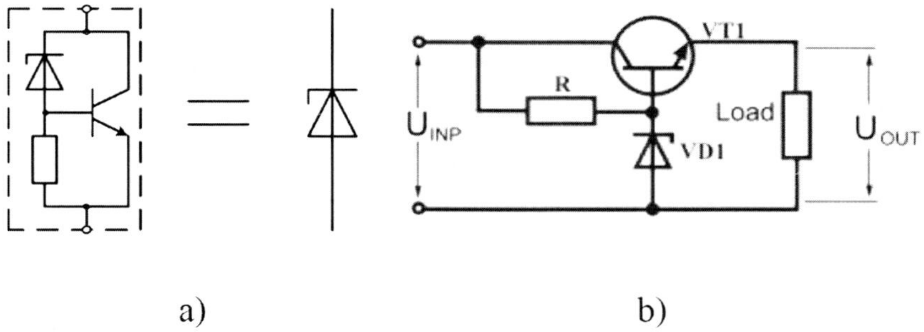

FIGURE 4.44 Power parametric stabilizer with transistor amplifier
a – equivalent circuit diagram; b – practical circuit.

By using the Zener diode to supply the base current to a power transistor, the power rating of load is limited only by the transistor. The maximum rated stabilized load current is equal to the product of the maximum current of the Zener diode and the gain of the transistor. This current may be implemented in practice with the condition that the transistor is selected in respect to the corresponding power dissipation. In this circuit it is not recommended to use a Darlington transistor as it adds to the Zener diode rated voltage an additional 1.5 – 1.7 V, and this can change the device's operational mode.

And finally, one can get a powerful adjustable parametric voltage stabilizer by means of combining several of the technical methods mentioned above (Fig. 4.45).

Let us now examine several variants of alternating voltage regulators. A typical circuit diagram of the simplest single-phase voltage regulator on one thyristor and one triac is shown in Fig. 4.46.

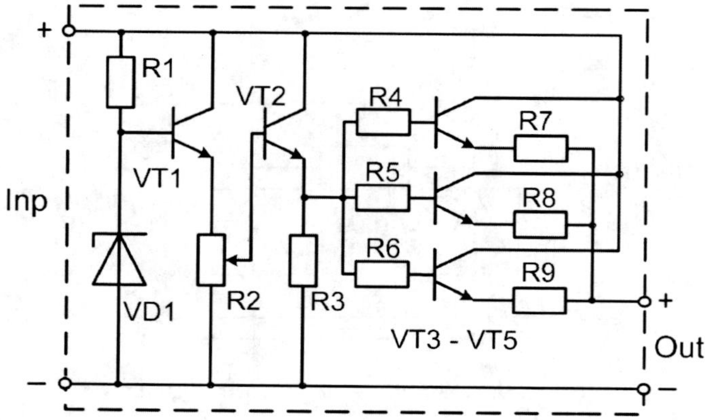

FIGURE 4.45 High power adjustable parametric stabilizer.

To provide a regulator working on both half-waves of sinusoidal voltage one may use a triac or thyristor connected in the diagonal of a rectifier bridge. The operating principle of these regulators is based on shifting the phase of the firing signal applied to thyristor gate with respect to the beginning of the sinusoid of power supply by means of circuit R1R2C (see Fig. 1.42). The average output voltage on the load changes at shifting the moment of the thyristor turning on relative to the voltage sinusoid beginning by means of adjusting the potentiometer R1.

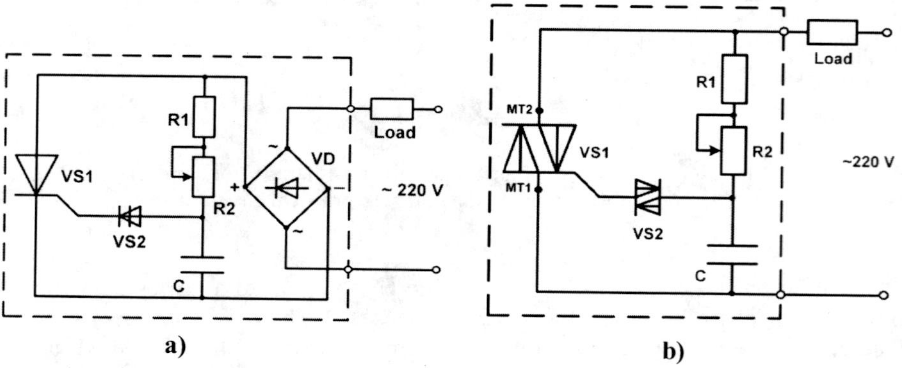

FIGURE 4.46 Typical single-phase voltage regulators.

In the circuit diagram 4.46a a diac is used as element VS2 forming the firing pulses, and in circuit 4.46b, an alternistor (triac) is used. For 220 VAC this element may be chosen for voltages of 30 – 40 V; C – 0.1 μF, 1200V; R2 – 470 kOhm; R1 – 4.7 kOhm. One should bear in mind that all elements of these devices, including the potentiometers, are under the full potential of the main network and this is why one cannot touch the potentiometer handle outside of the insulation body without taking safety measures. In the regulator shown in Fig. 4.45a only positive voltage is applied to the thyristor VS1 during operation and its natural turning off takes place close to the zero point of the current's sinusoid. In case of a load with a large inductive component, the current through the thyristor may not be reduced to zero and as a result the thyristor with the diode bridge may lose controllability.

In Fig. 4.47 another circuit diagram of a single-phase, two half-wave thyristor voltage regulator is shown.

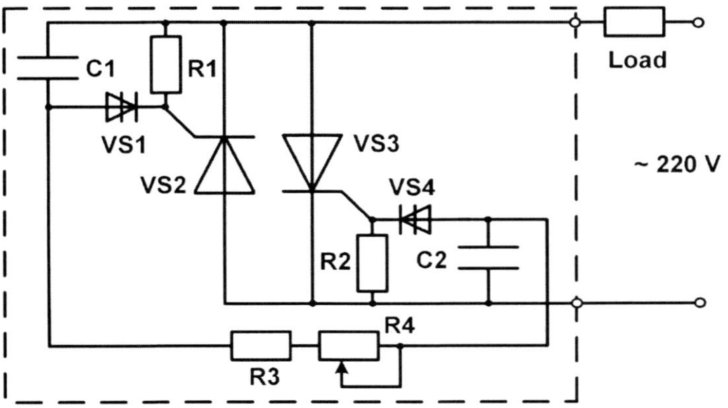

FIGURE 4.47 Two thyristor single-phase voltage regulator.

This device is more preferable than the triac regulator when it is needed to adjust high power because two separate thyristors dissipate power and withstand large overloads much better than a single triac. On the other hand, this circuit has advantages over the regulator shown in Fig. 4.46a, because it does not lose controllability in the event of high inductive loads. One may use the circuit elements with the following parameters: C1, C2 – 0.25 μF, 1200V, R4 – 220 kOhm; R1, R2 – 100 Ohm can be used when voltage is 220 V.

Three-phase voltage regulators are much more complex than single-phase ones, because one has to trace in them the phase angle of all three phases and synchronize regulating elements of all three phases to each other.

These problems are solved, as a rule, by using a microprocessor or a complex control system on microcircuit chips. At the same time, it is not always needed to implement the deep regulation of three-phase voltage. In many cases the matter concerns only compensation of voltage deviations in the supply mains for providing the normal functioning of consumers especially sensitive to voltage deviations. Such consumers include, for example, illuminating incandescent lamps, the service life of which very much depends on the mains voltage supply. As even maximum deviations of voltage level in 220 VAC network are limited to within 30 – 40%, regulators and stabilizers with limited regulating range may be used, and they are much simpler. Also voltage regulating in all three phases often times is not necessary. Rather it appears that the problem is one of providing stable voltage to a powerful one-phase consumer supplied from three phases. For these purposes one may use the simple regulator-stabilizer shown in Fig. 4.48. In the usual three-phase regulators, as was mentioned above, controlling (firing) pulses are delivered to thyristors in each phase separately with determined phase-shift, but in this regulator the gates of all three thyristors are united and the

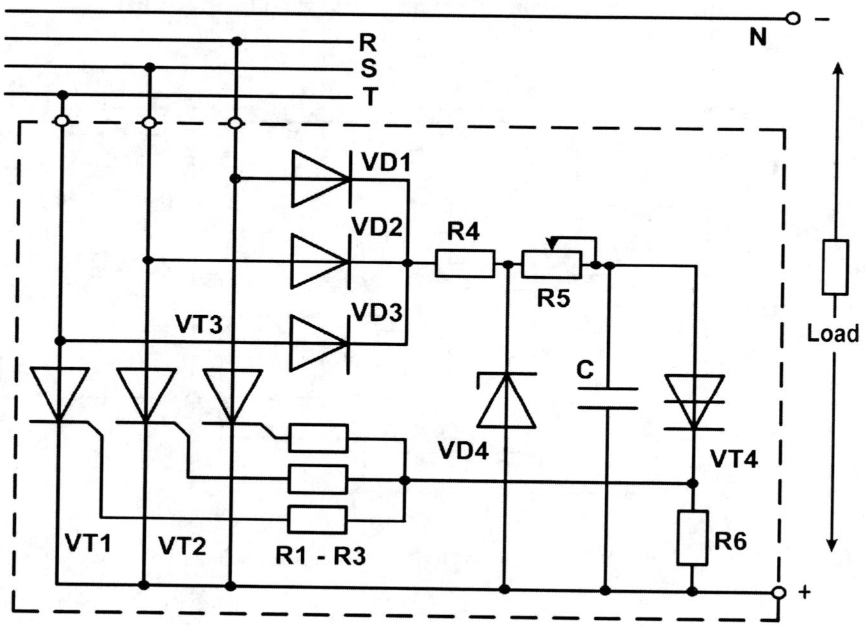

FIGURE 4.48 Voltage regulator-stabilizer with three phase supply.

firing pulses are delivered to them simultaneously with the common fixed phase-shift equal to approximately one third of supply voltage period. At this point, not all thyristors receiving the firing pulses are turned on, but only one, to which in a given moment of time the positive anode voltage is applied (see Chapter 1). Adjustment of output voltage is realized in the range of firing pulse phase change of 30° – 150° relative to corresponding phase voltage. This ensures a rather wide range of voltage adjustment; however, a shift of phase of the pulses by more than 150° in regards to the phase voltage may lead to the thyristors turning on at the beginning of the next half-wave of supply voltage, that is, to disturbances in the regulator functioning properly. Adjustment of this angle by the potentiometer R5 is limited, and this is why its resistance cannot be higher. The angle value is limited by the angle of thyristors turning on where a greater angle would exceed 150°.

The voltage regulator with a six-pulse rectifier appears to provide more qualitative rectification of alternating current (Fig. 4.49). Unfortunately, the rectification quality improvement is accompanied by constriction of the adjustment range: in this device it is as much as three times narrower than in the previous one. Nevertheless, even such limiting range is sometimes called for in practice. Device operating principles, however, do not differ from the one examined above.

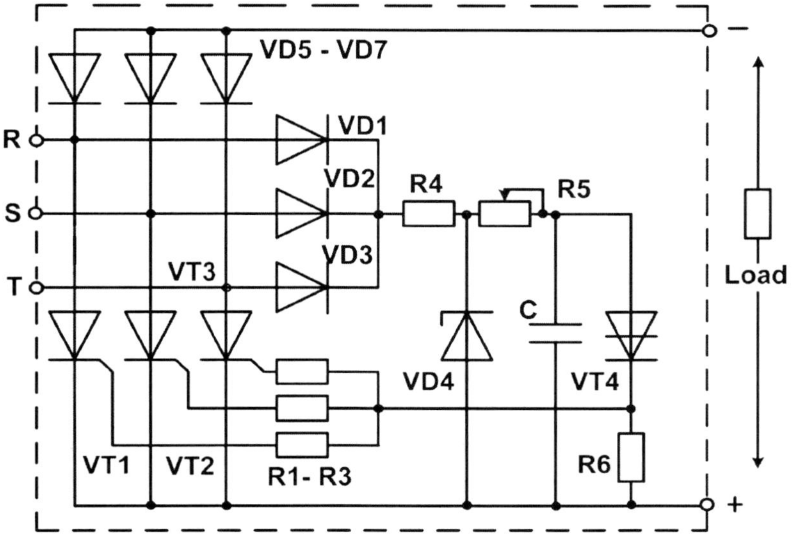

FIGURE 4.49 Voltage regulator-stabilizer with six-pulsed rectifier.

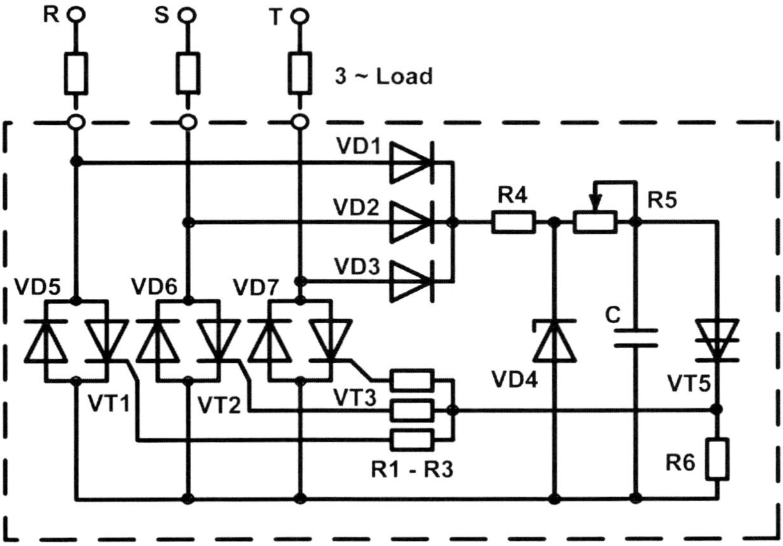

FIGURE 4.50 Three-phase voltage regulator.

One may use the same control principle, i.e., simultaneous delivery of firing pulses to the thyristors of all three phases, in a three-phase regulator of alternating current without a rectifier as shown in Fig. 4.50. This control principle allows creating mostly simple three-phase voltage regulators, which, in comparison to analogous manufactured regulators, are remarkably simple. One may use such a regulator to compensate voltage deviations in the network within 20 – 25%.

4.7 OTHER FUNCTIONAL MODULES FOR AUTOMATIC DEVICES

In this section we describe several functional modules that are certainly of interest, and are not related to any of the groups described thus far.

The circuit diagram in Fig. 4.51 depicts a device enabling the actuating sequentially on several different loads or a great number of stages in a single device by means of a single Set contact KS and a single Reset contact KR.

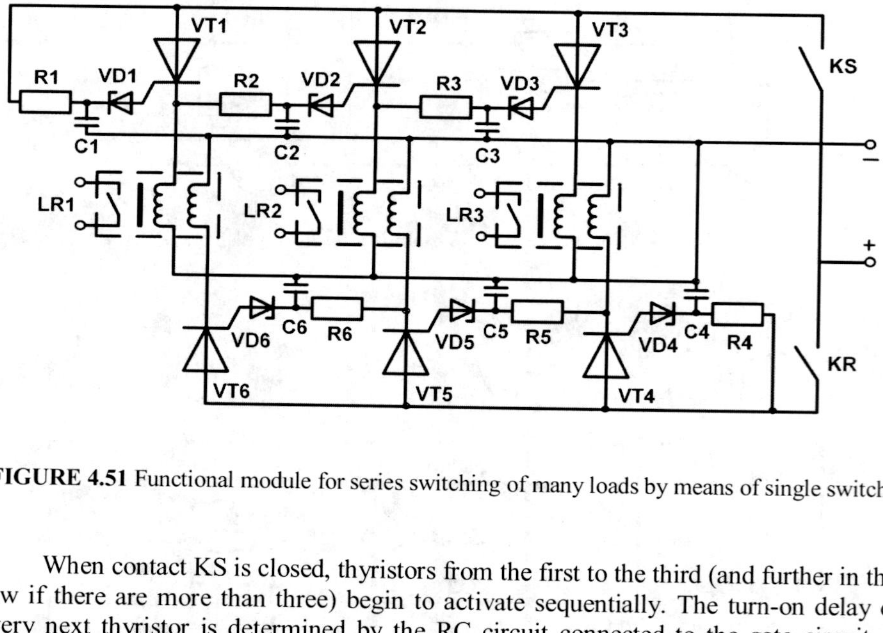

FIGURE 4.51 Functional module for series switching of many loads by means of single switch.

When contact KS is closed, thyristors from the first to the third (and further in this row if there are more than three) begin to activate sequentially. The turn-on delay of every next thyristor is determined by the RC circuit connected to the gate circuit of every thyristor. With activation every one of these thyristors delivers a power supply to the coils of the corresponding latching relay LR. Relays LR are pick ups one after another and they switch the circuit load. Release of the latching relays in a stepwise fashion returns the loads to their initial state beginning at the closing of Reset contact KR.

The duration of closed position of KS and KR switches determines the position of the load circuit. Thus, by means of only two control channels (two relays with contacts KR and KS) one may implement numerous switches in the load circuit. This is also important when control and executive circuits are under a high potential difference and contacts KR and KS are the contacts only of two high-voltage relays.

To indicate a device position parallel to the coils of the latching relays one may connects LEDs. It is advisable to connect resistors with resistance of several tens of kOhm in parallel with capacitors C1-C6.

A device for changing the voltage polarity applied to load is shown in Fig. 4.52.

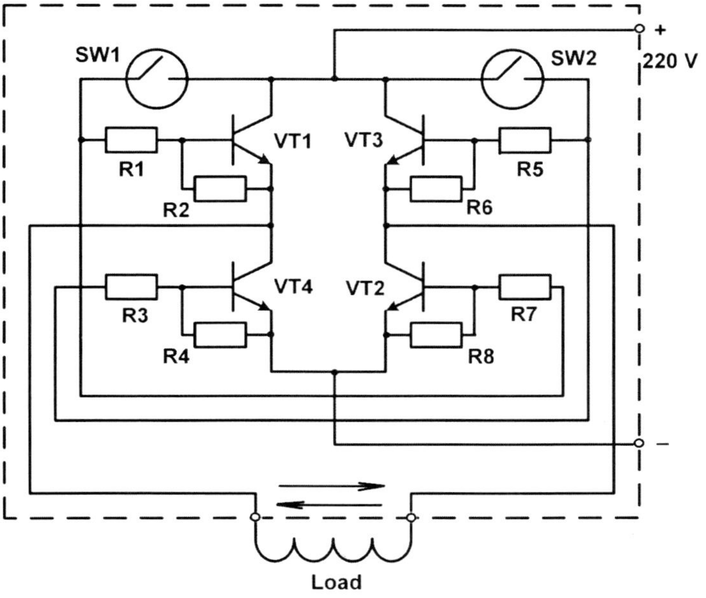

FIGURE 4.52 Polarity changer.

The device is based on a bridge circuit and its purpose is to periodically change the power supply polarity, as needed. For one polarity SW1, VT1, and VT2 elements must be activated. The other elements: SW2, VT3, and VT4 must be activated when other polarity is needed.

One use of the device is to change the direction of electromotor rotation.

In any case a blocking device preventing simultaneous activation of both starting switches during which short circuiting is possible is needed.

Instead of reed switches one may use transistors or opto-couplers.

In the device of an analogous purpose, shown in Fig. 4.53, one does not need any synchronizing elements preventing simultaneous activation of both switches. The change of polarity is implemented by means of SW1, and power-up by means of SW2. Thus, the load condition is determined by the position of both reed switches.

High-voltage, varistor-protected Darlington transistors should be used if the device is meant for powerful load control.

A circuit diagram of short pulses expander is shown in Fig. 4.54. The device allows energizing a load that is reliable for a period lasting from fractions of a second to several tens of seconds by the single closing control contact SW for a very brief time (fractions of milliseconds).

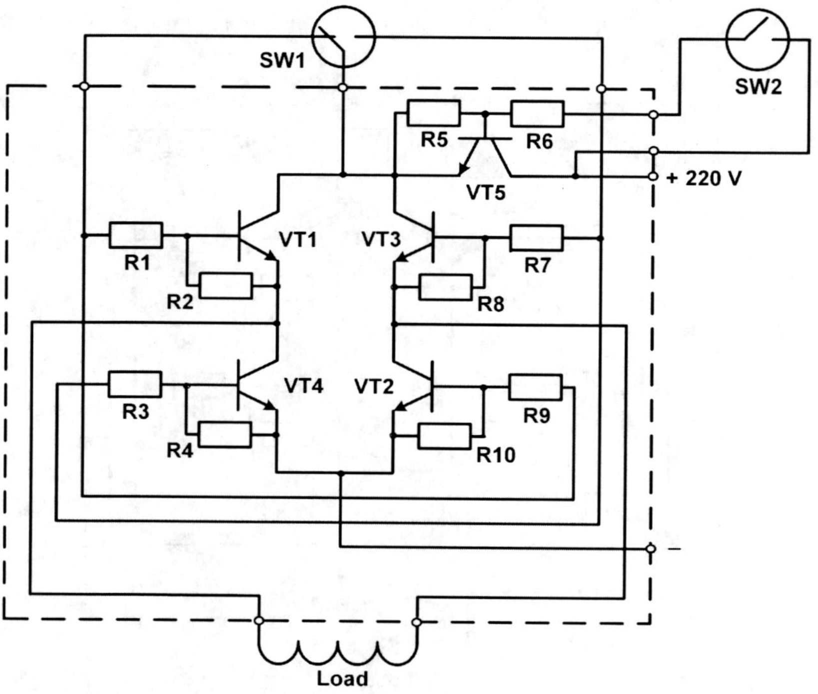

FIGURE 4.53 Polarity changer which does not need synchronization.

To charge capacitor quickly resistance R3 should not be high (20 – 50 Ohm), and thyristor VS1 should be high-voltage allowing current pulses of up to 10 – 15A. In order to ensure correct device functioning one should make the current holding of thyristor VS1 higher than the current flowing through resistor R4 (i.e., the current through the resistor should be lower than the current necessary for the thyristor holding in the turned on condition).

Transistor VT1 should have a high gain (about 1000) in order to reduce the current consumed by the base circuit from capacitor for keeping it turned on and for increasing the R4 resistance.

This is why transistor ST901T with high gain (1500) and high permissible emitter-collector voltages (350 V) is used. However, to ensure necessary voltage reserves this voltage is reduced (only for this transistor, as it is the only one that does not work with high current) to 50 – 100 V by means of the Zener diode VD1.

For correct device functioning one does not need to completely discharge the capacitor (this is why the device does not contain a circuit for its quick discharge). For sharp threshold of load de-energizing in series with resistor R4, a low-power Zener diode for 20-50 V may be connected; however, in this case the output pulse width decreases.

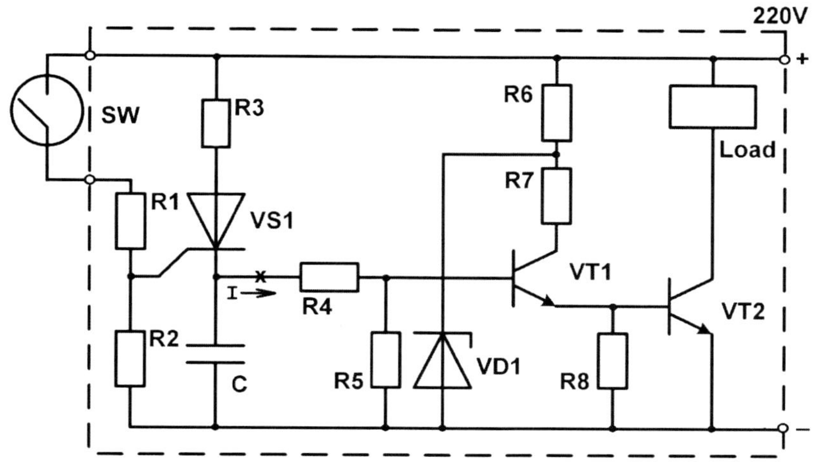

FIGURE 4.54 Single pulse expander.

Single pulses may not only be extended, but also transformed into two separate independent signals by means of the device shown in Fig. 4.55. This device is able to produce two single, galvanic-insulated signals first at the beginning and the second at the end of input pulse of random duration. The time difference between output signals is equivalent to input pulse duration.

The forward front of the input pulse turns on transistors VT1 and VT2 simultaneously. The first transistor shunts VT3 transistor base preventing its turning on. Transistor VT2 turns on first optocoupler Opt1 and thyristor VS1. At this, the signal appears at the first output of the device and at the same time LED1 illuminates to indicate device actuation. From the turned on thyristor the positive potential appears on the transistor VT4 base. This transistor forms, together with transistor VT3, a logical AND. The second output optocoupler Opt2 and LED2 remain inactive because transistor VT3 remains turned off.

Transistors VT1 and VT2 are turned off, but thyristor VS1 remains in conducting condition at the end of the input pulse. At this, the transistor VT1 shunting influence on the base circuit of transistor VT3 disappears. It is turned on and at the output of logical AND element a current appears which turns on the second output opto-coupler Opt2 and LED2.

As the device is low-voltage, low-voltage transistors of BC337 type and thyristor 2N5061 are employed. As output elements in the device high-speed transistor optocoupler of PVA3055N type are used.

Resistors' values: R1, R2 – 1 k; R3, R4 – 20 k; R5, R6 – 500; R7 – 300; R8 – 100; R9 – 230; R10, R11 – 30 k; R12, R13 – 5 k; R14 – 1.5 k.

The device shown in Fig. 4.56 is able to register current pulse flowing through it, i.e., remember the fact of the current flow having exceeded some determined threshold, keep and provide information about this condition.

In this device coils L1 and L2 are wound on two ferrite rings with inner diameters not less than 25 mm which are put together. A wire (cable) by which the current is being controlled is passed through the ferrite rings.

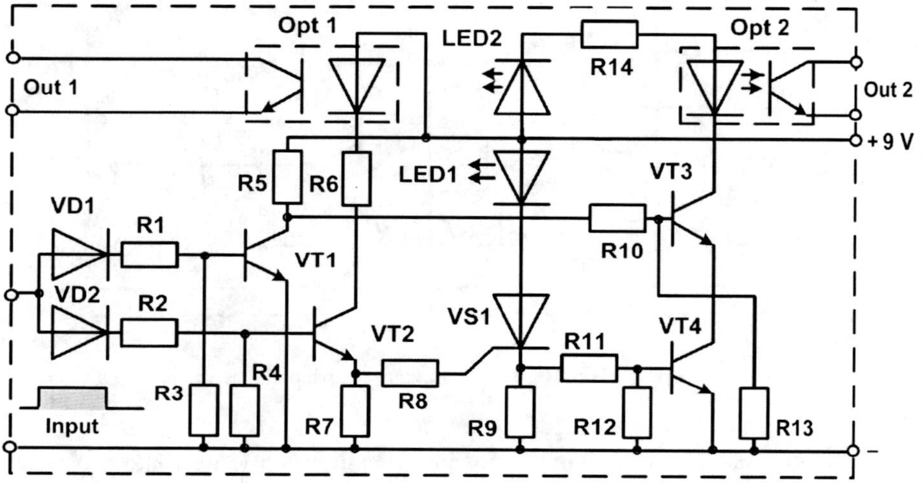

FIGURE 4.55 Converter of single pulse to two independent signals.

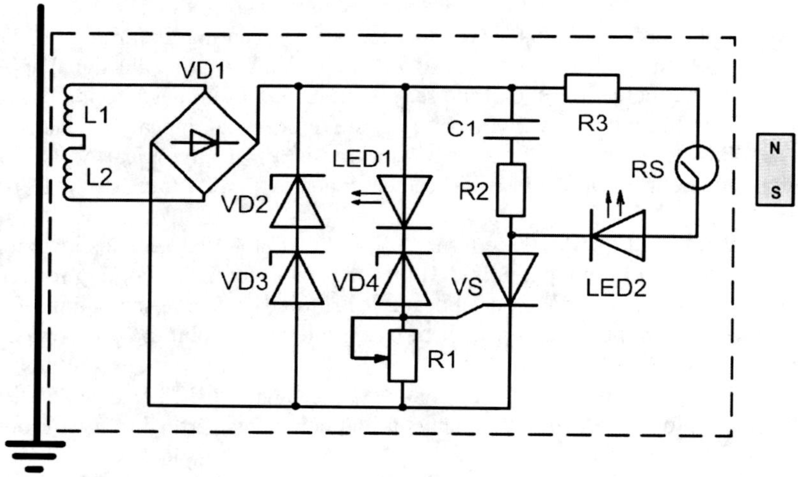

FIGURE 4.56 Memorized pulse current indicator.

Zener diodes VD1 and VD2 are meant for protection of electronic components against over-voltage in case a high current (for example, short-circuit current) passes through the controlling wire (cable). These Zener diodes limit voltage within 20 – 25 V. Capacitor C1 should be selected for the voltage of not less than 50 – 100 V.

Resistor R1 is intended for setting and adjusting the device's actuation threshold.

The minimal current in the wire (cable) by which the device can be actuated is 1.5 – 2 A. Because of this, the device may be desensitized considerably, up to hundreds of amperes. Its surge withstanding current can achieve kilo amperes.

LED1 serves as an indication of damage if high current sometimes flows through the controlled wire (cable). LED2 displays damage when cable is de-energized (turned off) by the protection apparatus. Information about high current that has been flowing through the wire (cable) is kept for several days. To do this capacitor C1 should be of high quality (with low leakage current). To check the condition of the device one connects the permanent magnet to reed switch RS. With this, LED1 flashing for the period of several seconds will mean that current flowing through the controlled wire (cable) has exceeded some pre-determined threshold. The permanent magnet should be installed at the end of the insulation rod if the device is intended for use in high-voltage applications.

5

Simple Protective Relays on Discrete Components

5.1 UNIVERSAL OVERCURRENT PROTECTIVE RELAY

Maximum current protection relays are the basic components in a majority of types of powerful electrical and electronic devices and also have use in power engineering. Analysis of the trends in relay technology development shows that the major relay developers do not share a solid and consistent direction for simple current relay design improvement. For instance, some experts reason that certain families of electric-mechanical relays currently in mass production need to be replaced by static IC relays. At the same time, they claim that regular electric-mechanical relays (including the simple maximum current relays) are most reliable and affordable for many electric utility companies, and will, therefore, be used in the majority of control and protection systems. It is important to note that the modern electric-mechanical relay is a high speed device, which is insensitive to pulse and high-frequency interference and surge voltage. It exhibits a very robust behavior in overload modes and has a satisfactory reset ratio. One has to agree, though, that electric-mechanical relays usually consist of many high precision expensive components, the production of which becomes inefficient for the relay manufacturer.

A dynamic measuring system with exposed contacts reduces the required relay reliability in a dust and gas intensive environment under a constant vibration factor. Besides, the need to clean and adjust the contacts contributes to increased labor intensive maintenance.

Static relays, however, have a lower complexity and a better assembly factor, since they consist of standard electronic components mounted on PCBs. They require zero maintenance and are decently robust when subjected to environmental and mechanical impact.

At the same time, the threshold components, such as IC triggers and comparators as well as the transformer by means of which ICs are connected to the high current circuits, cause an entirely new range of problems related to the interface immunity issue. The threshold components happen to be extremely sensitive to high-frequency signal interference, pulse peak interference coming through the feeding circuitry, etc. Therefore, it is difficult to filter out the useful signal in the wide spectrum noise background for these components.

In compliance with the recommendations of the IEC, static relays are to be subjected to obligatory special noise immunity tests. At the same time, manufacturers of such relays do not recommend to perform tests that include the application of electronic relay inputs having high-voltage pulses and powerful high frequency signals. Moreover, it is not recommended even to use megohmeter for insulation tests of such relays. Similar tests are resolved only for term, when all relay inputs are connected together. In this case, there is no sense in such tests!

The input relay transformer – an interface between the highly sensitive electronic module and the high current circuit – transforms the useful signal as well as the noise. Besides, in many cases, the transformer itself becomes a source of the interference. While this issue is very critical for static relays, the electric-mechanical relays are well compatible with transformers. In this way, the transformed-based interfaces have found a wide range of applications in relay protection engineering.

We have used the above considerations as a basis for the new methodology in current relay design to combine the advantages of the two relay concepts (electric-mechanical and static). The basic guidelines of our methodology are as follows:

- the threshold element, which in principle is a measuring organ, has to have an electric-mechanical structure to ensure interference immunity;
- it is expedient to use a reed switch equipped with a special module to move the former relative to the control coil;
- to ensure the compatibility of the reed switch specifications with the output commutation component, an interface unit should be implemented with the discrete electronic components (not ICs!) having wide current and voltage margins; the number of these components has to be minimal and their schematics should not result in a threshold circuitry (such as a trigger, comparator, monostable vibrator, etc.).

Relay designers are well familiar with the technical characteristics of these reed switches: high level of protection from environmental impact, extreme reliability, a large communication resource, zero maintenance requirements. It is less known, however, that their reset ratio in the AC magnetic field is about 0.8…0.9, and the pickup ratio (not to be confused with its statistical variance) is relatively stable, and its adjustment function has been technologically resolved. The problem of the statistical variance of the magnetomotive force becomes irrelevant with the introduction of the reed switch adjustment module. The initial value of the relay can be defined at the manufacturing stage.

The mentioned principles have been implemented in a whole new family of relays including the universal maximum current relay, arc protection relay, short circuit indicator, current relay with a non-transformer HV interface, etc. Some of these developments are described below.

The "Quasitron" is a multipurpose protection relay, based on a hybrid (reed-electronic) technology, with very high noise immunity (Fig. 5.1, 5.2).

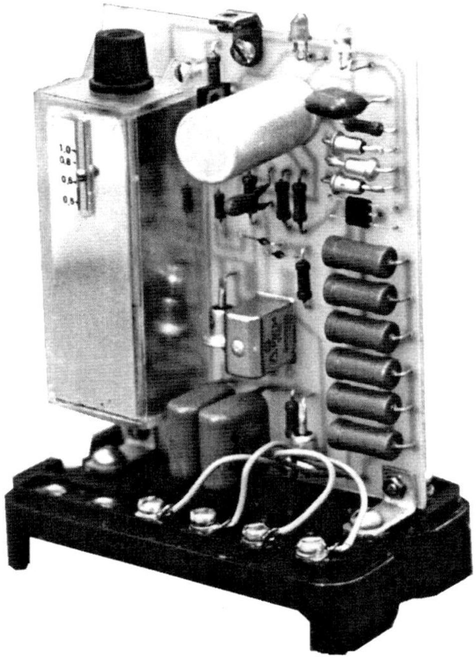

FIGURE 5.1 Hybrid over-current protection relay "Quasitron" series (without protection lid).

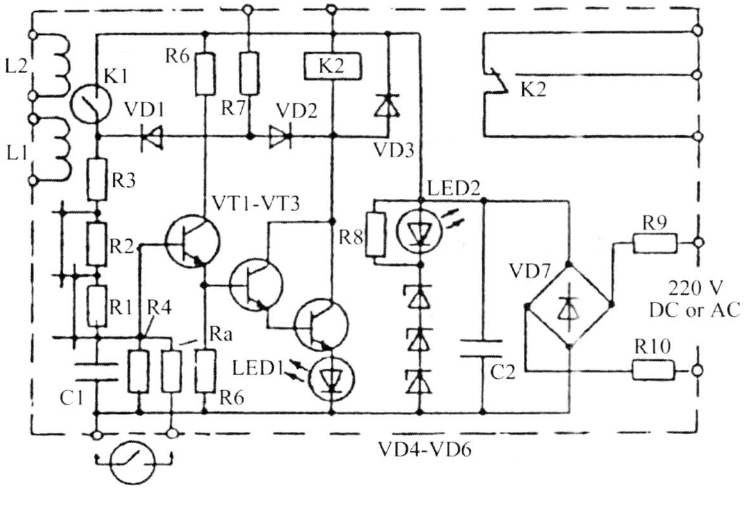

FIGURE 5.2 Circuit diagram of "Quasitron" relay.
K1 – reed switch; L1, L2 – input current coils; K2 – output auxiliary relay.

One relay unit may be used simultaneously with different current sensors: low voltage (above) and high voltage, each of which has a different current trip value.

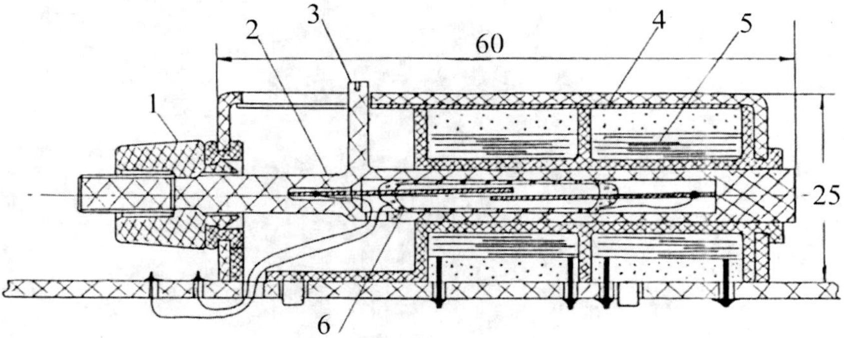

FIGURE 5.3a The "Quasitron" current sensor with adjustable current trip level. 1 – limb; 2 – movable dielectric capsule; 3 – level indicator of current trip; 4 – ferromagnetic screen; 5 – coil; 6 – reed switch.

A current sensor may be mounted into the relay unit (as shown in Fig. 5.1) or mounted outside the relay unit on an additional plate (Fig. 5.3).

FIGURE 5.3b External low voltage current sensors to "Quasitron" relays. 1 – cutting-circuit sensor type "1" for current trip 0.01 to 100 A; 2 – sensor type "2" for bus bar and cable installation (30 to 10.000 A).

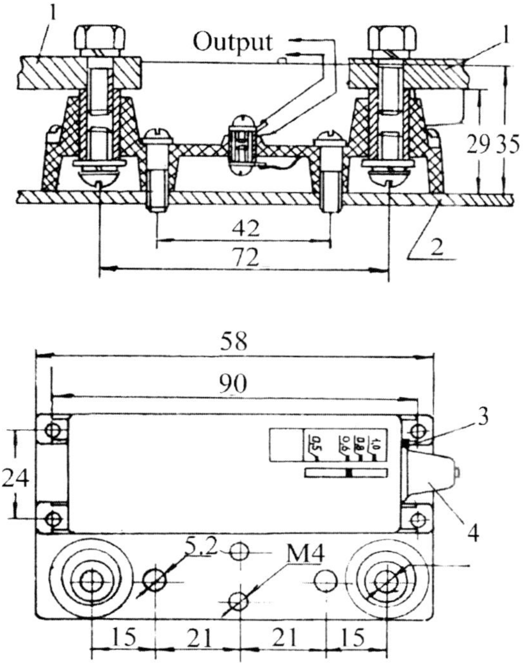

FIGURE 5.3c Outside dimensions of external low voltage current sensors. 1 – external wires of current circuit; 2 – plate; 3 – fixative element; 4 – limb; "output" is connected to "Quasitron" relay input.

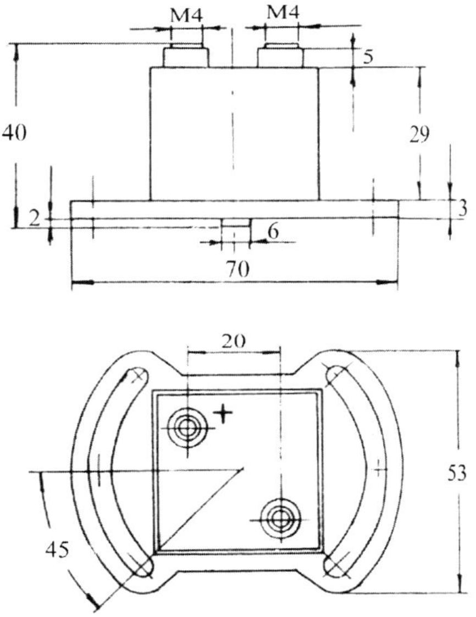

FIGURE 5.3d Outside dimensions of sensor type "2" for bus bar and cable installation.

The relay unit has three time/current characteristics (T1 – T3), Fig. 5.4, one of which can be selected by a customer by means of jumpers on resistors R1, R2 (see Fig. 5.2). For this purpose one (or two) jumper(s) can be cutting.

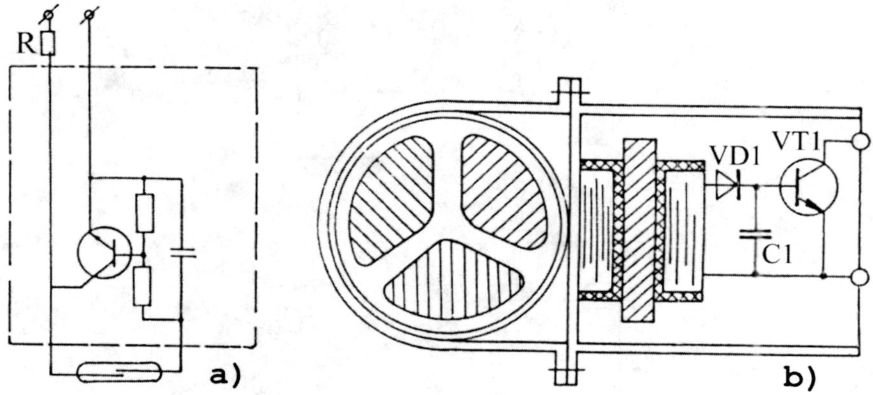

FIGURE 5.3e Circuit diagrams of type "2" sensors
a – for current level 100 A and more; b – for low current levels.

All sensor outputs are connected to the relay unit via low voltage wire.

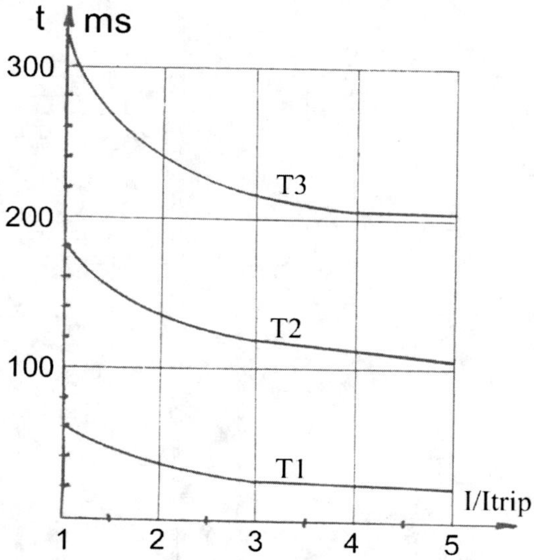

FIGURE 5.4 Time/current characteristics of "Quasitron" relay.

As the circuit configuration (Fig. 5.2) implies, it does not contain ICs; its active solid-state components (transistors) do not constitute a threshold element and are merely used as an amplifier. An interface between the electronic circuit and the outside network bus is implemented via an insulated interface, based on reed switch K1, which also plays the role of threshold elements and starts vibrating with the double network frequency when the relay trips. The contact erosion-free capacity of the reed switch (about $10^6 - 10^8$ operations) along with the short period of the maximum current relay's on-state, ensure the required commutation resource of the relay.

The amplifying module of the base circuit (Fig. 5.2) is nothing else than a compatibility link between the integrating couple L1-L2C1, and the output auxiliary relay K2 provides for the stability of the on-state of the relay under the K1 vibration conditions.

The feeding voltage of transistors (2N6517, 2N5657 or similar series) does not exceed 50...70 V, the nominal operating range being 350 V. The control coil (K2) current under the tripping relay condition is as low as 15 mA, while the maximum collector current limit for these transistors is 500 mA. This magnitude of the current and voltage margins ensure a high level of the relay's operational reliability.

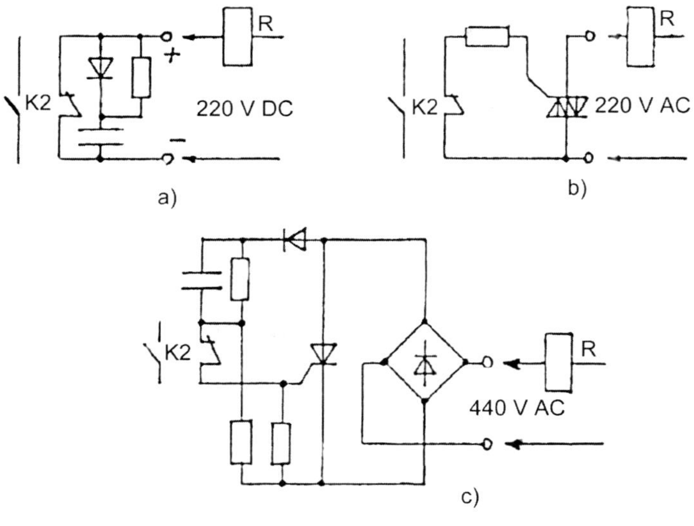

FIGURE 5.5 Output modules for "Quasitron" relays.
K2 – contact of output auxiliary relay, mounted on PCB in relay unit; R – load of RL-type;
a – with spark protection, for DC load with large inductance;
b – with power amplifier, for power AC load (up to 500 VA);
c – for AC load, connected to power supply with voltage more than switching voltage of output auxiliary relay.

The high frequency and short pulse interference at the relay input cannot migrate to the electronic module, since K1, being the interface link, does not react to the high frequency control signals due to the inherent inertia.

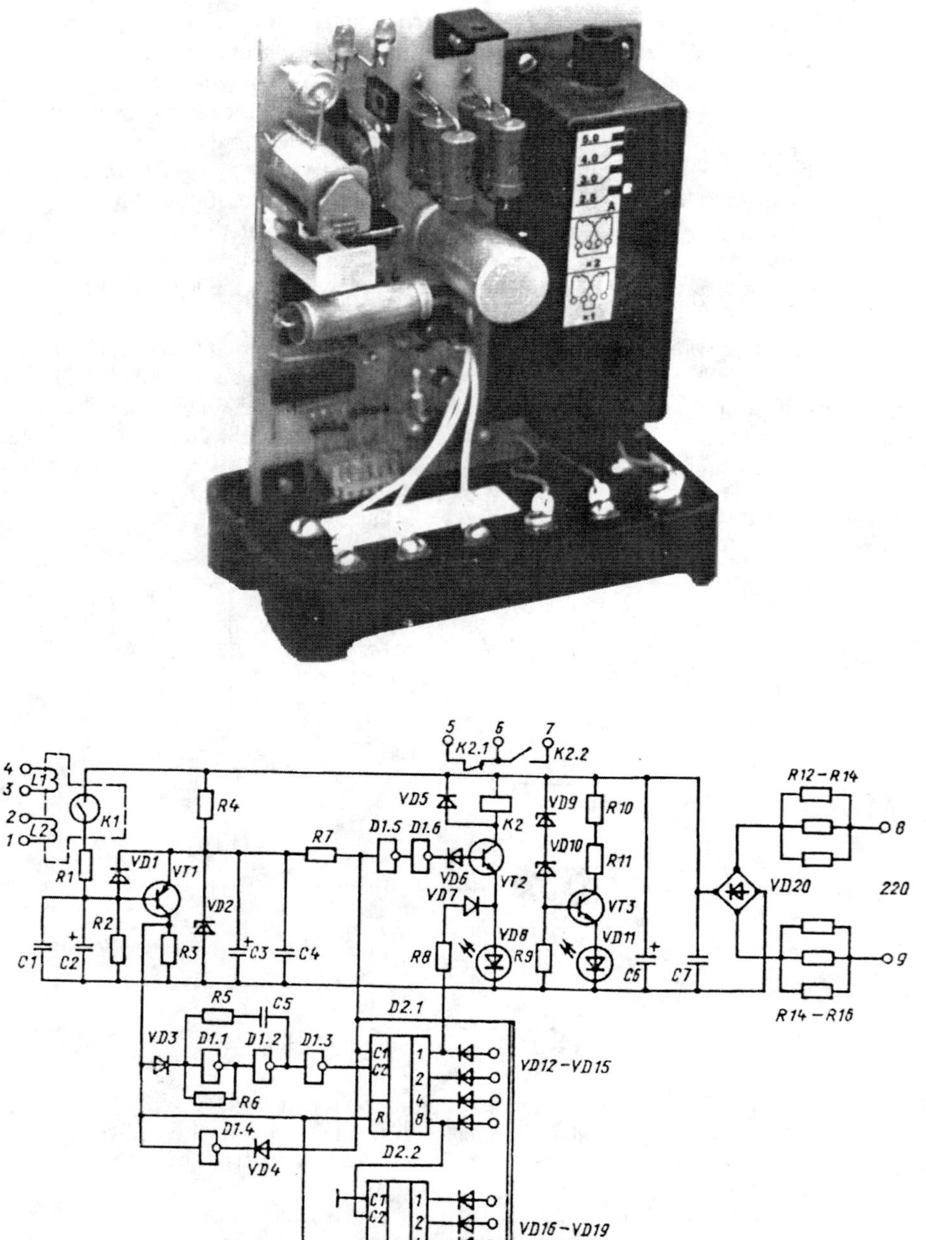

FIGURE 5.6 Overload protection relay "Quasitron-T" type with the built-in timer.

Neither does it respond to the transient interference from the power circuit commutation. Therefore, the whole relay becomes very robust to the power circuit pulse interference.

The effect of the magnetic component of the dissipation fields can be neutralized by introducing the ferromagnetic screen into the relay design (see Fig 5.1). The 1.5 mm screen shields the reed switch in the fields with an intensity much higher than that of the dissipation fields under the actual operation conditions.

For different applications of the "Quasitron" device, several types of output modules are available (Fig. 5.5).

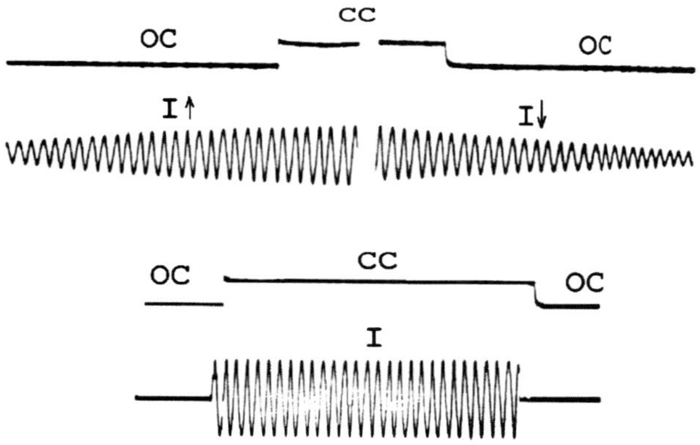

FIGURE 5.7a Some oscillograms, received during test process. Sudden increasing and decreasing of control current (**I**).
OC – opened contact; CC – closed contact; **I↑** – slow increasing of control current; **I↓** – slow decreasing of control current

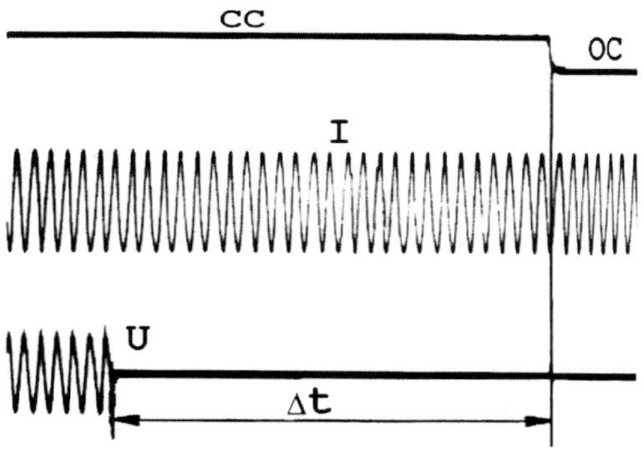

FIGURE 5.7b Some oscillograms, received during test process. Unset of relay after tripping in case of sudden disconnecting from power supply with voltage U ($U = U_{NOMINAL}$; frequency – 50 Hz).

Perspective improvement of this device is a current protection relay with the built-in timer (Fig. 5.6).

Time dial is adjustable in the range 0.1 to 25 (± 2.5%) seconds with a grade of 0.1 s. Dimensions of all modifications of relay units "Quasitron" and "Quasitron-T" device are 110 x 65 x 150 mm.

All types of relays were tested. Some oscillograms are shown in Fig. 5.7.

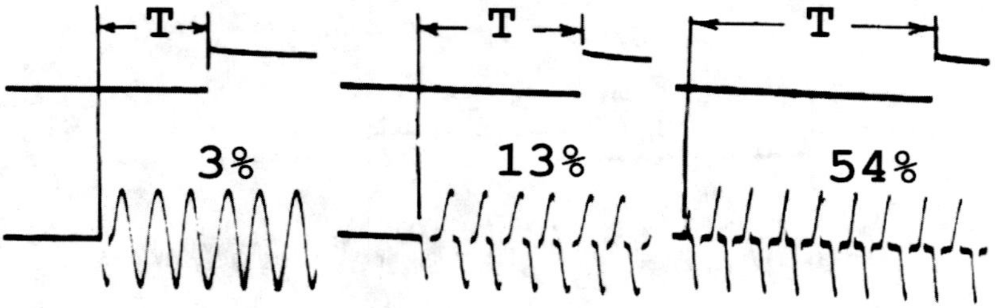

FIGURE 5.7c Some oscillograms, received during test process. Influence of distortion of the measured current in relay input (simulation of overload of current transformer) on a response time (T) of the relay.

5.2 SIMPLE VERY HIGH-SPEED OVERCURRENT PROTECTION RELAY

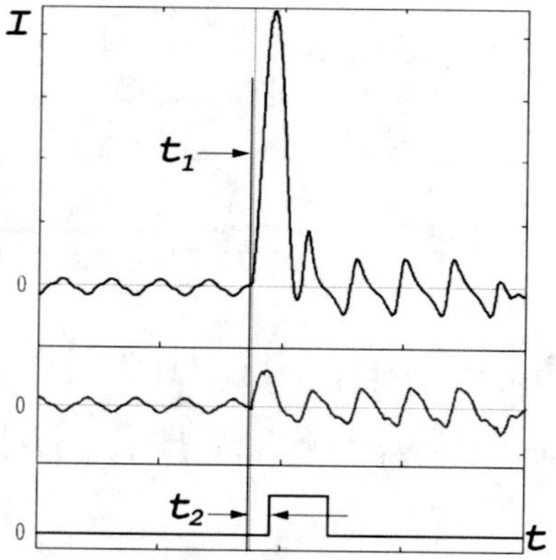

FIGURE 5.8 Oscillogram of operation of a high-speed microprocessor relay of the SEL-487B type. According to promotional materials its operation time is less than 20 milliseconds.

Overcurrent and overload protection functions for both low-voltage and high-voltage consumers of electric power (and also electric networks) are usually realized on current relays with dependent or independent time delay characteristics, or on high-speed differential relays or impedance (distance) relays (for power line protection). In some situations, however (at close short circuits and high-power sources), the multiplex overcurrent passing through the protected object is capable of causing destruction of the object, even when it is protected with one of the above-mentioned protection relays.

For such cases special very high-speed relays are stipulated. Usually the time delay of such high-speed relays, both electromechanical (for example, KO-1, produced by ABB) and microprocessor-based (for example, SEL-551C from Schweitzer Engineering Laboratories, BE1-50 from Basler Electric, RCS-931A/B from NARI, etc.) is within 20 to 40 milliseconds (as stated by manufacturers). In addition, electromechanical protection relays with instantaneous pick-up characteristics frequently provide even higher speeds (18 – 25 мс), than microprocessor-based relays.

Promotional materials may sometimes be found which claim that an especially constructed high-speed microprocessor relay is capable of operating with a time delay of less than one period (less than 20 мс) (Fig. 5.8).

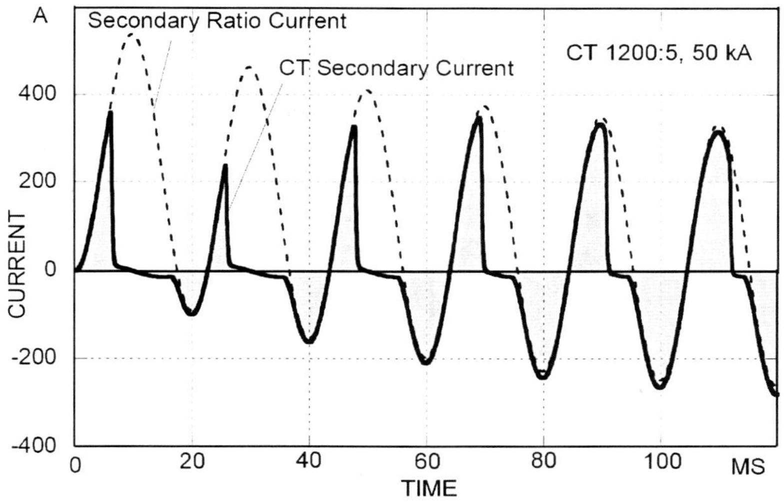

FIGURE 5.9 Shows relation between the CT secondary current, applied to input of the relay and secondary ratio current at close short circuit mode.

Such small operating time delays really can be realized sometimes for microprocessor relays with injection of high current with an artificially fixed phase for the first half-cycle (as on the oscillogram, Fig. 5.8). Unfortunately in practice such extreme artificially created conditions are rarely achieved, therefore such unique operation times look more like an advertising gimmick than a parameter provided under real operating conditions.

Many companies are engaged in development and production of actual high-speed relays. The analysis of real transients of short circuits with high DC components and strong CT saturation has brought some researchers to the conclusion that it is impossible to provide relay protection for operating times of about one half-cycle (10 ms). These researchers offer a new algorithm based on measurement of first *(di/dt)* and even second *(di2/dt)* current derivatives.

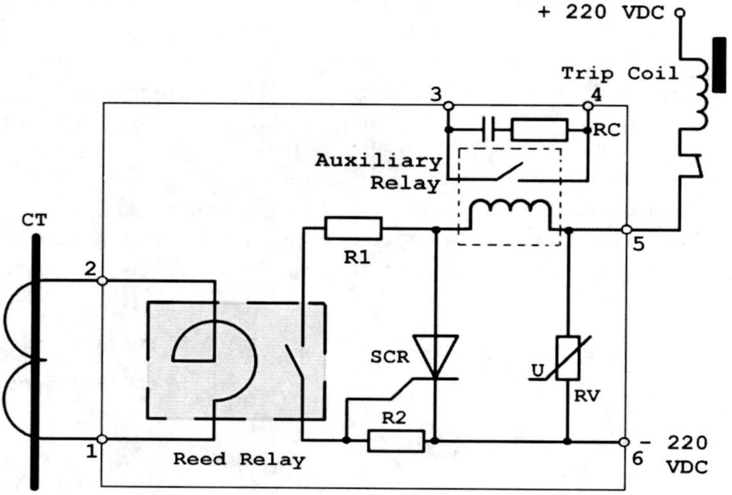

FIGURE 5.10 Basic circuit diagram of a simple very high speed overcurrent relay on reed switch.

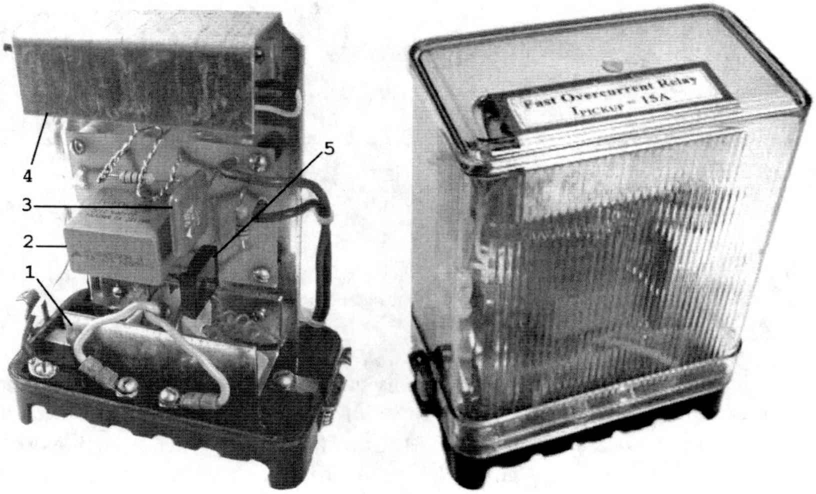

FIGURE 5.11 Construction of very high speed overcurrent relay.
1 – reed relay module; 2 – capacitor; 3 – varistors; 5 – ferromagnetic screen of auxiliary reed relay; 5 – thyristor.

In reality, experimental oscillograms of transients (Fig. 5.9) confirm the stability of such parameters as a current derivative (speed of change of current, or in other words an inclination angle of the front of the first pulse of a current at short circuit) even with high DC components contained in the current. On the basis of these researches one of the Israeli companies has developed a microprocessor relay with this algorithm.

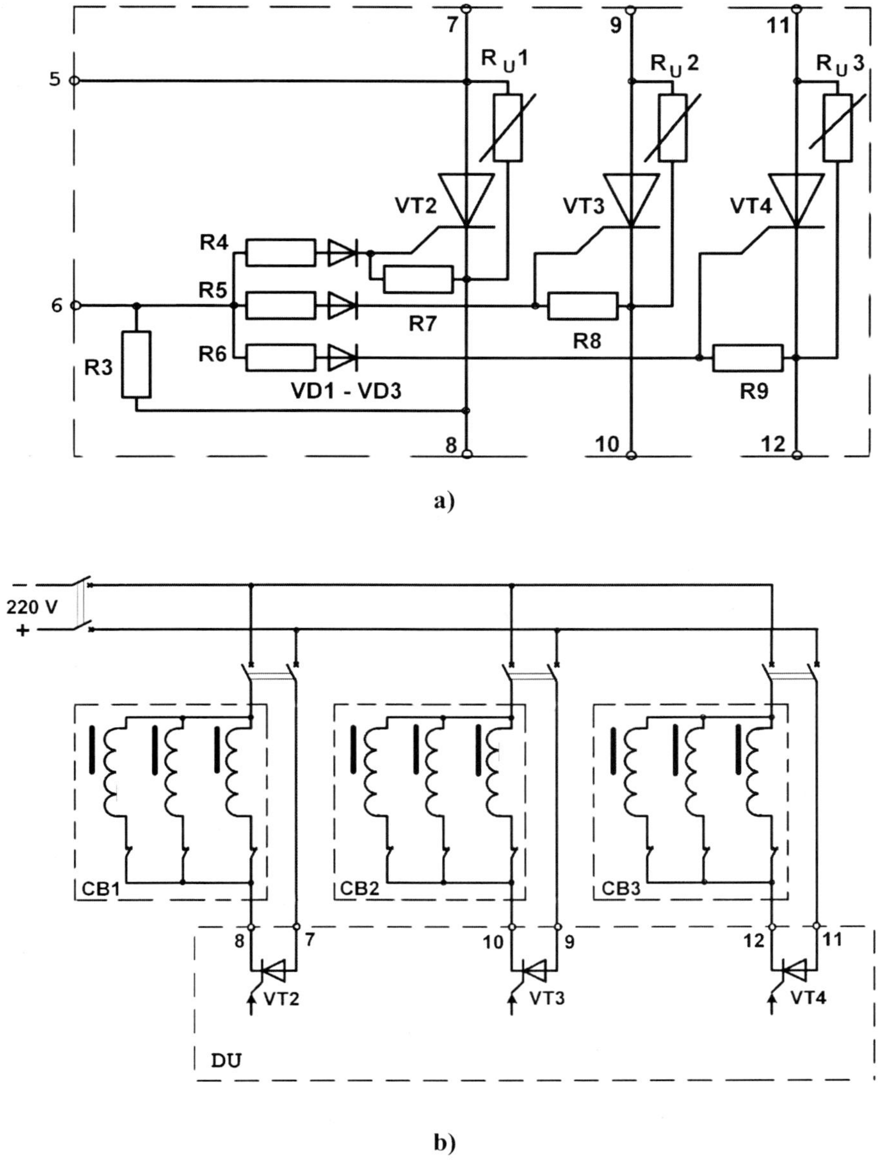

FIGURE 5.12 Demultiplexing unit (a) and connection diagram (b).

Thus the relay has turned out to be relatively complex because measurement of only the second current derivative is insufficient for realizing necessary relay stability. Inserting special elements for blocking of excessive relay operations is required because of the excessive sensitivity of the relay to some operating modes, as revealed. In addition, as the current derivative depends on a relation between an initial current before failure and a pickup current at failure, it appeared that the relay does not always work properly if relative high load current is preceded to failure mode, and vice versa, excessive relay operations sometimes take place for great current changes (from zero value up to high values, but less than pickup value).

Inserting special elements for blocking of excessive relay operations is required because of the excessive sensitivity of the relay to some operating modes, as revealed. In addition, as the current derivative depends on a relation between an initial current before failure and a pickup current at failure, it appeared that the relay does not always work properly if relative high load current is preceded to failure mode, and vice versa, excessive relay operations sometimes take place for great current changes (from zero value up to high values, but less of pickup value).

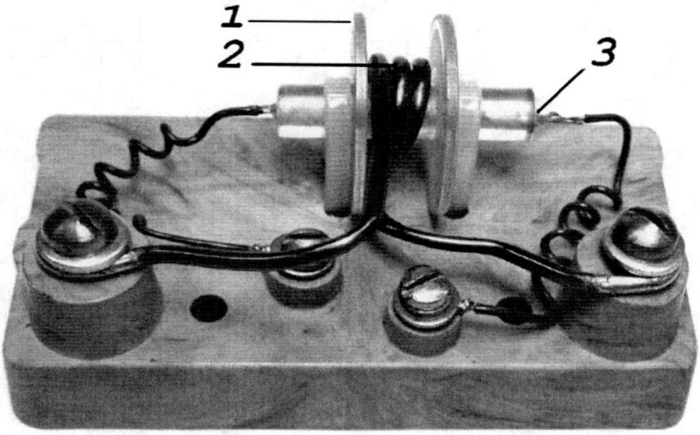

FIGURE 5.13 Unit of reed switch, submitted on tests with rated pickup current of 10A.

Despite some technical problems, preliminary tests of the relay prototype have confirmed its high speed. For the most difficult cases the time delay displayed was 8.4 MS, which is much less than any microprocessor relays existing today in the market. The EMI compatibility and some other important relay parameters have not been investigated yet; nevertheless, the possibility of creation of the overcurrent microprocessor relay with an operation time of about a half period has been confirmed.

The author offers an alternative variant of a very simple and low-priced high-speed overcurrent relay with an algorithm based on measurement of instantaneous value of a current. The relay is so simple that it can be produced by the staff of power systems. The offered overcurrent relay is based on a reed switch with a high-voltage thyristor as an electronic amplifier, Fig. 5.10, 5.11.

The basic sensitive element in this device is the reed switch, which begins to vibrate at a pickup with frequency of 100 Hz. Its first pulse opens a powerful thyristor SCR, which energizes a circuit breaker trip-coil.

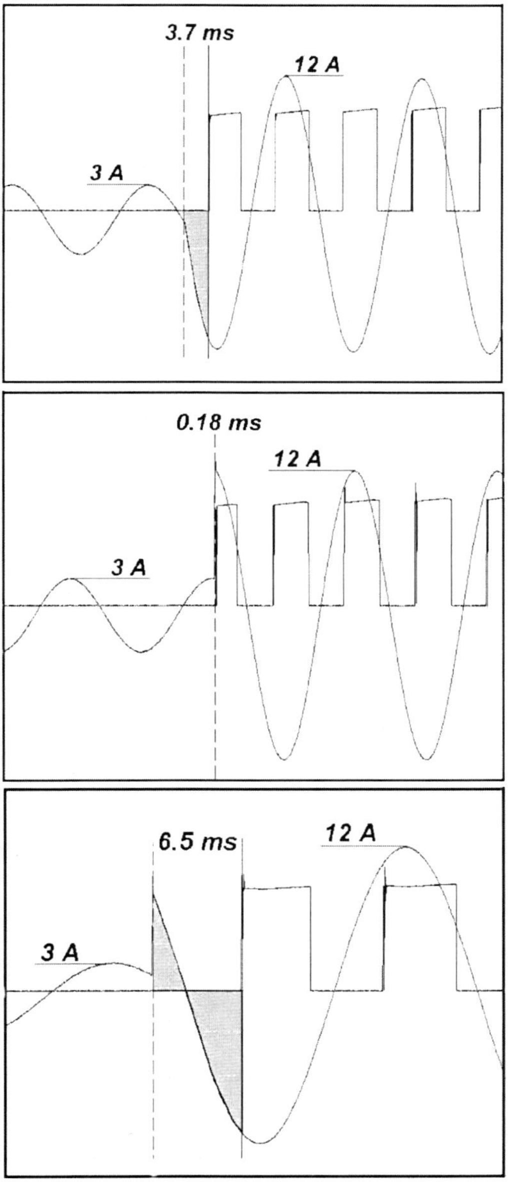

FIGURE 5.14 Some oscillograms of operation of reed switch unit at instantaneous changes of current with various phases of current transition. Non-operating zones of the relay are marked.

The thyristor remains in the conductive condition, despite reed switch vibration, so long as the circuit is not turned-OFF by auxiliary contact of the circuit breaker. An addition of an auxiliary relay with a low impedance current coil and a spark protected power reed switch may be used for energizing of external electromechanical relays of automatic or signal systems.

Subminiature high-voltage vacuum reed switches of the MARR-5 (Hamlin) or MIN-21 (Binsack Reedtechnik GmbH) types, with withstanding voltage of 1.5 – 2 kV and turn-ON times of not more than 0.6 – 0.7 ms, are used as metering elements that provide high reliability of the relay.

A thyristor of the 30TPS16 type was also chosen with a large reserve for current (30A) and for voltage (1600 B), making it possible to choose the varistor RV for protection from overvoltages with large reserve (clamping voltage of about 800 VDC) regarding rated voltage (220VDC), providing both higher reliability and longer durability.

The reed switch module can be provided with different methods of pickup adjustment: by means of a moving reed switch inside the coil, or by using different modules with different fixed values of pickup current. The last variant is quite acceptable, as this module is very simple and low-priced. After adjustment of the reed switch position in the coil, it must be fixed by means of silicon glue.

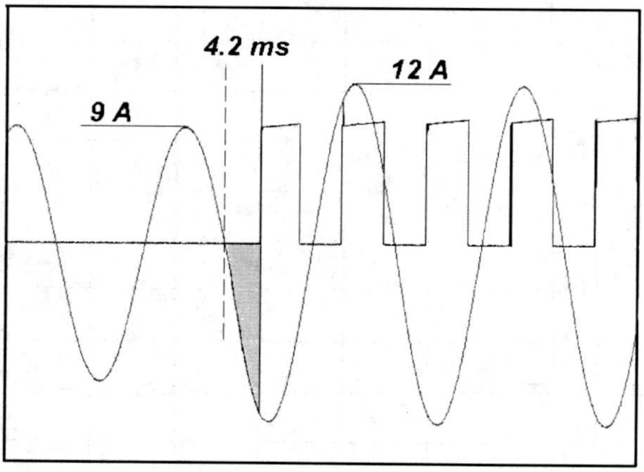

FIGURE 5.15 Oscillogram of operation of relay with previous non-operating current (9A), near to a pickup (10 A).

Output auxiliary relays are also made as reed relays (without adjustment) because their winding is not standard relay winding, but is designed as current winding (80 – 100 turnings) for current values suitable to the trip coil of circuit breakers. For such purpose power reed switches can be used, for example R14U, R15U (Yaskawa Electric America); MKA-52202 (Russia); GC 1513 (Comus Group); DRT-DTH (Hamlin), provided with spark protection (RC-circuit).

As usual, high voltage CB has separate drivers and separate trip coils for each phase. All these coils may be connected in parallel. In an actual emergency mode the single protective relay must disconnect not just one, but several CB simultaneously.

Otherwise the current of all trip coils may exceed maximal permissible current for a single thyristor. In this case a demultiplexer with single input is used, connected with trip relay contact, and several separate outputs for circuit breakers, as shown in Fig. 5.12.

The prototype model of a 10A pickup reed switch module (Fig. 5.13) without a thyristor amplifier (thyristor switch-ON time is less than 10 μs, which does not affect in any way the general time delay of the device) and without an auxiliary relay, has been submitted to tests.

Tests were performed by artificial simulation of various modes on a current by means of a Power System Simulator F2253 (DOBLE Engineering), and also by injection in the module, by means of the same simulator, of real secondary currents of short circuit transients restored from COMTRADE files of the real failures in a 160 kV power network, extracted from microprocessor-based transient recorders.

In the first series of experiences operation time of the module was measured at instant change of current on an input of the module in a range from 0.2 – 0.8 I_{PICKUP} up to 1.2 – 5 I_{PICKUP}, with various random phases of current transition and also with a zero phase of current sinusoid (Fig. 5.14). The tests verified that the lower limit predetermined current value preceded to pickup current does not affect operating time (Fig. 5.15), as against microprocessor based relay reactions to current derivatives.

Research also affects harmonics (contained in a current) on operating time at different phase transitions of a current, Fig. 5.16, and verified that even the high harmonics content does not affect operating time.

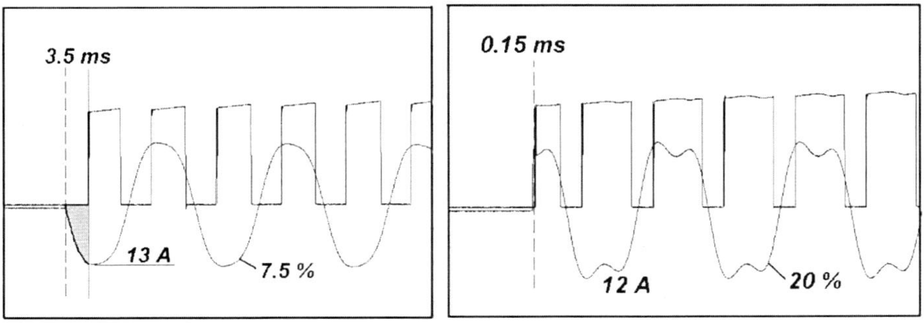

FIGURE 5.16 Oscillograms of operation of relay at high harmonic content in current (for contents of the third harmonic of 7.5 % and 20 %).

The main factors are still the phase and magnitude of a current transient. For the most difficult case, that is at small current $I = 1.2I_{PICKUP}$ and with switching current phase at close to 45°, maximal operation time can reach 7 – 8 ms.

Heavier testing appealed for real secondary currents of short circuit transients containing a high DC component, causing displacement of a sinusoid of current concerning an axis (Fig. 5.17). The maximal operation time fixed at these tests reaches 9.4 ms. In addition, in some experiences with high DC components, pickup current decreased by as much as 0.7 of the rated pickup current.

This occurred when the relay pickup phase occurred at the moment corresponding to the maximal displacement of the first half wave of a current sinusoid.

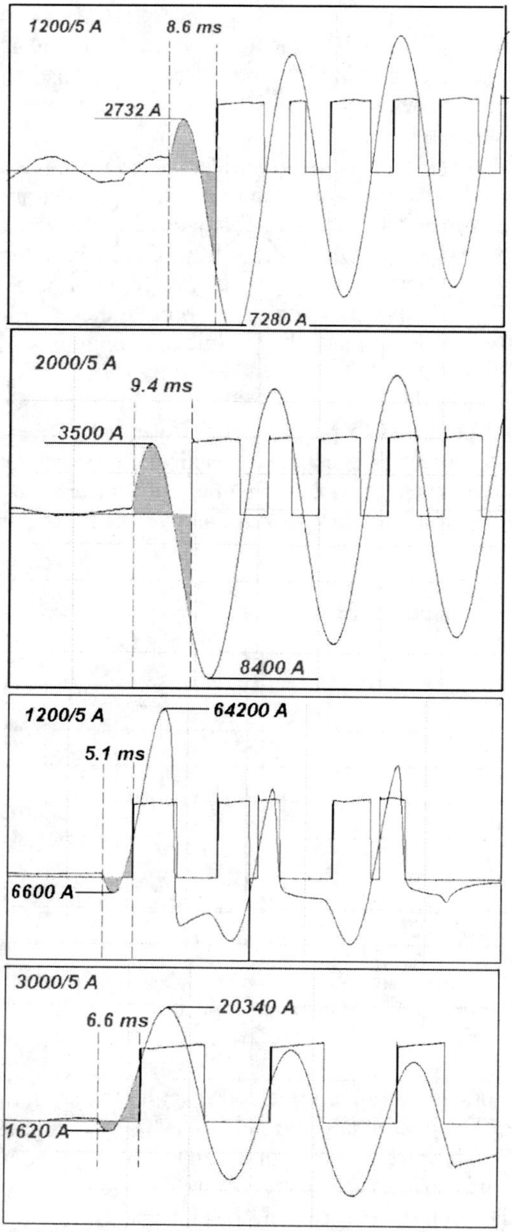

FIGURE 5.17 Oscillogram of operation of the relay for actual short circuit transients containing a high DC component.

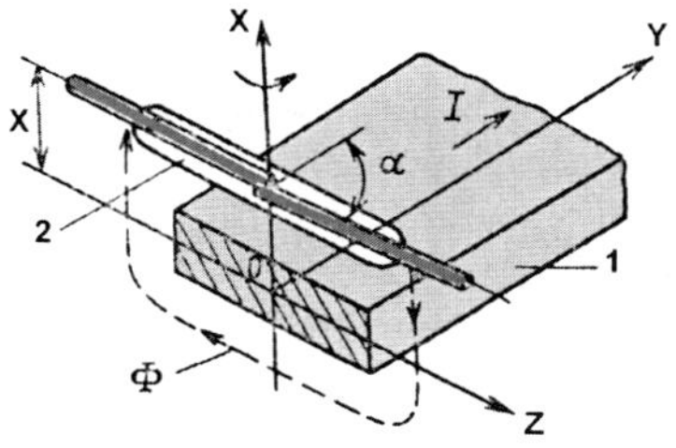

FIGURE 5.18 Principle of winding-free overcurrent reed relay. 1 — current carrying bus bar; 2 – reed switch with pickup depends on distance X from the bus bar and on angle α for longitudinal axis Y.

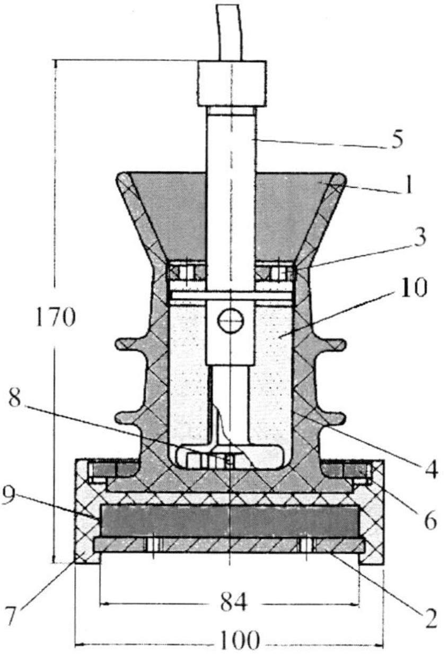

FIGURE 5.19 Design of high-voltage overcurrent reed relay not requiring CT. 1 – main insulator; 2 – fixative plate; 3 – inside nut; 4 – semi-conductive cover; 5 – bushing; 6 – fixative nut; 7 – fastener; 8 – reed switch; 9 – high-voltage bus bar; 10 – epoxide compound.

For such conditions relay picked up at much smaller current than at a normal sinusoid in the continuous mode. In our opinion, this phenomenon is not so essential, as the basic purpose of such high-speed relays is not exact current measure, but only detection of the presence of a dangerous short circuit for acceleration of action of basic relay protection. In other words, at adjustment of the relay for a primary current, for example,

20 кA, it is possible to achieve pickup in some cases at a current of 14 кA that also specifies a dangerous short circuit, as well as pickup of a 20 кA current. Nevertheless, in some cases this phenomenon can limit application of reed relays.

For such cases we have developed reed relays which do not demand use of the CT (the CT is reason for pickup relay decrease). Such relays can be applied for both low-voltage and high-voltage (up to 24 kV) electric networks.

By a principle of action it is a winding free relay, sensitive to a magnetic component of the current carrying the bus bar (Fig. 5.18). The reed switch has been placed in an internal cavity of the polymer isolator in the form of a glass (Fig. 5.19) and filled with epoxide compound for direct installation on the high-voltage bus bar (Fig. 5.20).

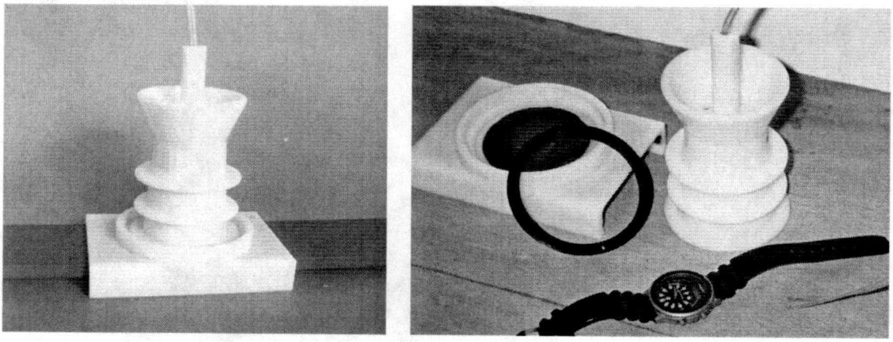

FIGURE 5.20 External view of high-voltage (24 kV) overcurrent reed relay not requiring CT.

Due to use of the reed switch as a sensitive threshold element, the high speed overcurreent relay developed is not only very simple, low-priced, and accessible to manufacturing even by technicians, but also highly steady to external electromagnetic influences: to distortions of a current, to voltage spikes, to powerful high-frequency radiations, etc. Such sensitive elements on a reed switch, adjusted on operation at the high rate of a current, can be built-in also in various microprocessor protection relays (or can be connected to them outside, through a separate input) as the bypassed element of the microprocessor for accelerator tripping of the circuit breaker.

The winding free reed relay can be used also as self-sufficient over current relay. For this purpose the device must be equipped with a simple electronic converter. The magnetic field of the current carrying bus excites the reed switch whose pulses are then converted into a standard binary signal compatible with the relay protection devices. Fig. 5.21 shows the solid-state converter circuitry with reed switch as a triggering component.

The sensitivity of this module is directly proportional to the sine of the angle α between the longitudinal axes of the reed switch and the HV bus, and is inversely proportional to the distance h between these axes. Keeping in mind that for a reed switch operating in the magnetic field of the current carrying bus, the operative (F_o) and the release (F_r) magneto-motive forces are respectively adequate to the operative and the release currents in the bus, one can say :

$$I_o = \frac{F_o}{K_h \sin\alpha};; \tag{5.1}$$

$$I_r = \frac{F_r}{K_h \sin\alpha} \tag{5.2}$$

where

I_o, I_r – are the values of the current in the bus causing the triggering and the release of the reed switch, respectively;

K_h — remoteness factor accounting for the distance between the longitudinal axes of the reed switch and the bus.

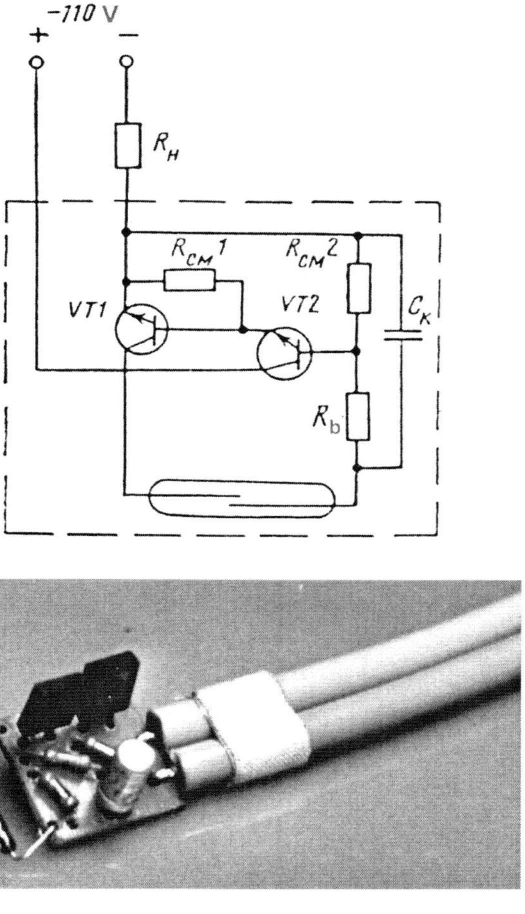

FIGURE 5.21 Circuit diagram and external view of the built-in converter.

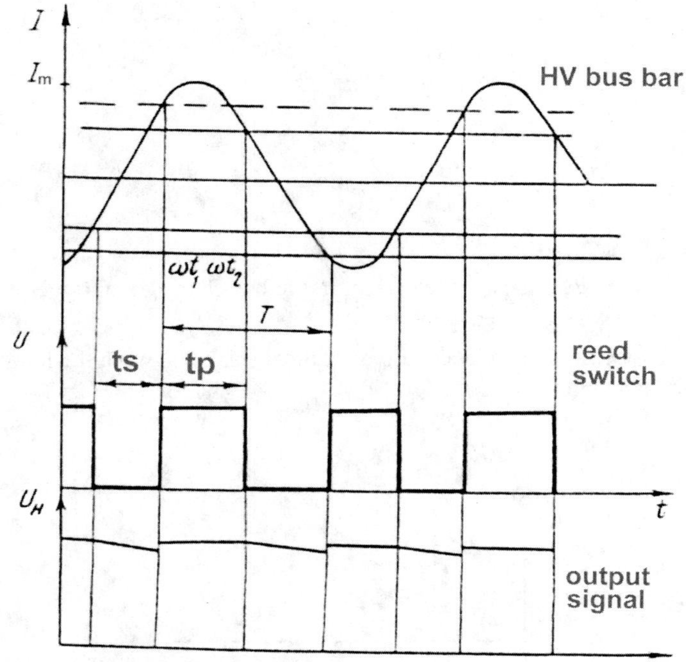

FIGURE 5.22 AC operational oscillogram.

Therefore, by rotating the current transducer with respect to its longitudinal axis, one can set up the operative current at differential values.

The solid-state converter is based on a transistor filter with a peculiar parameter vector, which is explained below.

Excited by AC magneto-motive force, the reed switch generates rectangular pulses, Fig. 5.22, the duration of which is t_p and space is the t_s. Capacitor C_k is supposed to gain the full charge during period t_p, i.e., the full charge time must satisfy the following condition:

$$\pi R_H C_k < t_p ,$$

which defines the capacity:

$$C_k \leq \frac{t_p}{\pi R_H} . \tag{5.3}$$

During the discharge period of C_k, the transient voltage free component attenuation on the R_b resistor is not supposed to exceed a given load pulsation factor K_p:

$$K_p = U(t_s)/\,U,$$

where: U, $U\ (t_s\ )$ – nominal voltage and voltage drop on the load at the end of the pulse space, respectively. (If the load R_H is represented by a control coil of an auxiliary relay, then this pulsation factor can be expressed through the relay reset ratio). The above requirement can formally be expressed as:

$$K_p \geq \exp\left(-\frac{t_s}{R_b C_k}\right), \tag{5.4}$$

where $\tau = R_bC_k$ – the discharge time constant.

The resistance of R_b should, therefore, satisfy the condition:

$$R_b \geq \frac{t_s}{-\ln K_p C_k}. \tag{5.5}$$

The oscillogram in Fig. 5.22 shows that the duration of pulses generated by the reed switch is:

$$t_p = \omega^{-1}\ (\ \omega t_2 - \omega t_1\ )\ , \tag{5.6}$$

where the current phases ωt_2 and ωt_1 are, respectively, given by

$$\omega t_1 = \arcsin\ (I_o\ /\ I_m\ ) \tag{5.7}$$

$$\omega t_2 = \pi - \arcsin\ (I_r\ /\ I_m\ ) \tag{5.8}$$

where I_m – the current amplitude.

Based on (5.1), (5.2), (5.7), and (5.8), the reed switch pulse duration takes a form of:

$$t_p = \omega^{-1}\ [\ \pi - \arcsin\ (\ I_o\ /\ I_m\ ) - \arcsin\ (I_r\ /\ I_m\ )]. \tag{5.9}$$

The pulse duration and space are related to each other by

$$t_p = T - t_s$$

where T is the pulse period, which under the sinusoidal form of the current in the bus bar is equal to π .

Taking in to account (5.9):

$$t_s = \omega^{-1}\,[\arcsin\,(I_o / I_m) + \arcsin\,(I_r / I_m)]. \tag{5.10}$$

By incorporating (5.9) and (5.10) into (5.3) and (5.5), we finally get:

$$C_k \le \frac{\pi - \arcsin(I_o / I_m) - \arcsin(I_r / I_m)}{\pi\omega R_H}$$

(5.11)

$$R_b \ge \frac{\arcsin(I_o / I_m) - \arcsin(I_r / I_m)}{C_k\omega(-\ln K_p)}$$

Thus, expressions (5.11) define the parameter vector for the above solid-state converter.

5.3 THE NEW GENERATION UNIVERSAL PURPOSE HYBRID REED – SOLID-STATE PROTECTIVE RELAYS

In the past few years small-sized standard case TO-247 and TO-220 thyristors and transistors intended for soldering on the printed-circuit-board for the switching of current of tens of amperes at voltages of 1200 – 1600 V have appeared.

Various companies manufacture miniature high-speed (fractions of milliseconds) vacuum reed switches that can withstand voltages of 1000 – 2500 V which can serve as precision threshold (pickup) elements in the protective relays. The Japanese company Yaskawa and its branches manufacture a series of middle size powerful reed switches for switching currents of up to 5 A at a voltage of 250 V (Fig. 2.14).

When using reed switches it should be kept in mind that their high reliability will be guaranteed only when observing the restrictions imposed by the switching ability determined in the technical specifications. As in semi-conductor switches, the reed switches quickly fail when the allowed switching parameters are exceeded even for a short time. At the same time, even though modern reed switches are electromechanical elements, their reliability and number of switching cycles is closer to that of semi-conductor elements, and so are many of their parameters, such as withstanding electromagnetic interferences, surge capability, etc. It should be pointed out they considerably surpass semi-conductors in withstanding surges. Because of extraordinary features of the reed switch relays, not possessed by usual electromechanical relays, such as high speed, precise and stable pickup value, and high release factor on an alternating current, etc., many devices for protection and automation systems in the industry, power engineering, and military techniques have been developed on their basis.

The combination of the reed switches with magnetic circuits and semiconductor elements opens new avenues in the development of interesting and promising devices distinguished by simplicity and low cost.

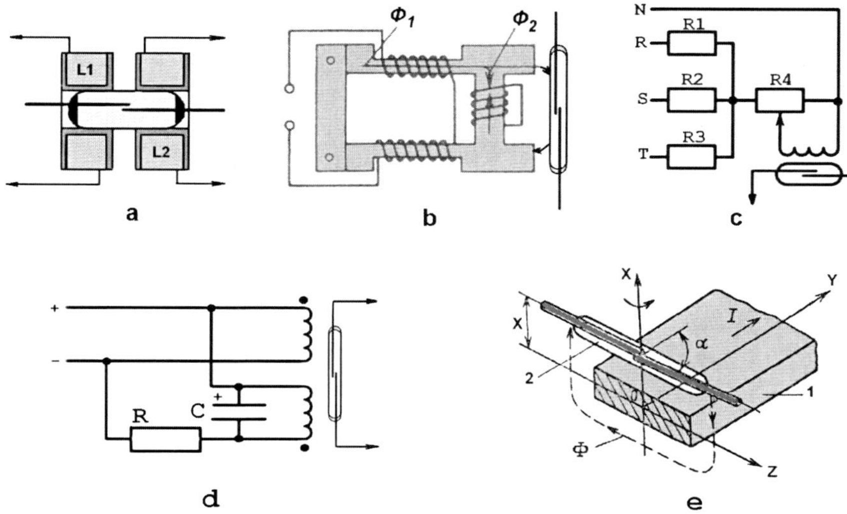

FIGURE 5.23 Examples of various applications of reed switches in the protection devices.

For example, a simple device such as a reed switch with two operating coils (Fig. 5.23a) can be a basis for creation of the differential protection, logic elements, threshold summing element, etc. A reed switch with a special magnetic circuit (Fig. 5.23b) appears to be insensitive to the DC (aperiodical) component of the current in the coil. The reed switch, connected to a simple circuit, Fig. 5.23c, responds to the voltage asymmetry. In the circuit, Fig. 5.23d, the reed switch picks up only at rapid changes of current (voltage) in a control circuit which is distinctive for emergency modes and does not respond at slow changes of the current, related to the changes in load, and as described above. The reed switch is also directly responsive to the magnetic field of the current passing in the bus bar without additional windings (Fig. 5.23e).

Let us consider concrete examples of the most widespread kinds of protective relays based on the suggested technology.

Instantaneous current relay (Fig. 5.24). The over-current relays without time delays are widely used for the protection of electric networks and electric equipment against overloads. This version of the relay is intended for directly energizing the trip coil of the high-voltage circuit breaker (CB).

The sensitive threshold (pickup) element of the device is the Rel1 relay made with a miniature high-speed vacuum reed switch. Its coil contains 2050 turns of 0.16 mm wire. At pickup this reed switch starts to vibrate at double the current frequency. Upon the initial closing of the circuit by the reed switch, thyristor VT will turn-ON and energize the CB trip coil. The thyristor only switches this coil ON; it is switched OFF by the own auxiliary-contact of the CB. Rel2 is an auxiliary relay, intended for signaling or blocking circuits and it uses a medium capacity reed switch such as GC1513. Its coil has very low resistance and it is designed for the short-term carrying of a direct current in a range from 0.5 up to 15 A (typical currents of CB trip coils of various types) at which this reed switch operation is reliable.

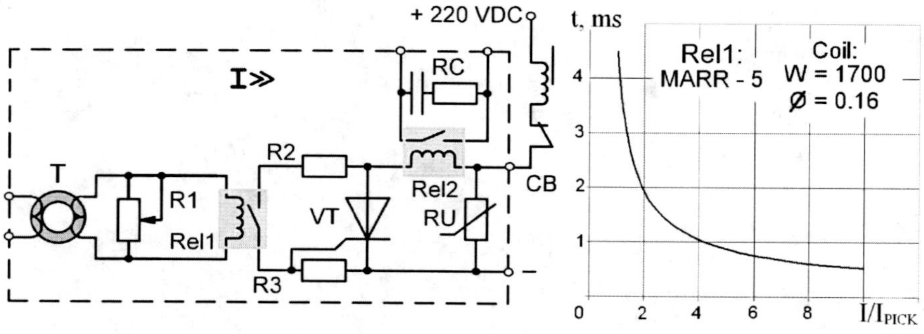

FIGURE 5.24 The simplest hybrid protective relay: instantaneous current relay. The basic circuit diagram and experimental time-current characteristic curve. I/I_{PICK} – multiples of pickup setting.

Adjustment of pickups (coarsening the relay) is carried out with the help of potentiometer R1. In the relay a thyristor such as 30TPS12 (in case of TO-247AC) is used with rated current 30A and the maximal withstanding voltage of 1200V and a miniature vacuum reed switch such as MARR-5. The input CT is made on a low-frequency ferrite ring with the external diameter of 32 mm.

RC-circuit serves for protection of the auxiliary contact (reed switch) from spark erosion at switching of inductive loads. Varistor RU such as SIOV-Q20K275 protects the device from spikes in the DC circuit. Its clamping voltage does not exceed 350-420V DC. This voltage level should be higher than the rated voltage of a DC network, but lower than the maximal withstanding voltages of the thyristor and the reed switch. As shown in the experimental time-current characteristic curve, the relay speed is higher than that of the electromechanical, static, or microprocessor-based devices, it does not need power supply, is insensitive to high-frequency interferences and spikes in a current circuit, and remains reliable at strong distortions of current.

Instantaneous current relay with high release ratio, (Fig. 5.25). In this relay design a powerful reed switch, such as R15U (Yaskawa), is used as a contact of tripping output relay Rel2 instead of power thyristor VT. The second CT (T2) serves as an energy source necessary for the operation of the power reed switch.

Closing and opening reed switch (Rel1) with double the current frequency at energizing does not suits the operation of relay Rel2. Therefore a special active filter can be used between pickups element Rel1 and output relay Rel2. The filter is formed with capacitor C2 (22 μF), resistors R2, R3, and transistor VT, which can be any low-power transistor for voltages not less than 100V with current gain (h_{FE}) not less than 100, for example, such as ZTX753, or ZTX953. With a low-power Darlington transistor (for example, such as ZTX605), as shown in Fig. 5.25, in the capacity of the capacitor C2 can be considerably reduced. By means of this filter the current pulsation in the reed switch circuits of Rel1 will be transformed to a stable current in the coil circuit of relay Rel2.

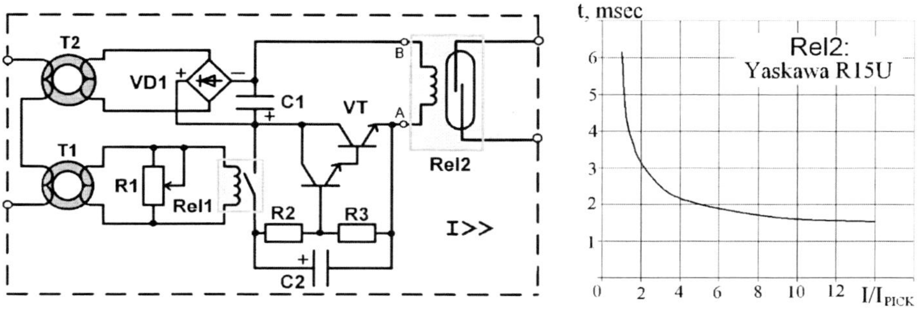

FIGURE 5.25 Instantaneous current relay with high release ratio: the basic circuit diagram and experimental time-current characteristic curve.
I/I_{PICK} – multiples of the pickup setting.

The release ratio of reed switch Rel1 is close to 0.99 at alternating current. For a lower release ratio of the relay (0.7 - 0.6) it is sufficient to connect the Rel1 coil through a rectifier bridge, and to transfer capacitor C2 to a different location, in parallel to this bridge. Since the capacity needed to feed the powerful reed switch is much greater than for a miniature reed switch, CT (T2) is formed with two identical transformers, similar to transformer T1 in which the secondary windings are connected in parallel, and the primary winding – communicating, covers both ferrite rings. The total power consumed by the relay from the current circuit (at a current 5A) does not exceed 4 W. The winding of relay Rel2 consists of two coils placed on the reed switch and connected between them in series.

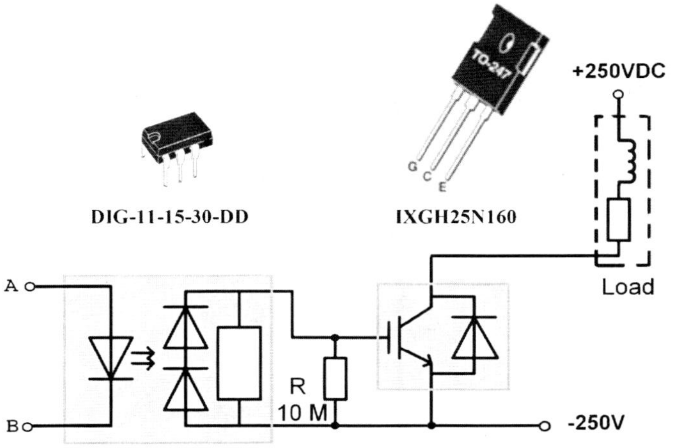

FIGURE 5.26 An embodiment of output switching unit of the relay based on the modern IGBT-transistor (IXGH25N160) and specialized driver with the dynamic discharging (DIG-11-15-30-DD) for this transistor control.

Each of them contains 7600 turns of a 0.08 wire. Experimental time-current characteristic curves (Fig. 5.25) were obtained for a series of consecutive pickups of the relay, during the time intervals between which the charge of capacitor C1 was kept unchanged. At the initial pickup of the relay with an uncharged capacitor the time delay is approximately twice as long. Such an acceleration of operation in case of repeated pickups at short circuit is a positive property of the protective relay. Even in view of increasing the operation time at the initial pickup, the relay speed still remains very high. Modern IGBT-transistors and complete modules used for their so-called "drivers" enable realization of a very simple switching output unit of the relay on a contactless basis (Fig. 5.26).

Current relays with independent and dependent time delay (Fig. 5.27). Similar to the above design, the relay contains two independent current transformers: the first one, T1, is used as a source of control value for the pickups module on the reed switch, Rel1, and the second, T2, for feeding the time delay unit.

When the micro-switch, S, is switched on, the Zener diode, VD3, is connected to the output of rectifier bridge, VD1, and provides a constant level of a voltage (on an input of the time delay unit time) which is independent of the input current in the current pickups range. In this case the relay works with the constant time delays which are determined by capacitance C2 and resistance R2. As this capacitance is charged to a certain voltage value, thyristor, VT1, is turned on and capacitance C2 is completely discharged through low resistance (81 Ohm) winding (2050 turns by a wire 0.16) of relay Rel2, activating the reed switch. In order to turn this device into a relay with time delay depending on the current it is necessary to turn the micro-switch, S, to OFF.

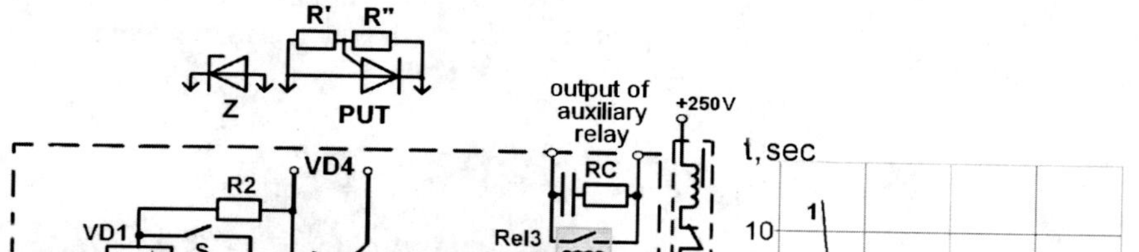

FIGURE 5.27 Universal protective current relay with the time delay: the basic circuit diagram and set of experimental time-current characteristic curves. For the relay with the dependent time-current characteristic curves the various values of capacity C2 (in μF) are as follows: 1 – 4400; 2 – 3200; 3 – 2200; 4 – 1000; 5 – 300.

In this way the voltage charge of capacitor C2 will depend on the input current level: the higher the current, the higher the voltage applied to capacitors C2 and the shorter is the time of their charging up to a voltage level at which thyristor VT1 turns ON, forming a typical time-current characteristic curve (Fig. 5.27), of a relay of this kind. If a second reed switch is removed from the center of the coil and mounted in the coil of relay Rel1 (so that its pickup will be 10 – 15 times higher than that of the first reed switch) and is connected in parallel to a reed switch of the relay Rel2, the device pickups will be instantaneous at high rates of the input current and energize the trip coil of the CB within 3 – 4 milliseconds. A turned ON thyristor VT1 was used as the threshold element VD4, and a standard Zener diode was used in the relay prototype; however the best results can be obtained with so-called "programmable unijunction transistor" (PUT), e.g., 2N6027, or 2N6028 types. This element of the structure and characteristics is similar to a thyristor with very low leakage current (microamperes) through a gate junction that allows more efficient use of capacitance C2. Its turn-ON voltage can be adjusted, i.e., "programmed," by means of resistors R' and R".

Relay of a power direction (Fig. 5.28). Even such a complex function as detection of power direction can be realized very simply by means of the hybrid technology.

As is known, the power direction is determined by the angle of phase displacement between the current and the voltage; therefore, actually, the power direction relay responds to the change of angle between the current and the voltage. It turns out that application of two equivalent phase-shifted voltages to two primary windings of the intermediate transformer T3 causes the output voltage on the third winding to depend very strongly on the phase displacement between these voltages (Fig. 5.28).

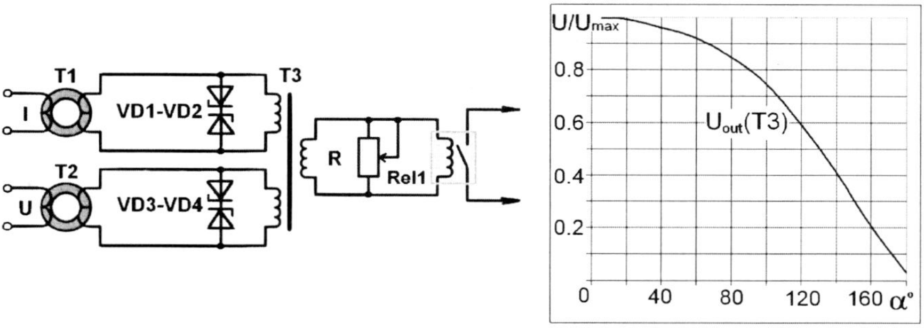

FIGURE 5.28 The relay of power direction: basic circuit diagram of measuring the threshold module and experimental dependence of an output voltage of transformer T3 on an angle shift between two voltages on its primary windings.

This is necessary only in order to prevent the effect of a change of the input voltage supplied from current transformer, T1, and voltage transformer, T2, at a level of the output voltage of transformer T3. The simplest solution of this problem is provided by means of two back-to-back connected Zeners as it is shown on Fig. 5.28. A pickup relay can be adjusted by means of potentiometer R.

Relay of differential protection (Fig. 5.29). The use of two current transformers (T1 and T2) connected to the input of the pickup module of any of the devices described above enables realizing a two-input relay of differential protection (Fig. 5.29a). Interesting opportunities are provided with the use of two separate auxiliary transformers with the secondary windings connected in series in this device.

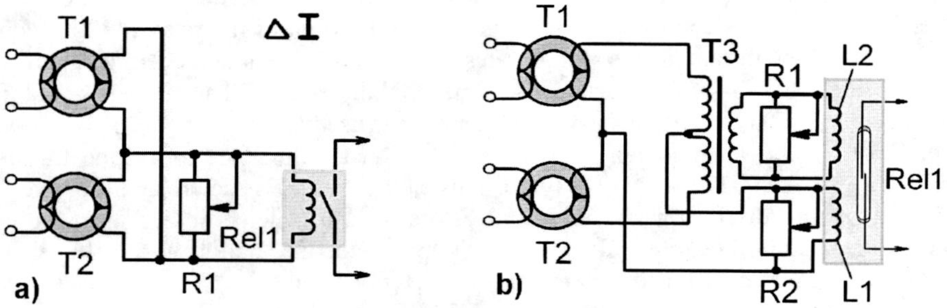

FIGURE 5.29 Measuring modules for the relay of differential protection: to the left the simplest embodiment, to the right – an embodiment with restraint.

To allow more complex functions, such as decreasing the relay sensitivity with increasing the current carried directly through protected object (so-called "restraint"), an auxiliary transformer, T3, is included in the relay. In addition, the output reed relay Rel1 consists of two windings: L1 – differential and L2 – restraint, which shift the working point of the relay proportionally to the current carried directly through the protected object (Fig. 5.29b).

Current relay with restraint. As is known, during the first moment after switching a power transformer on, almost all the current is expended in the magnetization of the iron core. This current is unipolar (Fig. 5.30) and leads to a large sinusoidal displacement of the alternating current (consumed by the transformer) relative to the zero value during the first moment after being switched on.

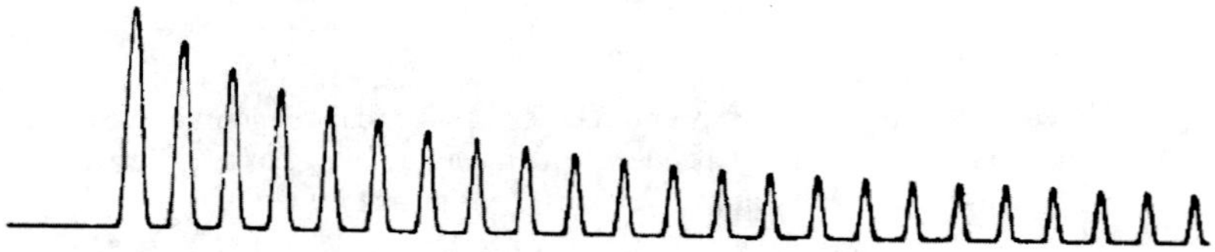

FIGURE 5.30 DC component of the magnetization current.

A similar sinusoidal displacement is caused in the aperiodical component of a transient in electrical power network when it is short circuited (Fig. 5.31).

Because of the high rate of the short circuit current, the line current transformer causes an additional deformation of this already displaced sinusoid because of its core saturation (Fig. 5.32).

It is obvious that the protective relay (current, differential) should not pickup an increasing current value caused by displacement of a sinusoid.

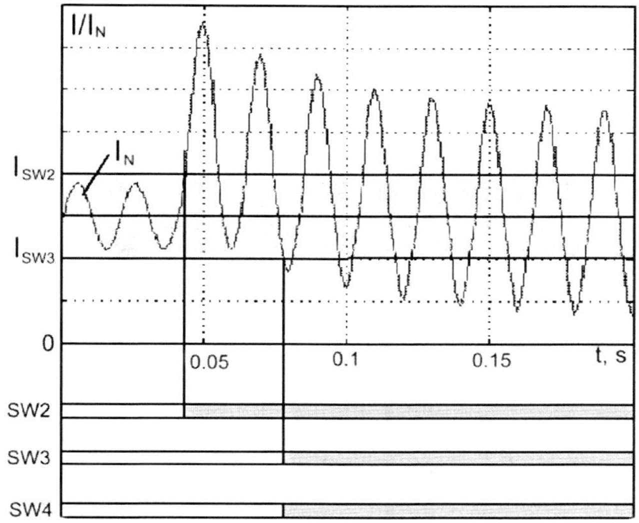

FIGURE 5.31 Displacement sinusoid of the short circuit current at aperiodical component and time diagram of the operating reed switches.
I_N – maximal working current in power line; I_{SW2}, I_{SW3} – picks up currents of reed switches SW2 and SW3, respectively.

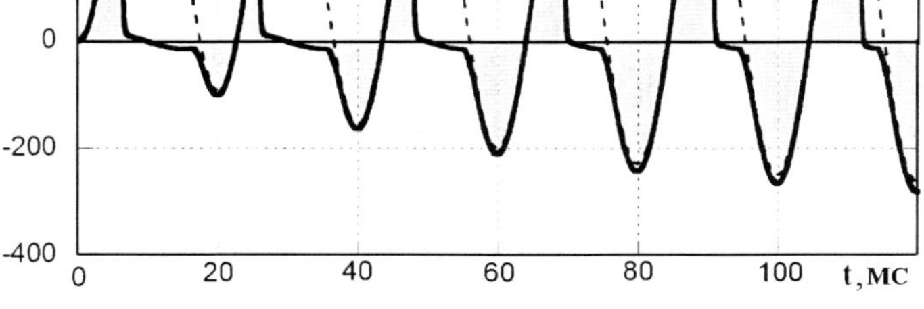

FIGURE 5.32 Deformation current sinusoid in conventional CT caused of core saturation.
I_1 and I_2 – theoretical and actual curves of the secondary CT currents.

In practice, various methods preventing false operations of the relay are used: from simple relay interlocking for some fixed time (time of transient), up to distinguishing a second harmonic of a current by means of filters with electronic amplifiers and the formation of a restraint signal from it. In microprocessor relays even more complex algorithms (which are carried out by the microprocessor) are used.

The hybrid relay (Fig. 5.33) employs another principle of the relay interlocking for the aperiodical component and inrush current. The relay consists of the three independent, high-speed current relays on reed switches: SW1, SW2 and SW3; one two-half-period rectifier VD1 and two one-half-period rectifiers (separately for positive, VD4, and negative VD5 half waves); two current transformers: CT2 (as a source of an input signal) and CT1 (as a power supply of transistors) based on low-frequency ferrite rings; the filter on the capacitor C2 and the Darlington transistor VT1; two-input logic elements, transistors VT2 and VT3; and also an output switching unit on the basis of high-voltage reed switch, SW4, and the high-voltage thyristor, VS, with a working voltage 1200V for directly switching the circuit breaker trip coil on.

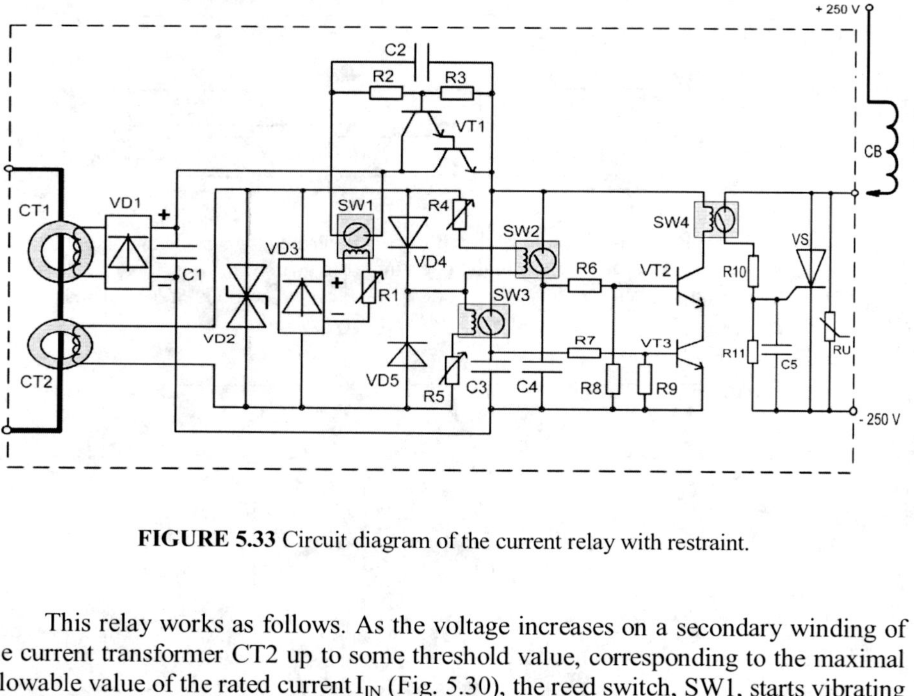

FIGURE 5.33 Circuit diagram of the current relay with restraint.

This relay works as follows. As the voltage increases on a secondary winding of the current transformer CT2 up to some threshold value, corresponding to the maximal allowable value of the rated current I_{IN} (Fig. 5.30), the reed switch, SW1, starts vibrating at a frequency of 100 Hz. The voltage on the transistor VT1 base, acted upon by SW1, remains stable (due to capacitor C2) which leads to the unlocking of this transistor. VT1 switches on the power supply of the logic elements of transistors VT2 and VT3, putting them into a ready state mode. Each of the reed switches, SW2 and SW3, controls only one half-wave of the current, which diodes VD4 and VD5 extract for each reed switch. The pick-up values of both reed switches are identical and correspond to the amplitude of the established value of a short circuit current (or close to it). At the initial moment of a short circuit (when positive and negative half-waves of a current have different amplitudes because of the displacement of the current sinusoid relative to the zero value) only one reed switch is picked up, namely the one whose polarity corresponds to the half-wave having greater amplitude relative to zero. It is, in fact, the reed switch SW2 (in Fig. 5.31) corresponding to a positive half-wave of the sinusoid displaced upwards.

Therefore transistor VT2 switches on and keeps conducting at vibration frequency of the reed switch SW2 due to capacitor S4. Reed switch SW1 at this time does not work, as the amplitude of a negative half-wave of the sinusoid (displaced upwards) is not yet sufficient for the picking up of this reed switch. The output reed switch, SW4, remains open. In the process of attenuation of the aperiodical component the sinusoid of a current gradually becomes symmetric, and thus the amplitude of the negative (bottom) half-waves increases (Fig. 5.31). When it reaches a pick-up threshold of the reed switch, SW3, it starts to vibrate at a frequency of 50 Hz and unlocks transistor VT3. Now both transistors, VT2 and VT3, enter a conducting condition that leads to the switching on of the output stage with relay SW4 and thyristor VS and energize trip coil of the circuit breaker. The device is supplied with a standard spike protection element: VD2 and RU, and some other auxiliary elements. Requirements for solid-state elements and reed switches are the same, as for other hybrid relays described above. Adjustment of the relay is carried out by potentiometers R1, R4 and R5.

Descriptions of quite interesting and promising devices based on the suggested elements could be continued. However, the purpose of this publication is not to present the advantages of reed switches, but to prove that on the basis of a combination of modern reed switches and modern power semi-conductor elements a new generation of hybrid protective relays, not including complex mechanisms, can be easily created and can replace the out-of-date electromechanical relays at the upper level while retaining their high noise and surge stability, maintainability and other positive features. The use of the new generation of the relays would allow sparing considerable financial expenses connected with the necessity of purchasing the expensive microprocessor-based protective devices. Thus, further perfection of automatic control systems in electric networks through equipping them with microprocessor recorders of emergency modes, optical communication systems, and other modern systems can be gradually accomplished, during accumulation of financial resources independently of the relay protection. In the author's opinion, the above examples of the protective relays developed and tested by the author can support our conclusion.

5.4 AUTOMATIC HIGH-VOLTAGE CIRCUIT BREAKERS

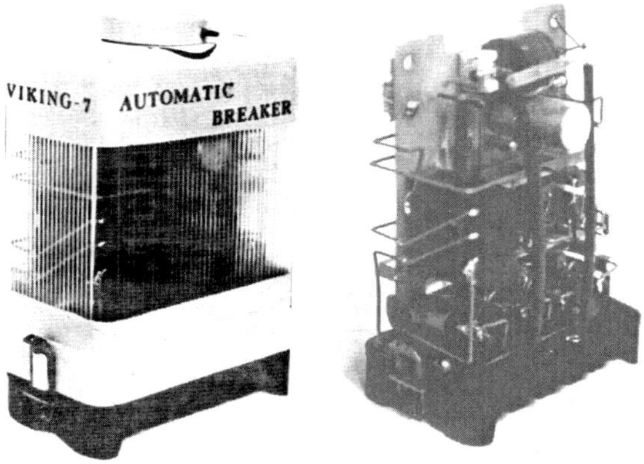

FIGURE 5.34 Automatic high-voltage one channel circuit breaker "VIKING-7."

The "VIKING" series of HV automatic circuit breakers, Fig. 5.34, 5.35 (ACB) is intended to be used as part of standard insulation testing equipment that applies high voltage to electric devices under test, and checks their insulation parameters through the current leakage levels

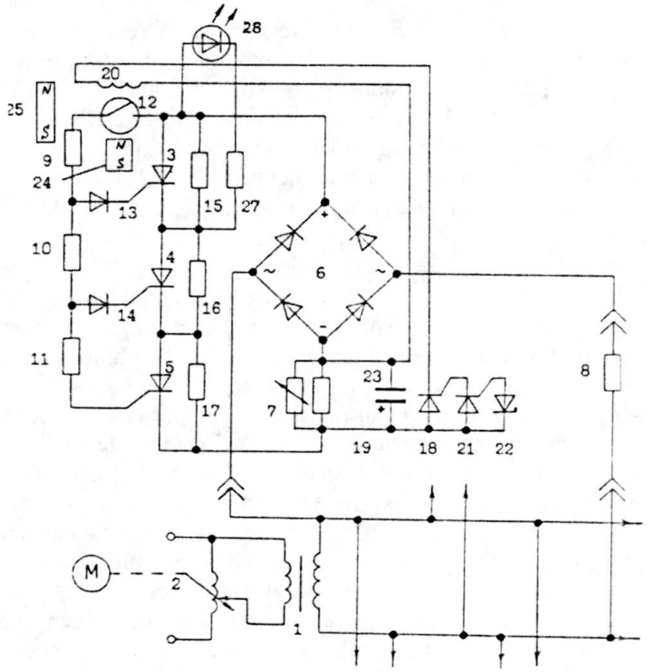

FIGURE 5.35 Circuit diagram for one channel automatic HV circuit breaker "Viking-7." 1 – HV transformer; 2 – variac with motor drive; 3…5 – HV thyristors; 6 – HV rectifier bridge; 7 – trip regulator; 8 – object under test; 9…11, 15…17 – resistors; 18…23 – elements of triggering circuit; 24, 25 – permanent magnets; 28 – trip indicator (LED).

The "Viking-7" series is designed based on a hybrid reed-thyristor technology (Fig. 5.35). "Viking-7" is provided with a manual-reset module, in which a permanent magnet 25 is used (Fig. 5.36).

Table. 5.1 Main specifications of ACB "Viking-7":

Parameters	Value
Switching Voltage, V AC (rms)	100 to 3000
Maximum Interrupted Current, A (rms)	5.0
Dimensions, mm	140 x 100 x 60
Mass, kg	0.5
Number of Operations	10^6

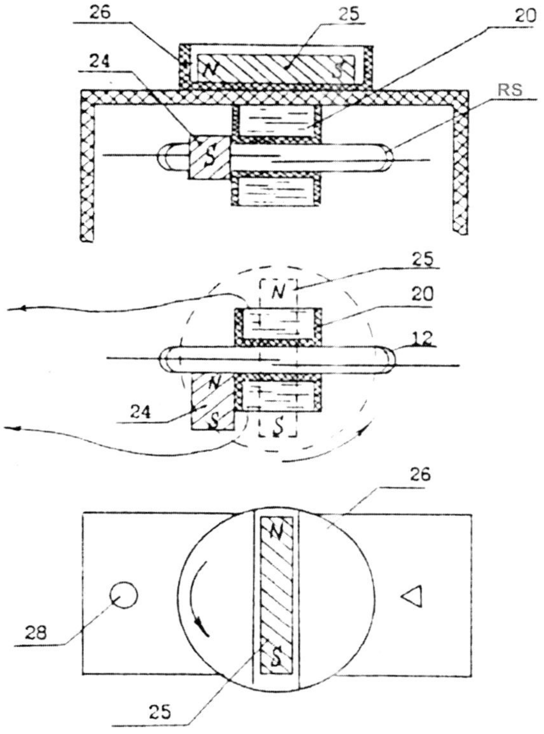

FIGURE 5.36 Construction of a manual-reset module in ACB "Viking-7."
12 (RS) – HV reed switch; 20 – trip coil; 24 – latching magnet; 25 – reset-magnet; 26 – revolving limb in upper part of ACB body; 28 – trip indicator (LED).

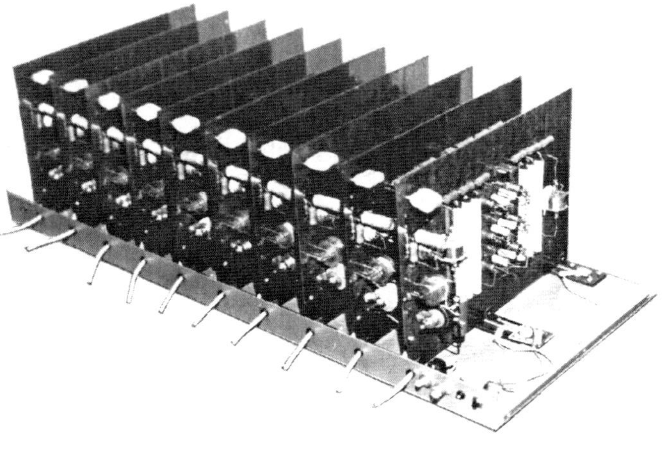

FIGURE 5.37 Multi channel automatic HV circuit breaker "Viking-10" (without lid).

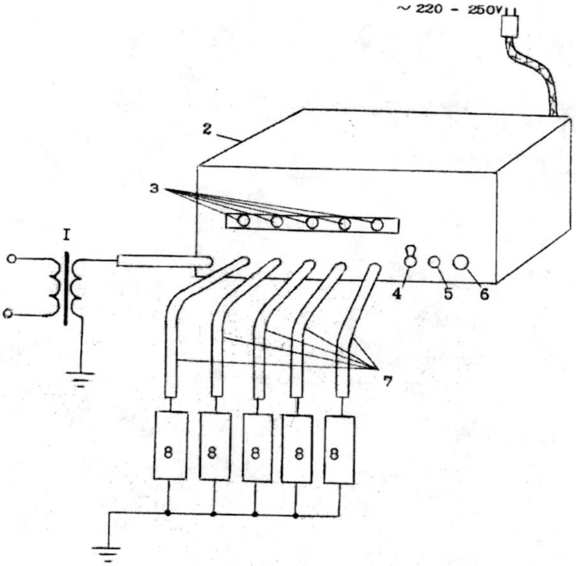

FIGURE 5.38 Block-diagram for outside connections of ACB "Viking-10."
1 – high voltage power supply; 2 – ACB "Viking-10"; 3 – trip indicators (LEDs); 4, 5, 6 – main switch and indicators; 7 – ouputs (high voltage wires); 8 – objects under testing.

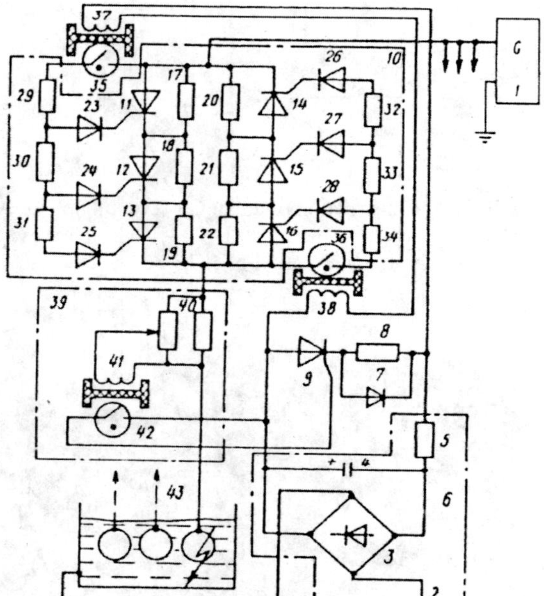

FIGURE 5.39 Circuit diagram for one channel of a multi channel automatic circuit breaker "Viking-10."
1 – HV power supply; 10 – hybrid reed-thyristor commutating unit; 37, 38 – HV reed switch relay; 39 – trip unit; 41, 42 – RG-relay; 43 – objects under test; 44 – reset switch.

For manual-reset of a reed switch, a limb 26 with magnet 25 must be turning. In its initial position, an axis of the permanent magnet is arranged perpendicularly to the axis of the reed switch, and therefore, has no affect on the reed switch. After tripping and disconnection a test object from the HV power supply (when the current running through it exceeds the admissible leakage current value), the reed switch is held by magnet 24 in the open position.

Multi channel automatic HV circuit breaker "Viking-10," (Fig. 5.41; Table 5.2) enables the insulation testing equipment to test simultaneously many devices from a single HV power supply, Fig. 5.42.

Thus, the testing can be fully automated and much more efficient. The HV circuit disconnection is accomplished without causing any electric arc during the first alternating current passing though the zero value and is not followed by overvoltage in the devices under test. When the current leakage through a channel exceeds a level trip, then the "Viking-10" disconnects that channel and turns ON the corresponding light indicator (LED), Fig. 5.39.

The current level trip of each channel can be preset individually.

Table 5.2 Specifications of multi channel HV automatic circuit breaker "Viking-10."

Parameter	Value
Switching Voltage, AC, V (rms)	100…3000
Maximum Interrupted AC current, A (rms)	25.0
Range of current trip regulation, A	0.01…1.0
Dimensions, mm 5 – channel breaker 10 – channel breaker	 430 x 325 x 260 760 x 325 x 260

5.5 HIGH SPEED VOLTAGE UNBALANCE RELAY

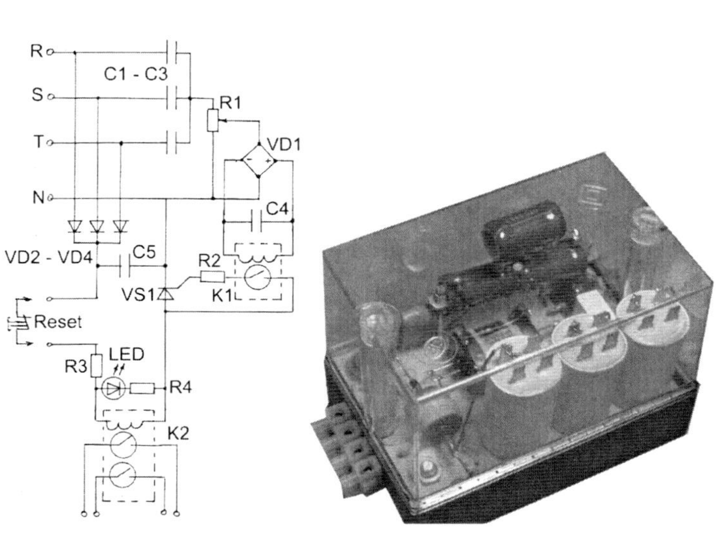

FIGURE 5.40 The high speed voltage unbalance relay, v. 1.

Some types of industrial complex electronic power systems are quite sensitive to electrical energy quality. For example, repeated occurrence of emergency conditions with strong over-currents in 500 V power reversing thyristor drive of powerful coal lifting mechanism engines at thermal power stations have been quite common.

The thermal power station staff found out the relation between these conditions and single-phase short circuits occurring in a 161 kV powering circuit used to power the drive (via transformer). Highly distorted signals arriving at the drive elements sensitive to current and voltage angles cause failures in the drive control circuit and unpredictable power thyristors activation.

In order to prevent emergencies we have developed two versions of High Speed Voltage Unbalance Relays instantly acting on the trip coil of the power switch in case of voltage unbalance (Fig. 5.40)

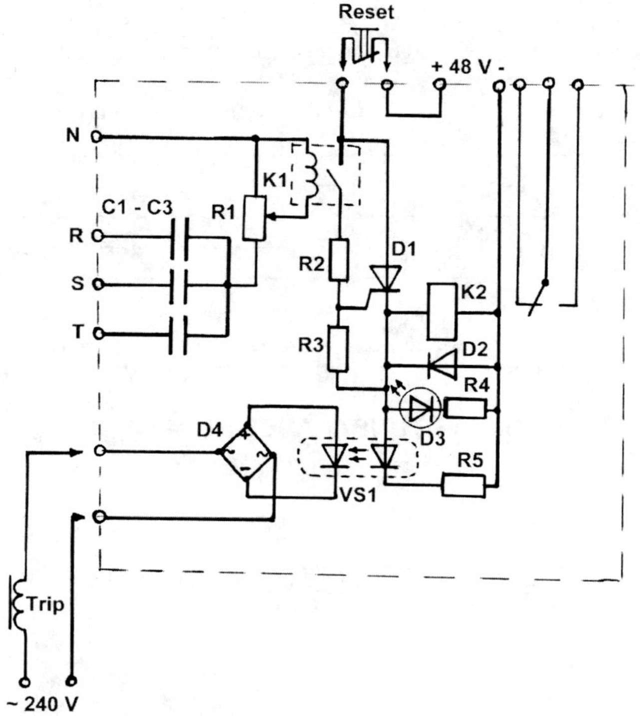

FIGURE 5.41 Circuit diagram of the high speed voltage unbalance relay, v.2.

In both types of relays a zero sequence voltage filter on capacitors C1-C3 is used. Also, a high-speed reed switch relay K1 is used as a sensitive element at the filter output.

The two versions differ in the output switching elements (reed switch or thyristor in the first and the second version respectively) and in different powering circuits. The first version of the relay has a built-in diodes VD1-VD3 rectifier, which is insensitive to a phase loss.

The second version (Fig. 5.41, 5.42) is more stable and quick, and has an external source of DC voltage and additional K2 auxiliary relay for remote indication.

Laboratory pilot tests for ver.2 Unbalance Relay on test-set EPOCH-10 (Multi-Amp Corp.) proved action time 0.0008 sec (0.8 ms) and minimal voltage unbalance – 15 %.

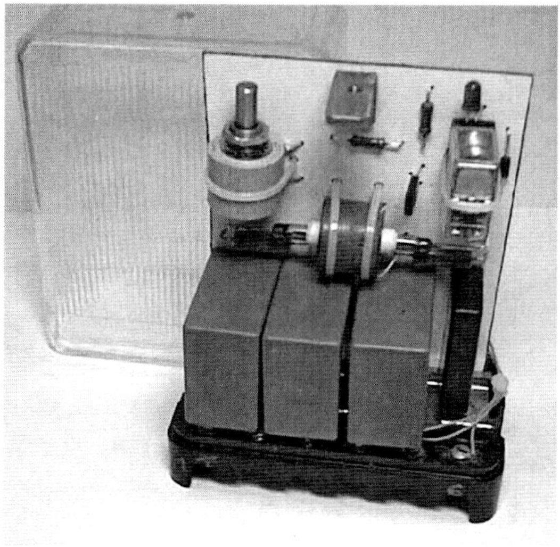

FIGURE 5.42 Experimental sampler of the high speed voltage unbalance relay (Ver. 2). Dimension: 65 x 52 x 115 mm.

5.6 IMPULSE ACTION PROTECTIVE RELAY

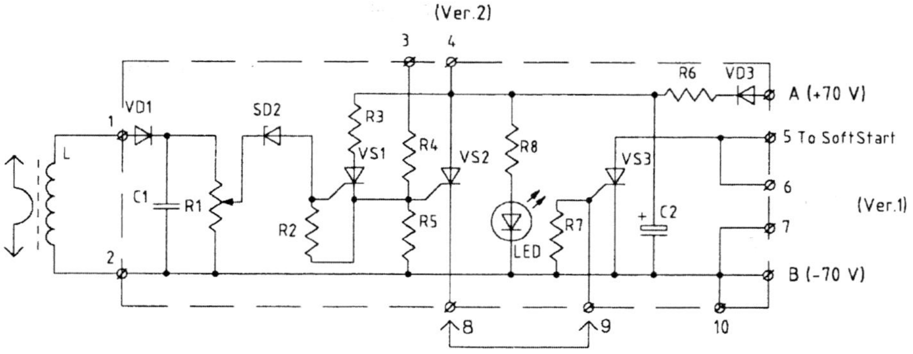

FIGURE 5.43 RG-PLS device circuit diagram.

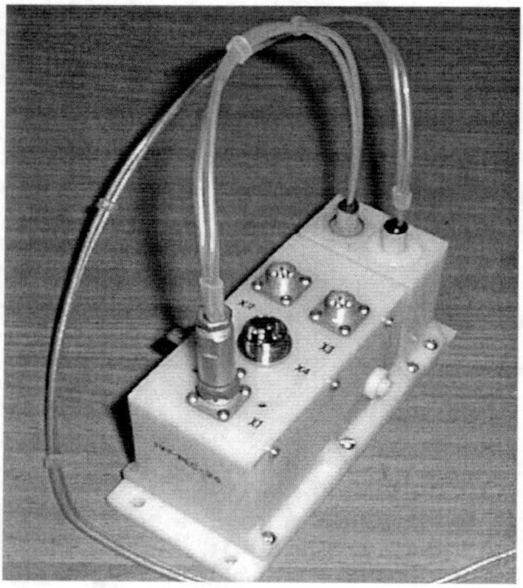

FIGURE 5.44 The impulse action RG-PLS-25 type device on common base plate with RG-25 device (see Chapter 3).

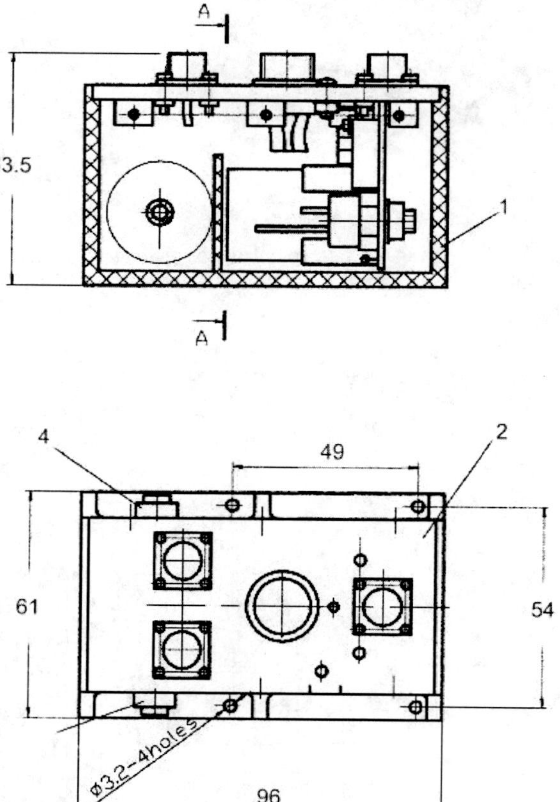

FIGURE 5.45 RG-PLC-25 type device drawing.
1 – insulated box; 2 – upper panel; 3, 4 – bushing for HV wire.

RG-PLS type devices (Figs. 5.43, 5.44, 5.45, Table 5.3) are based on impulse action, and unlike the RG devices (see Chapter 3) are insensitive to the DC current value flowing through them.

An RG-PLS device is operated by a drastic current change: di/dt. Moreover, a built-in filter in the RG-PLS device prevents its operation through short modulation current pulses occurring in the controlled circuit. This type of device can be adapted to different response current ranges (from fractions of amperes to tens of amperes) and have a built-in trip level controller.

An RG-PLS interface can be connected to any other RG type relay (to RG-25, for example, as shown in Fig. 5.44) and either share a common output circuit that generates a strong current pulse on operation of any interface part, or have two separate outputs (an external RG relay can be connected to 3 – 4 input or 5 – 6 inputs, see Fig. 5.43).

In the latter case, the output reed switch of the RG controls the thyristor contactor in the LV circuit, whereas the pulse RG-PLS output controls both the LV thyristor contactor and the HV thyristor crowbar protection device which causes short circuiting of the HV output. As a relatively slow current increases in the controlled circuit, the RG is activated and disconnects the low voltage thyristor contactor, whereas at fast current jumps with a steep front edge (typical for internal breakdowns), RG-PLS is operated and simultaneously triggers control signals to the LV contactor and the HV short circuiting device.

Table 5.3 Main parameters of the RG-PLC devices.

	SUBTYPE	
RG-PLS-	**25**	**50**
Nominal Voltage, kV DC	5	50
Test Voltage, kV DC	35	70
Max. Continuous Operating Current, A (rms)	10	
Over Current in Operating Circuit for 1 sec, A	50	
Maximal Time Delay, microsecond	10	
External Power Supply: voltage, V DC current, A DC	 70...120 0.01	
Internal Resistance of Operating Circuit, Ohm	≈ 0	
Dimensions, mm	95 x 60 x 70	

6

Improvement of Microprocessor-Based Protective Relays

6.1 POWER SUPPLY OF MICROPROCESSOR-BASED PROTECTIVE RELAYS AT EMERGENCY MODE

As is known, both auxiliary AC and DC voltages are used at power substations. Use of DC auxiliary voltage increases the essential reliability of relay protection due to use of a powerful battery, capable of supporting the required voltage level on the crucial elements of the substation at emergency mode with the AC power network disconnected.

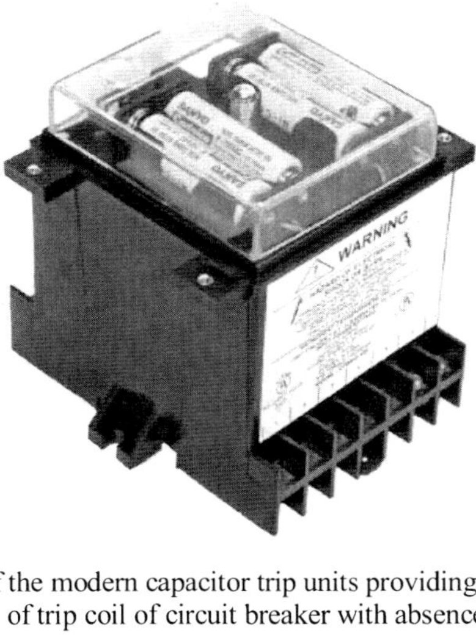

FIGURE 6.1 One of the modern capacitor trip units providing accumulation and long storage of energy for a feed of trip coil of circuit breaker with absence of an auxiliary voltage.

However, this increase of reliability comes at the cost of an essential rise in price of the substation and its maintenance. On the other hand, electromechanical relays of all types do not demand an external auxiliary power supply for proper operation, as their operation requires input signals only.

There may be some problem when it is necessary to energize the trip coil of the high-voltage circuit breaker at loss of auxiliary voltage in the emergency mode, but this problem has been solved for a long time and simply enough through use of a storage capacitor. It is constantly charged at the normal operating mode from the AC auxiliary power supply through a rectifier and provides a power current pulse to the trip coil on operation of the protective relay in the emergency mode.

A modern capacitor trip unit contains, in addition, little nickel-cadmium cells and a low-power solid state inverter for an output voltage of 250V, through which the main capacitor is constantly recharged from a battery while auxiliary voltage is disconnected.

The power capacity of the inverter makes mill watts which are spent only for compensation of self-discharging of the capacitor. Such compact devices (Fig. 6.1) are issued by many companies and allow keeping the capacitor charged for several days. Clearly, in such conditions, sufficient reliability of relay protection, even on an operative alternating current, is provided. For this reason, the operative alternating current is applied very widely.

The situation began to change with the introduction of microprocessor-based relays and the mass replacement of electromechanical relays by them. To the many problems caused by this transition,[1] one more problem was added.[2] As is known, the internal switching-mode power supply, admitting use as auxiliary AC and DC voltages, has an overwhelming majority of microprocessor-based protective relays (MPR). Therefore, at first sight, there should be no reasons to interfere with the use of an auxiliary AC voltage on substations with MPR. The problem arises when there is not enough power for normal operation of an overwhelming majority of MPR and only the presence of corresponding input signals (as for electromechanical relays) and also requires a feed from an auxiliary supply. How will the MPR behave at loss of this feed at failure mode when the hard work of the microprocessor and other internal elements is required? How will the complex relay protection (containing some of MPR, incorporated in the common system by means of the network communication when there are also losses of auxiliary feed) function? How will the MPR behave during voltage sags (brief reductions in voltage, typically lasting from a cycle to a second or so, or tens of milliseconds to hundreds of milliseconds) during failure? We shall try to understand these questions.

The internal switching-mode power supply of the MPR contains, as a rule, a smoothing capacitor of rather large capacity, capable of supporting the function of the relay during a short time period. According to research which has been led by General Electric[3] for various types of MPR this time interval takes 30 – 100 ms.

[1] From Gurevich, V, *Electrical Relays: Principles and Aplications*. – Taylor & Francis Group, London – New-York, 2005, 704 pp.

[2] From Gurevich, V, "Nonconformance in Electromechanical Output Relays of Microprocessor-Based Protection Devices Under Actual Operation Conditions," *Electrical Engineering & Electromechanics*, 2006, vol. 1, pp. 12 – 16.

[3] From Fox, Gary H, "Applying Microprocessor-Based Protective Relays in Switchgear with AC Control Power," *IEEE Transaction on Industry Applications*, vol. 41, No. 6, 2005, pp. 1436 – 1443.

In view of that time of reaction, the MPR for emergency operation lies in the same interval and depends on that type of emergency mode, it is impossible to tell definitely whether protection will have sufficient time to work properly. At any rate, it is not possible to guarantee its reliable work. It is a specially problematic functioning of protection relays with the time delay, for example the distance protection with several zones (steps of time delay, reaching up to 0.5 – 3.0 s). Also it is possible to only guess what will take place with the differential protection containing two remote complete sets of the relay, at loss of a feed of one of them only.

Voltage sags are the most common power disturbance. At a typical industrial site, it is not unusual to see several sags per year at the service entrance, and far more at equipment terminals.

These voltage sags can have many causes, among which may be peaks of magnetization currents, most often at inclusion of power transformers. Recessions and the rises of voltage arising sometimes at failures and in transient modes are especially dangerous when coming successively with small intervals of time. The level and duration of sags depend on a number of external factors, such as capacity of the transformer, impedance of a power line, remoteness of the relay from the substation transformer, the size of a cable through which feed circuits are executed, etc. MPR also have a wide interval of characteristics on allowable voltage reductions. Different types of MPRs keep working their capabilities at auxiliary voltage reduction from the rated value of up to 70 – 180 V.

Table 6.1 Parameters of capacitors with large capacity and rated voltage of 450 – 500V.

Capacity, µF	Rated Voltage, V	Dimensions (diameter, height), mm	Manufacturer and capacitor type
6000	450	75 x 220	**EVOX-RIFA** PEH200YX460BQ
4700	450	90 x 146	**BHC AEROVOX** ALS30A472QP450
10.000	450	90 x 220	**EVOX-RIFA** PEH200YZ510TM
4000	500	76.2 x 142	**Mallory DuraCap** 002-3052
4000	450	76.2 x 142	**CST-ARWIN** HES402G450X5L
6900	500	76.2 x 220	**CST-ARWIN** CGH692T500X8L

Thus MPR with a rated voltage of 240 V supposes a greater (in percentage terms) voltage reduction than devices with a rated voltage 120 V. It is also known that any microprocessor device demands a long time from the moment of applying a feed (auxiliary voltage) to full activation at normal mode. For a modern MPR with a built-in system of self-checking this time can reach up to 30 sec. It means that even after a short-term failure with auxiliary voltage (voltage sag) and subsequent restoring of voltage level, relay protection still will not function for a long time.

What is the solution to the problem offered by the experts from General Electric? Fairly marking, that existing capacitor trip devices obviously are not sufficient to feed MPR, as reserved energy in them has enough only for creation of a short duration pulse of a current and absolutely not enough to feed MPR, the author comes to the conclusion that it is necessary to use an uninterrupted power supply (UPS) for feeding the MPR in an emergency mode.

The second recommendation of the author – to add an additional blocking element (a timer, for example, or internal logic of MPR) will prevent closing of the circuit breaker before the MPR completely becomes activated. Both recommendations are quite legitimate. Here only usage UPS with a built-in battery is well known as a solution for maintenance of a feed of crucial consumers in an emergency mode. This solution has obvious foibles and restrictions (both economic and technical).

Use of blocking for switching-on of the circuit breakers can be a very useful idea which should be undoubtedly used, however it does not always solve the problem as failures of voltage feeding connected to operation of the circuit breaker is always a possibility.

In our opinion, a more simple and reliable solution of the problem is use of a special capacitor with large capacity connected in parallel to the feed circuit of every MPR instead of UPS usage. High-quality capacitors with large capacity and rated voltage of 450 – 500 V are sold today by many companies under the price, approximately, €150 - 200 and are not deficient (Table 6.1).

Use of such capacitor for auxiliary voltage of 220 V AC requires, naturally, a rectifier and some more auxiliary elements (Fig. 6.2).

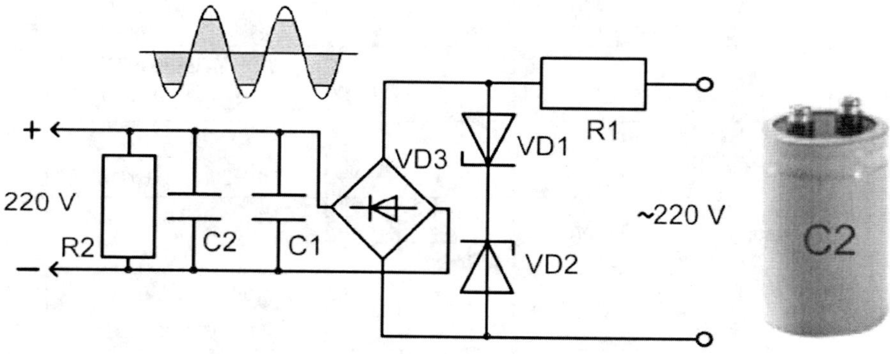

FIGURE 6.2 The device for reserve feed of MPR at emergency mode with AC auxiliary voltage.

In this device a capacitor of large capacity is designated, such as C2. The C1 is auxiliary not-electrolytic capacitor with capacity in some microfarads serves for smoothing pulsations on electrolytic capacitor C2. It is possible to include also in parallel to C1 one more ceramic capacitor with a capacity of some thousand picofarads, for protection of C2 against the high-frequency harmonics contained in mains AC voltage. A R1 (200 – 250 Ohm) resistor limits the charging current of C2 at a level near 1A. The same resistor also limits pulse currents proceeding through back-to-back connected Zener diodes VD1 and VD2. Resistor R2 has high resistance and serves to accelerate the discharging capacitor up to a safe voltage at switching-off of the auxiliary voltage. Zener diodes are intended for the maximal value voltage limits of capacitor C2 at a level of 240 V. Without such limitations on the device output voltage would reach a value of more than 300 V due to the difference between r.m.s. and peak values of voltage. That is undesirable both for MPR and for C2. The Zener diodes slice part of a voltage sinusoid in which amplitude exceeds 240V, forming a voltage trapeze before rectifying. As powerful Zeners for rating voltage above 200 V are not at present on the market, it is necessary to use two series connected Zeners with dissipation power of 10 W and rating voltage of 120 V, for each of Zeners (VD1, DD2 – for example types 1N1810, 1N3008B, 1N2010, NTE 5223A, etc.).

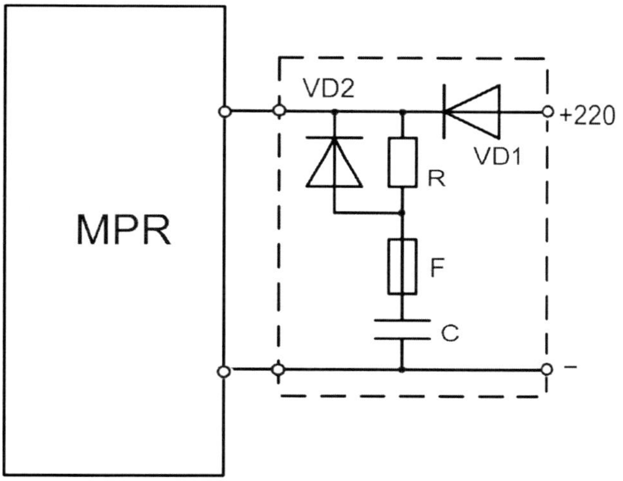

FIGURE 6.3 The device for reserve feed of MPR at emergency mode with DC auxiliary voltage.

As further research of this type of situation is clarified, the problem of maintenance of reliable feed MPR is relevant not only for substations with AC auxiliary voltage, but also for substations with DC voltage. Many situations where the main substation battery becomes switched-off from the DC bus bars are known. In this case nothing terrible occurs, as the voltage on the bus bar is supported by the charger. However, if during this period an emergency mode occurs in a power network, the situation appears to be no better, since use of an AC auxiliary voltage as charger feeds from the same AC network. Usually electrolytic capacitors with high capacitance for smoothing voltage pulsations are included on the charger output. Since not only many MPRs, but also sets of other consumers are connected to charger output it is abundantly clear that this capacity is not capable of supporting the necessary voltage level on the bus bars during the time required for proper operation of the MPRs. Our research has shown that such high capacitance as 15,000 μF does not provide proper functioning of MPRs at consumption from charger reaches up to 5 – 10 A.

For maintenance of working capability of MPRs in these conditions it is possible to use the same technical solution with the individual storage capacitor connected in parallel to each MPR feeding circuit. Now the design of the device will be much easier, due to a cut-out from the circuit diagram of Zeners and rectifier bridge (Fig. 6.3). The resistor R (100 Ohm) is necessary for limiting of the charging current of the capacitor at switching-on auxiliary voltage with a fully discharged capacitor. Diode VD1 should be for a rated current of not less than 10A. High capability quick blow fuse F (5A/1500A, 500V) is intended for protection of both feeding circuit of MPR and external DC circuit at damaging of the capacitor.

The prototype of such device with the capacitor 3700 μF (Fig. 6.4) has shown excellent results at tests, with the various loadings which simulate MPRs of various types with different power consumption at nominal voltage 240V (Fig. 6.5).

FIGURE 6.4 The prototype of the device for reserve feed of MPR.

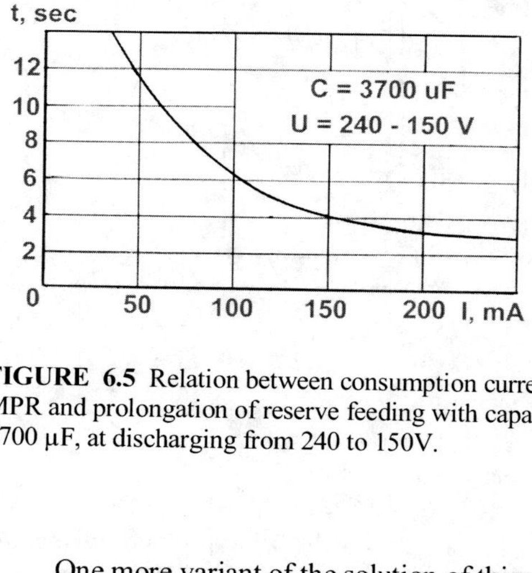

FIGURE 6.5 Relation between consumption current of MPR and prolongation of reserve feeding with capacitor 3700 μF, at discharging from 240 to 150V.

One more variant of the solution of this problem for substations with DC auxiliary voltage is to not use an individual capacitor for each MPR, but rather a special "supercapacitor" capable of feeding a complete relay protection system set together with conjugate electronic equipment within several seconds. Such supercapacitors can already be found on the market under brand names such as: "supercapacitors," "ultracapacitors," "double-layer capacitors," and also "ionistors" (for Russian-speaking technical literature). There are electrochemical components intended for storage of electric energy. On specific capacity and speed of access to the reserved energy they occupy an intermediate position between large electrolytic capacitors and standard accumulator batteries, differing from both in their principle of action, based on redistribution of charges in electrolyte and their concentration on the border between the electrode and electrolyte.

Today, supercapacitors are produced by many Western companies (Maxwell Technologies; NessCap; Cooper Bussmann; Epcos; etc.) and also some Russian enterprises (Esma; Elit; etc.). The capacity of modern supercapacitors reaches hundreds and even thousand of Farads, however the rated voltage of one element does not exceed, as a rule, 2.3 – 2.7 V. For increasing voltage of the supercapacitor its internal separate elements connect among themselves in parallel and series as consistent units (Fig. 6.6). Unfortunately, supercapacitors are not so simply incorporated among themselves as ordinary capacitors, and demand leveling resistors at series cells connection and special electronic circuits for alignment of currents at parallel cells connection. As a result, such units turn out to be rather "weighty," expensive, and not so reliable (there could be enough damage to one of the internal auxiliary elements to cause failure of the entire unit).

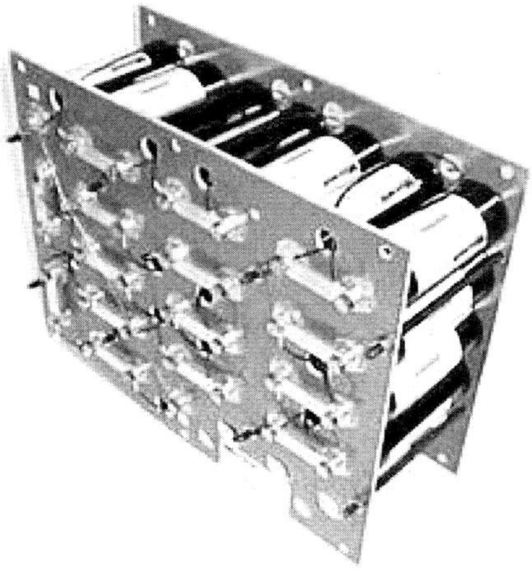

FIGURE 6.6 Internal design of high-voltage (ten voltages) supercapacitor, assembled from number of low-voltage elements.

FIGURE 6.7 High-voltage supercapacitor modules ESCap 90/300 type (Table 6.2).

Table 6.2 Main parameters of the ESCap90/300 type supercapacitor.

Rated Voltage, V	300
Capacitance, F	2.0
Max. Power, kW	75
Max. Energy, kJ (at 300 V)	90
Internal Resistance, Ohm	0.3
Dimensions, mm	Dia. 230 x 560
Weight, kg	35
Temperature, °C	-40 +55
Price per Unit (for 2006), $	1000.00
Rated Voltage, V	300
Capacitance, F	2.0
Max. Power, kW	75
Max. Energy, kJ (at 300 V)	90
Internal Resistance, Ohm	0.3
Dimensions, mm	Dia. 230 x 560
Weight, kg	35
Temperature, °C	-40 +55
Price per Unit (for 2006), $	1000.00

For example a combined supercapacitor manufactured by the NessCap firm, with a capacity of 51 F and voltage of 340 V, weighs 384 kg!

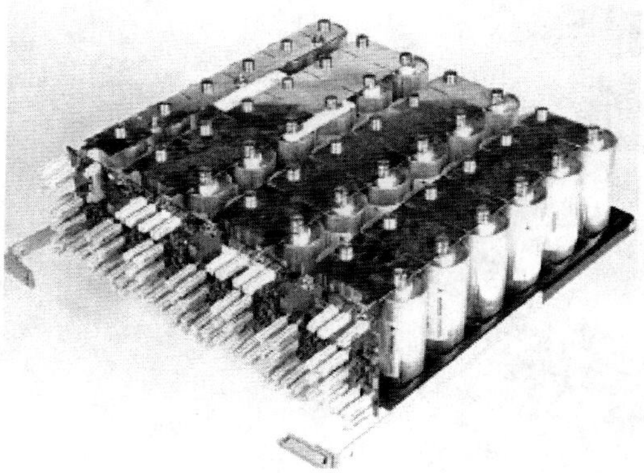

Sitras®

Nominal voltage – DC 750 V
Number of Ultracapacitors in single module –1344
Energy stored – 2,3 kWh
Energy saving per h – 65 kWh/h
Max. power – 1 MW
Capacitor efficiency – 0,95
Temperature domain –20 to 40 °C

FIGURE 6.8 High-voltage supercapacitor module Sitras® series from Maxwell Technologies.

Some other companies also produce combined supercapacitor modules for high voltages (Fig. 6.7). Such ESCap90/300-type supercapacitors, (see Table 6.2) meet our purposes quite well.

Another example is the supercapacitor module Sitras® series from Maxwell Technologies (Fig. 6.8). At use of supercapacitor SC, the feeding circuit of the protective relays should be allocated into a separate line connected to the DC bus bar through diode D (Fig. 6.9).

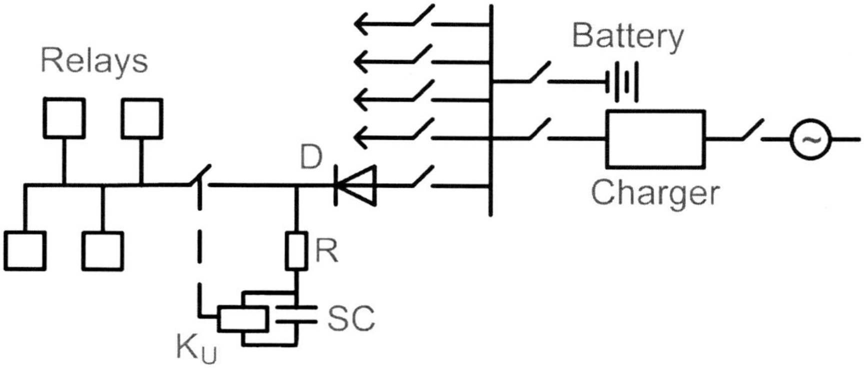

FIGURE 6.9 Example of usage of the supercapacitor as group power supply for protective relays at emergency mode with DC auxiliary voltage.
K_U – voltage relay; SC – the supercapacitor.

Due to the large capacity of the supercapacitor the voltage reduction on feeding input of MPR at emergency mode (with loss of an external auxiliary voltage) will occur very slowly, even after passage of the bottom allowable limit of the feeding voltage. From the personal experience of the author, cases of false operation of the microprocessor systems have been known to occur at slow feeding voltage reduction, below allowable levels.

This can be explained by the existence of different electronic components of a high degree of integration serving the microprocessor, having different allowable levels of voltage feeding reduction and stopping the process of voltage reduction serially, breaking the internal logic of the MPR operation. If such equipment is found in the MPR, used on the given substation, in parallel to the supercapacitor, it should be connected to a simple voltage monitoring relay K_U, which disconnects the supercapacitor at a voltage reduction below the lowest allowable level, for example, lower than 150 – 170V.

6.2 INCREASING RELIABILITY OF TRIP CONTACTS IN MICROPROCESSOR-BASED PROTECTIVE RELAYS

Whether this is good or bad may be open to debate (the advantages of microprocessor-based protection means over the traditional are far from being absolute or obvious) yet we must acknowledge that this is the general trend. While acknowledging this trend we must also note that microprocessor-based protection means do have several specific drawbacks. In this paper one of these problems will be discussed.

Microprocessor-based relay protection devices (MPR) with different functionalities (differential, distance protection, generators protection, capacitor banks protection, etc.), made by different leading companies in the world such as ABB, General Electric, Areva, Alstom, Cooper, and Crompton Instruments, were analyzed for compliance of output of the electromagnetic relays used in these MRPD, with the standard requirements and parameters set forth in the manufacturers' specifications and the actual operating conditions in the power systems.

It was established that in all types of MPR, electromagnetic relays of the same class were used as output elements: subminiature relays with one make or changeover contact enclosed in a sealed plastic box having dimensions of about 30 x 10 x 12 mm (Fig. 6.10). These are G2RL, RY6100 G6RN, RTE24012, ST2, JS, and similar relays made by the Schrack, Omron, Matshshita, and Fujitsu companies. Normally the ***maximal values*** of switched voltage and switched current are marked on the bodies of the subminiature relays, in contrast to the maximal switching power and the type of current to which these current values are related, which are usually omitted. This creates a problem when choosing the relay since maximal switched power is not equal to the multiple of maximal switched voltage and maximal switched current. For adequate evaluation of the switching ability of such relays the accompanying technical documentation needs to be analyzed. The results of our analysis of technical documentation accompanying these subminiature electromagnetic relays are presented in Table 6.3.

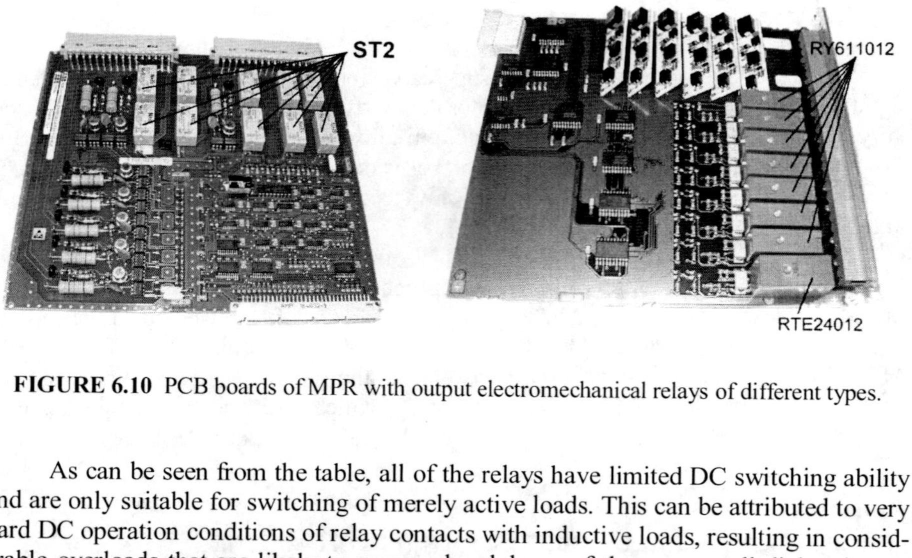

FIGURE 6.10 PCB boards of MPR with output electromechanical relays of different types.

As can be seen from the table, all of the relays have limited DC switching ability and are only suitable for switching of merely active loads. This can be attributed to very hard DC operation conditions of relay contacts with inductive loads, resulting in considerable overloads that are likely to cause a breakdown of the very small dielectric gap between the relay contacts which maintains arcing on the contacts, as well as nonoccurrence of periodic current zero crossing, characteristic of AC. Voltage across the contacts may become six-fold the value of the nominal voltage. When the voltage over the contacts exceeds 50 V a strong spark is generated at switching such a load that causes strong erosion of the contacts.

As the applied voltage is increased (to 100-150 V), the spark at the relay contacts is changed to a stable arc, which totally melts even powerful contacts (rated for nominal currents of 10-15 A) within 0.5-1.0 seconds at a current of 0.5-2 A. Therefore the switching ability of DC relays is much lower than that of AC relays (Fig. 6.11). However, according to the manufacturers of MPR, subminiature relays installed in MPR are designed for the direct making of tripping coils in high voltage circuit breakers – CB (for line protection) or for the making of lockout relays – powerful intermediate latching relays with a manual reset (for transformer protection). Such types of loads (inductive loads in 220 V DC circuits) are the heaviest duty for relays. What are these loads under actual operating conditions?

Table 6.3 Switching capability of subminiature electromechanical relays using MPR.

Relay Type (Manufac-turer)	Maximal Switching Power (for resistive load)		Rated Current & Voltage (for resistive load)		
	AC	DC	AC	D	for 250 V DC
ST series (Matsusita)	2000 VA	150 W	8 A; 380 V	5 A; 30 V	0.40 A
JS series (Fujitsu)	2000 VA	192 W	8 A; 250 V	8 A; 24 V	0.35 A
RT2 (Schrack)	2000 VA	240 W	8A; 250 V	8A; 30 V	0.25 A
RYII (Schrack)	2000 VA	224 W	8A; 240 V	8A; 28 V	0.28 A
G6RN (Omron)	2000 VA	150 W	8 A; 250 V	5 A; 30 V	-
G2RL-1E (Omron)	3000 VA	288 W	12 A; 250 V	12 A; 24 V	0.30 A

Table 6.4 includes the results of analysis of tripping coil parameters for different types of circuit breakers made in different countries. As can be concluded from the comparison of the abovementioned relay parameters (Table 6.3) and the parameters of the tripping coils of CBs (Table 6.4), the switching ability of subminiature relays for DC circuits (0.3 – 0.4 A) is not at all sufficient for the direct making of tripping coils of CBs (the required currents are 1 – 6 A). Connecting of lockout relays between output MPR relays and high voltage CB still does not provide a solution since the self-current consumed by the coil of lockout relay (2 A for the HEA type relay and 2.8 A for a modern HEA63 relay made by General Electric) falls into the same range of currents of tripping coils of CBs.

The situation becomes even more complicated because switching of ***the DC inductive load*** for these relays cannot be foreseen at all, so subsequently the use of these relays for direct switching of tripping coils of CB's, as well as auxiliary lockout relays, results in the generation of loads beyond those allowed. What do the standards and technical documentation related to MRPD say?

According to the ANSI/IEEE C37-90-1989 and IEEE Standard for Relays and Relay Systems Associated with Electric Power Apparatus [6.4], Part 6/7: Make and Carry Ratings for Tripping Output Circuits establishes, that the making current and the carry current accompanying it for 4 seconds that is provided by the contacts of the input relays controlling the tripping coil of CBs shall be at least 30 A. Why is this current value so high in comparison with the actual currents in disconnecting switches? Because the AC tripping coil (a solenoid with a movable core) has considerable starting currents (up to 10-fold) caused by a low initial impedance of a solenoid with an extended core. For devices with alternating operation current such requirements are quite justified.

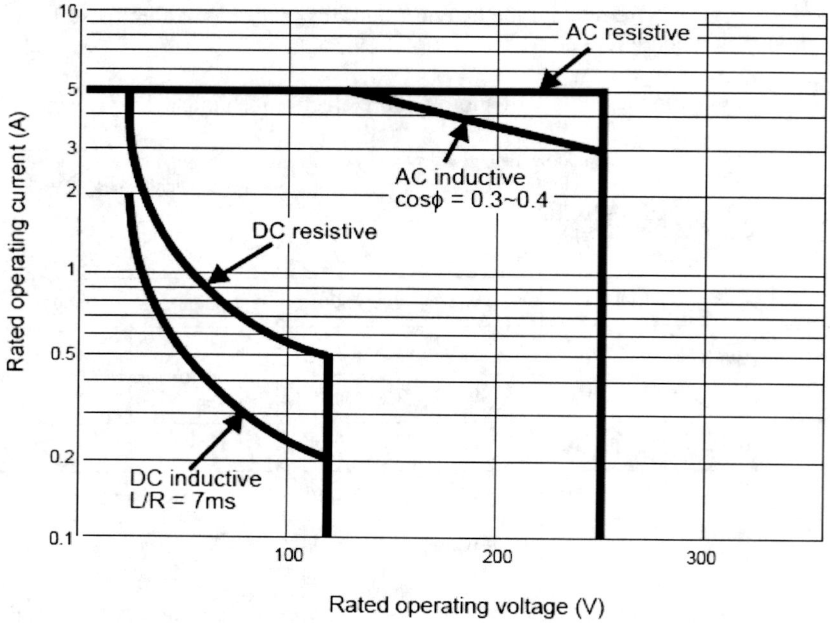

FIGURE 6.11 Typical relations between commutation parameters (voltage, current) and load characters for relay contacts.

It is reasonable that because of this requirement MPR manufacturers included this parameter in the standard provided by their MPR devices. In this way the MPR specification with regard to this parameter totally complies with the standard requirements. However the situation is different for the MPR themselves, since the specifications for specific types of output electromagnetic relays do not mention at all the capability to switch 30 A of current, even if it is AC.

Here we should be more precise and mention that specifications for some types of special extended power subminiature relays (not used in MPR) mention the ***inrush current,*** i.e. short duration making currents reaching values of up to 30 A. These relays are as if specifically designated for use in MPR. Maybe if MPR manufacturers were requested to use these relays this would provide a solution to the problem?

The issue turns out to be not so simple, as there are some other standards related to the relay switching modes. In particular the IEC 60947-4 standard, in which the switching modes of relays and contactors are divided into so called ***"categories of ap-***

plication," specifies requirements for contact relays operating in these modes. In particular the contacts designated for controlling the electrical magnets of other intermediate relays, contractors, solenoids, and valves are classified as AC-15 for AC and DC-13 for DC (Table 6.5).

Table 6.4 Parameters of tripping coils of high-voltage circuit breakers.

Circuit Breaker Model	Circuit Breaker Kind	Trip Coil	
		Rated Voltage, V DC	Rated Current, A
ELK SD14 (ABB)	SF_6, 170 kV, 4000 A	220	2.3
B3-S101 (ALSTOM)	SF_6, 170 kV, 2000 A	220	0.7
CPRG180/10-360 (AEG)	Generator CB for 13.8 kV	110	2.0
3AP1F1 (Siemens)	Oil CB, 245 kV, 3150 A	220	5.8
BBP-6-10/630 (Russia)	Vacuum CB, 10 kV, 630 A	100	5.0
ВБГ-35 (Russia)	SF_6, 35 kV	220	2.5
BBOA-15-14/12500 (Russia)	Air CB, 15 kV, 12500 A	220	4.5

From Table 6.5 it follows that increased (ten-fold) switching current of the relay at closing (making capacity), with respect to nominal current, is allowed only for AC. In switching of DC circuits this increase does not exceed 10%. This is accounted for by fact that not all making of relay contacts is terminated after the initial contacting of those contacts. Actually the making process is always accompanied by the contacts bouncing after their first closing. Relay contacts make several open-close cycles of contact bounce before coming to rest in the final state, Fig. 6.12. This contact-bounce interval is in addition to the relay's operate and release times, which can measure (upon the type of relay) from one millisecond for small relays to tens of milliseconds for larger relays. Therefore, all other factors being the same, the making power of the contacts of a DC load relay is much lower than that of an AC one. From the above it follows that short duration making currents of 30 A (3.75 I_N) for subminiature relays may be allowed only in AC circuits (even though this restriction for the use of relays is not pointed out in any of the specifications, for obvious reasons!).

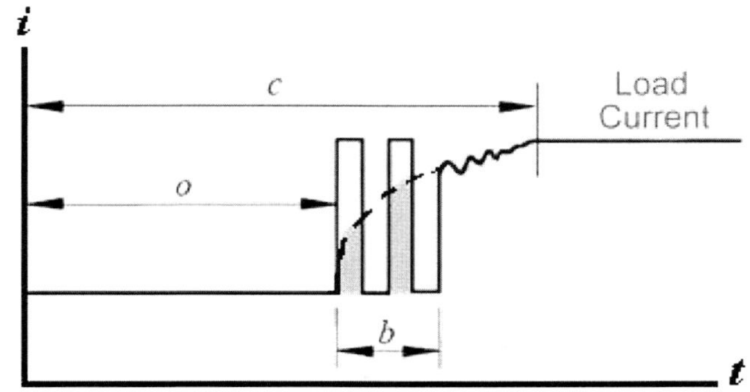

FIGURE 6.12 Oscillogram of relay making process with contact bounce (according to IEC 61810-7).
o – time period from coil energizing up to first contact closing; ***b*** – bounce time; ***c*** – time to stable closing.

That is quite reasonable since there are no DC making currents at solenoid and control coil activation. From here it also follows that when the tripping coil of the CB is controlled by DC 220-250V circuits, the allowed value of making current for contacts is only 110% of the nominal value, namely within a range of 0.35 – 0.45A, which is much lower that the actual currents.

In actual operating conditions output relays of MPR are operated relatively rarely (only in case of failures in the networks), which postpones detection of switching problems. This saves MPR manufacturers from customers' claims. Because of erosion that is intensified at each relay operation the contacts, surface condition is gradually deteriorated and their resistance and heating increased, which results in welding of the contacts during the next switching. In the course of the above research we have approached many manufacturers of subminiature relays with a request for an opportunity to use their relays for making without breaking currents of inductive load at a voltage 220V DC and we received the following answers:

a) the danger of welding of contacts may be very great because of bouncing;

b) the relay to be used only in the authorized modes specified in the technical specifications.

Moreover, in some cases, for example for accelerated (forcing) operating of the CB (used in some types of Siemens CBs, for example) a special circuit is used (Fig. 6.13) that provides higher making current of up to 75A. Direct connection of contacts of subminiature relays in such circuits is prohibited.

The problem becomes even more complicated since at infrequent protection operation the mentioned discrepancies in MPR are not detected at once. Under these conditions MPR can function well for a couple of years, during which the output relay contacts accumulate defects that eventually lead to a sudden failure, resulting in serious damage.

Table 6.5 Switching capacity of contacts depending on the type of load for control electromagnets, valves, and solenoid actuators.

Utilization Category IEC 60947-4	Type of current	Switching capacity of contacts in the mode of normal switching					
		Make (switching ON)			Break (switching OFF)		
		current	voltage	cosφ	current	voltage	cosφ
AC-15	AC	10 I_N	U_N	0.3	10 I_N	U_N	0.3
DC-13	DC	I_N	U_N	-	I_N	U_N	-
Switching capacity of contacts in the mode of infrequent switching							
AC-15	AC	10 I_N	1.1 U_N	0.3	10 I_N	1.1 U_N	0.3
DC-13	DC	1.1 I_N	1.1 U_N	-	1.1 I_N	1.1 U_N	-

I_N and U_N are rated values of currents and voltages of electric loads switched by relay contacts.

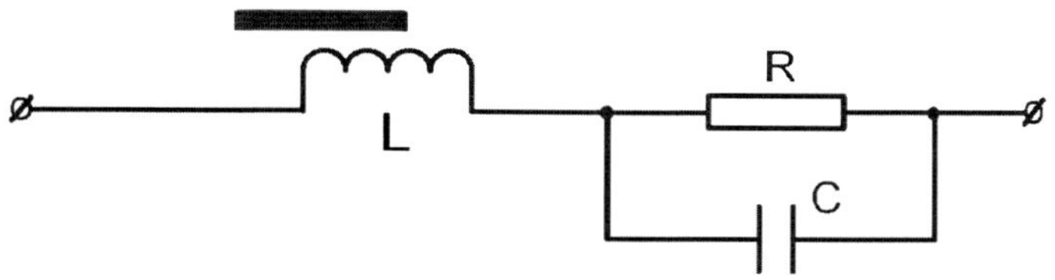

FIGURE 6.13 A circuit diagram for forcing switching of the CB tripping coil (L).

Provided the device can operate for several years without visible faults, it is difficult to present a claim to the manufacturers of MPR. It is also quite difficult to determine the exact working life of a relay operating under such conditions, and to predict when damage is due.

What can be done in this situation? The MPR manufacturer should be requested to install several output relays, complying completely with the standard requirements for industrial relays.

It should be noted that in the past this approach was very popular in semiconductor protection devices (Fig. 6.14), however at present it is not practicable, as this would require a major change in the MPR structure, and an increase in size. The problem could be resolved by having the user connect external power amplifiers between the MPR output relay and the tripping coil of the CB. This amplifier would have to be simple, fast-acting, jam-resistant, and highly reliable under actual operation conditions.

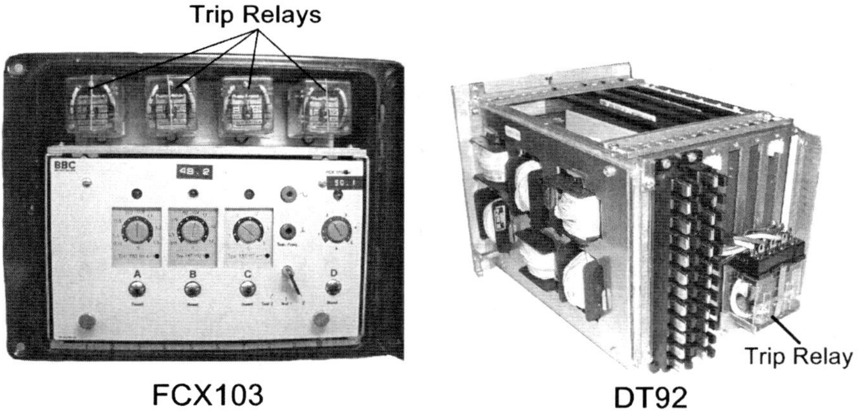

FIGURE 6.14 Static (electronic) protection devices with large-output electromechanical relays (trip relays) of industrial type.

We have analyzed the technical parameters of various strong-current solid-state relays (SSR) made by the leading companies in the world (ABB, Tyco Electronics, Crouzet, Teledyne, Magnecraft, Celduc, Crydom, Comus, etc.) and determined that each production sample SSR has at least one or more parameters that do not comply with the requirements essential for their use as a power amplifier of output MPR relays. Such parameters as maximal DC voltage withstood over the main electrodes in a cut-off state, which must be at least 1500 V; making current in DC circuits with an inductive load, which must be at least 5 – 10 A; operational suitability in DC circuits (many SSR can be operated only in AC circuits); and make-time which must be not more than 1ms.

Due to unavailability, in today's market, of power amplifiers conforming to the requirements of combined operation with MPR, we have developed an amplifier conforming to these requirements. Due to its simple structure (Fig. 6.15), in-house making of this amplifier by the electric power companies is quite possible. The main switching element of the device is a particularly small-sized thyristor VT designed for current up to 30A and voltage up to 1600V. It has additional protection against spikes by means of a varistor RV with clamping voltage of 1200 V (at nominal mains voltage of 250V this provides high reliability of the varistor). With the help of a special normally closed high voltage optical coupler Opt the thyristor is forcibly blocked in the OFF-state in order to prevent its accidental switching by induced voltage or noise signal.

The thyristor is switched ON by the control current flowing in the thyristor control circuit at closure of contact K of the output relay MPR. Capacitor C (0.01 μF 1600 V) is used as an additional filter preventing that the noises reach the thyristor. Unfortunately only a few of the thousands of electronic components available in the market comply entirely with the requirements. In the first place this relates to the thyristor VT (type 30TPS16, STMicroelectronics) and optical coupler Opt (type TLP4597G, Toshiba). In order to provide higher reliability and faster response of the device it is recommended to use only these elements.

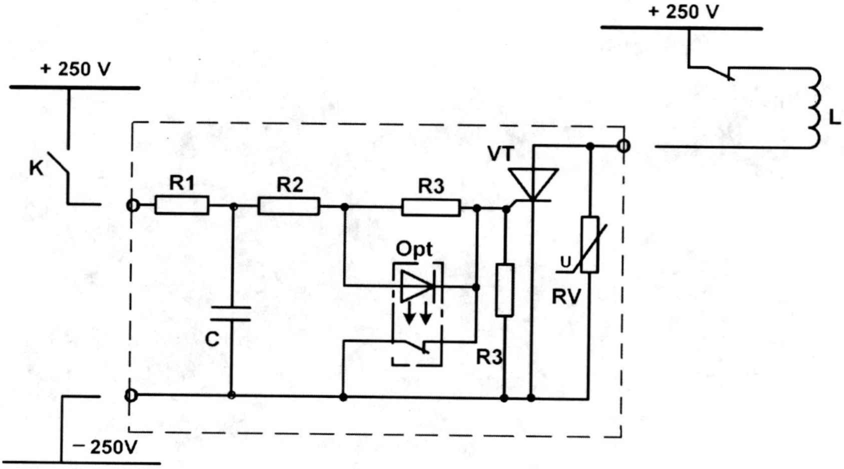

FIGURE 6.15 Switching amplifier for output MPR relays: K – contact of output MRPD relay; L – tripping coil of CB.

Compact electromagnetic relays intended especially for switching inductive load at 125 – 250V DC can be used for energizing of the CB tripping coil. For such ability the contact system of the relays incorporated a blow-out magnet, placed between contacts. Magnetic field of the magnet interacts with DC arc between contacts and rapidly repulses it away from the contacts at their breaking. Many companies produce such relays today (Fig. 6.16). Some of them are suitable for installation on the printed-circuit-board (PCB). For example, the JC2aF-H73 type, produced by Matsushita, IG2C-24VDCM type, produced by Kuhnke, etc. Application of such relays can appear effective not only for making of the CB trip coils used in usual schemes, in view of what was stated above, but much more for making of these coils in special schemes with the forcing capacitor, which creates a high current pulse through the coil (and contacts, of course) in the initial stage of the making process for its acceleration (see above).

In perspective, the decision should be, in MPR, to use hybrid devices containing a subminiature electromechanical relay (SER) contact and a solid state switching element, connected in parallel. One of these devices is suggested by the author in Fig. 6.17.

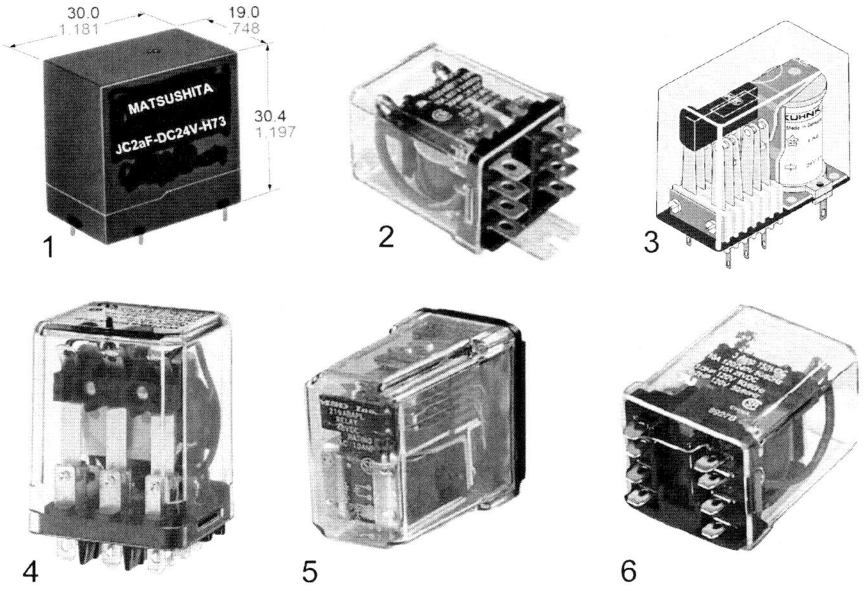

FIGURE 6.16 Compact relays with blow-out magnets suitable for switching inductive load at 125 – 250 VDC.
1 – JC2aF-DC24V-H73 type (Matsushita); 2 – 300 series (Magnecraft); 3 – IG2C-24VDCM type (Kuhnke); 4 – KUEP-3D17-12 (Potter & Brumfield); 5 – 219 series (Magnecraft); 6 – A283 series (Magnecraft).

In this device the control signal is put to a winding of the SER and simultaneously charges capacitor C, through a limit resistor (R1) and the gate circuit of thyristor VS. The thyristor instantly opens by means of this charging current (with a delay of some microseconds) and picks-up the CB trip coil L. Contact K closes and shunts the thyristor through 7 – 10 ms (the time of operation of the SER).

The current of trip coil L flows to a circuit of contact K. At the beginning of this process there is an opening of contacts at bouncing, increasing inter-contact resistance, and increasing the voltage on the contacts.

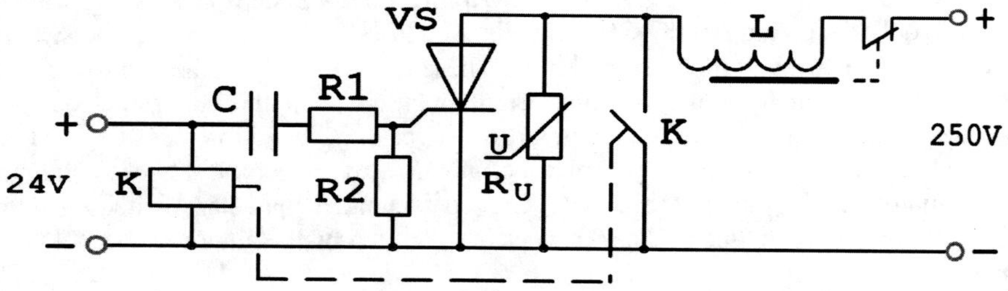

FIGURE 6.17 Circuit diagram of hybrid element intended for energizing trip coil at 250 VDC.

This voltage is put to thyristor VS. At an increase in this voltage of up to 5 – 7 V the thyristor again instantly opens as this time the capacitor C has not yet been completely charged and charging current continues to flow through the gate of the thyristor. The capacity C and resistance R1 leave this condition in order to guarantee a thyristor gate current of about 50 – 70 mA during 15 – 20 ms, that is, before the full termination of contact bouncing. Thus, during the contact making process there is no break of the trip coil circuit and there is no arc on the contacts. After full charge of the capacitor, the current in the thyristor gate circuit stops, is finally locked, and no longer influences the condition of the load (trip coil) circuit. At disappearance of the control signal at the input of the device the capacitor is discharged through the coil of the relay K.

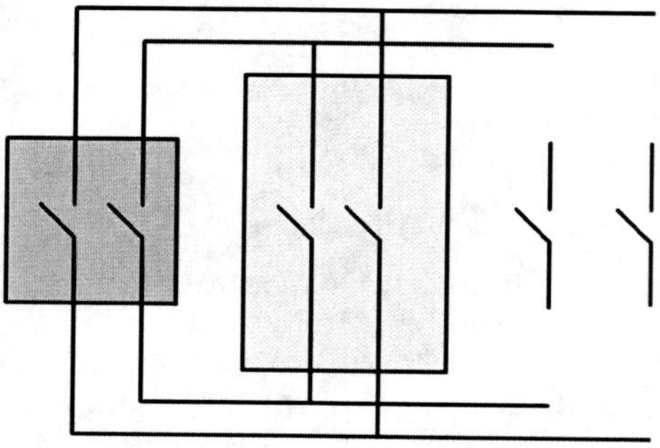

FIGURE 6.18 Connection in parallel-in pairs contacts from the relay of various types for excluding of bouncing.

As the basis of the problem of switching-on of the trip coil is the bouncing of contacts, accompanying short-term breaks of current during the making process, the natural solution of this problem could be to compensate these breakages with a combination of two contacts connected in parallel with the various parameters of vibration which are not conterminous on a phase (Fig. 6.18). It is abundantly clear that if contacts of relays of various types, with various mechanical properties of contacts, are connected in parallel-in pairs, we can predict with practically full confidence that it is possible to guarantee unconformity in the phase of bouncing that is preventing circuit breaks during the switching-on of inductive load. MPR manufacturers can take arms on this principle, using their usual SER products as described above, and SER manufacturers can produce relays containing, in the single case, two contacts with different rigidities or weights, in which vibration and bouncing at switching-on does not coincide on a phase, having connected both these contacts in parallel.

An alternative solution is the series connected contacts. Some relay manufacturers permit series connection of contacts for increasing of switching DC capability (Fig. 6.19).

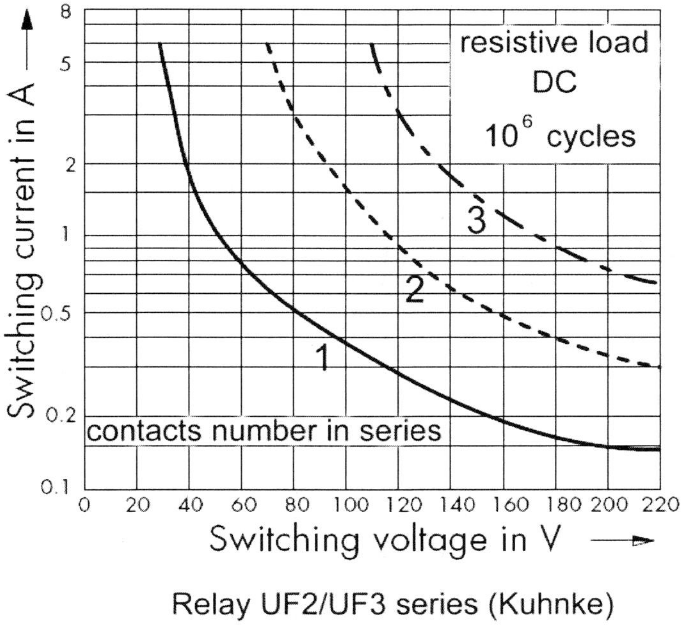

FIGURE 6.19 Increasing of DC switching capacity for series connected contacts.

Relay UF2/UF3 series is a compact industrial but not subminiature relay. It would be good if manufacturers of subminiature relays gave the same characteristics for the relays.

Power reed switches with increased switching capability have now appeared in the market (Fig. 6.20) and the relays on which they are based could be used with success as output contacts in MPD. Such reed switches with a rated current of 5A are capable not only to make, but also to break circuits with inductive loads at 230V DC. For example, they are capable of switching-off 0.4 – 0.6 ADC with a constant time of 40 ms. That it is quite enough for their usage as the auxiliary contacts of MPR intended for control by external auxiliary relays.

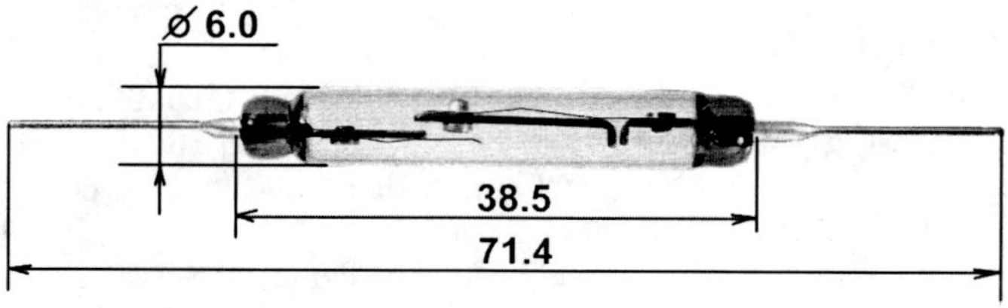

FIGURE 6.20 Power reed switch R14U and R15U (Bestact™, produced by Yaskawa) with two-stage contact system for switching inductive load.

The operating time of these reed switches does not exceed 5 ms, bouncing time not more than 2 ms; switching cycles number 50 – 100 million. It is obvious that when reducing the requirement of the number of switching cycles 10,000 times, we should expect an additional increase in their switching current (that should be confirmed with the manufacturer, of course). When choosing auxiliary output relays for MPR it is necessary to take into account that some of them will be used for switching of the coils of the external electromagnetic relays at 125 – 240 V DC, and some others for switching of low-voltage and low-current signals on logic inputs of other MPR. Usually that is not taken into account in any way by manufacturers of MPR; however, it is known that for switching powerful loads and for switching low-current, low-voltage signals contacts should have different properties and should be made from different materials. This is reflected in the technical specifications on SER. For example, a relay with powerful contacts will have limits on the lower threshold of a switched current and voltage, and this threshold frequently lies within the values used in practice for small level control signals. Therefore, two types of SER corresponding to two types of loads should be used as auxiliary output relays in MPR, as mentioned above.

Modern high-voltage IGBT transistors can be successfully used as powerful output auxiliary relays of MPR. The small sizes (plastic case TO-247 and similar types) have high values of collector current, high withstanding overvoltages, high power dissipation, and high allowable working temperatures of the crystal (see attachment), making such transistors rather attractive elements for switching inductive loads (coils of external auxiliary relays) with a consumption of 0.1–5 A at 250 VDC. For overvoltage protection when switching inductive loads, such transistors can in addition be protected by varistors with a clamping voltage of 500 – 700V. As is known, some problems may arise when using IGBT transistors represented by the correct organization of their control circuits; however, today these problems are successfully solved and there are numerous drivers for controls of IGBT transistors on the market which are made as small modules (Fig. 6.21). In such drivers, all necessary elements are contained inside for reliable switch-on and switch-off IGBT transistors.

Single modules of this type and two IGBT transistors, form analogues of high-quality changeover contacts, galvanically isolated from the internal control circuits of the MPR. There are in the market also completely assembled modules (solid-state relays on the basis of IGBT technology) ready to use at 250 VDC.

These modules have bigger sizes (58.4 x 45.7 x 22.9 mm) than the single IGBT with the driver but can be used in MPD of anyone's design as they do not demand the printed circuit board or any additional elements for their installation (Fig. 6.22).

Both these modules have a high current and overvoltage capability (75A, 1500 V - for 1, and 25A, 1200 V – for 2), that made them suitable for usage in MPR.

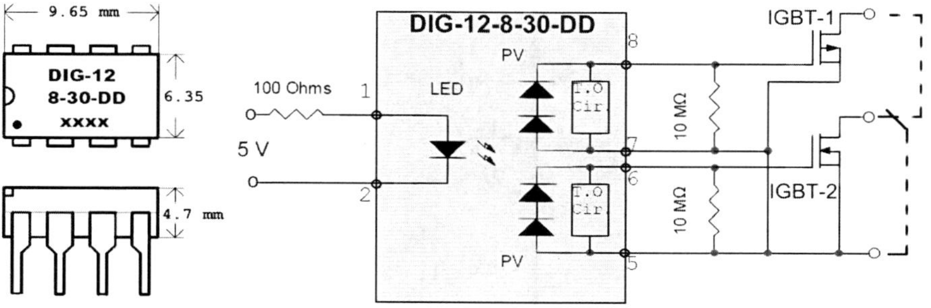

FIGURE 6.21 Modern galvanic isolated driver for controls of a pair of IGBT transistors, formatives single changeover contact.

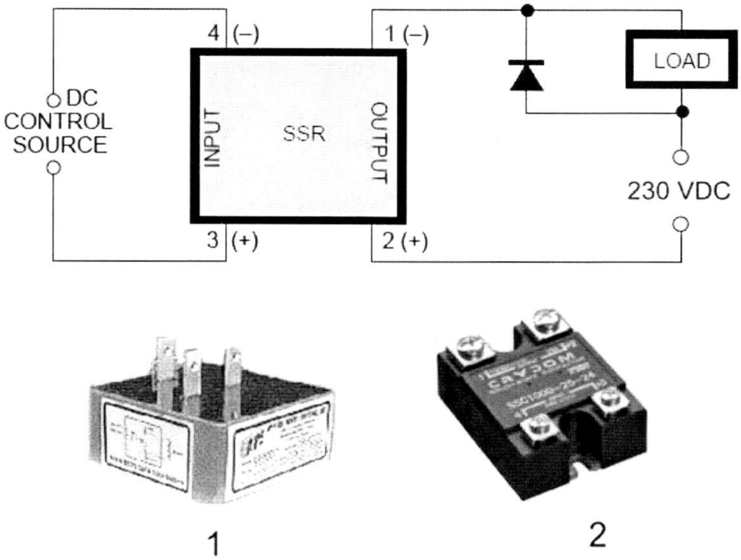

FIGURE 6.22 Solid-state modules for switching of inductive load at 250 VDC.
1 – APSW-DC75 type (Applied Power Systems);
2 – SSC1000-25 type (Crydom).

One elementary solution of the problem of MPD output contacts, including switching-on the CB trip coil, could be to use an external power amplifier of an elementary type, inserting it between the output contact of the MPR and the trip coil (Fig. 6.23).When the single output MPR contact must switch-on a group of trip coils belonging to different circuit breakers, it is possible to use the above mentioned power demultiplexer on thyristors connected to the output of the amplifier.

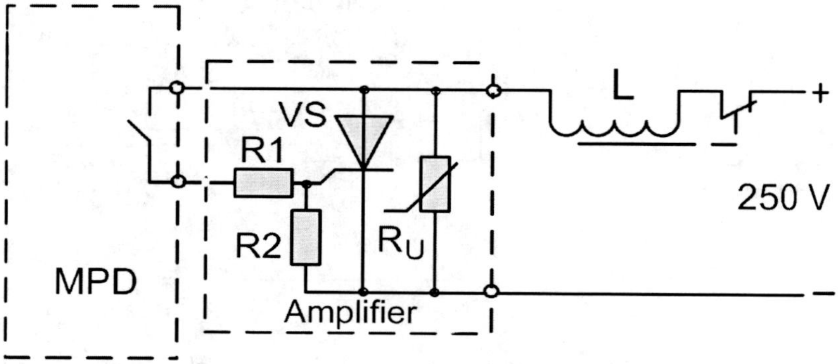

FIGURE 6.23 Simple switching amplifier on single thyristor for trip coil energizing.

For contacts of auxiliary relays (which require not only switching-on, but also switching-off the inductive load), arc protective modules of the passive type, connected in parallel to contacts of the relay, can be used: for example an RC-circuit of self-made or industrial types (Fig. 6.24), manufactured by many companies.

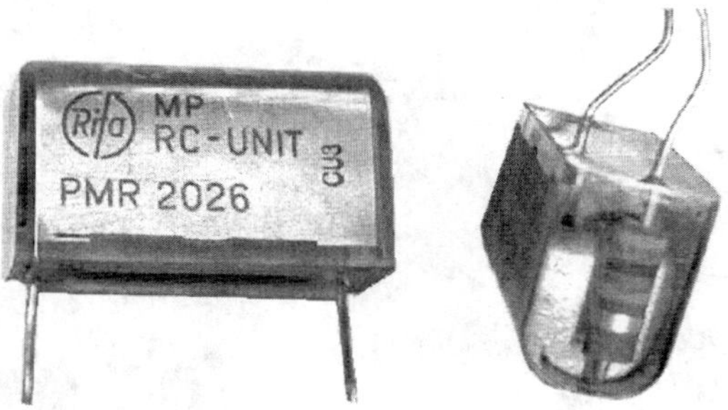

FIGURE 6.24 Passive arc-protective module contains series connected R and C elements (produced by RIFA).

More effective protection of relay contacts against an electric arc is provided by protective modules of the active type, containing semiconductor elements such as transistors (Fig. 6.25).

Naturally, modules of this type are much more complex and expensive than modules of the passive type. Even a more simplified version of such a module (USA Pat. 5703743) contains two transistors (IGBT and FET types), one triac, three diodes, and three Zeners. A more sophisticated updating (USA Pat. 6956725) consists of the current transformer, a rectifier bridge, and some capacitors and resistors in addition to the above-listed elements. Such modules are sold in the open market by Schweitzer Engineering Laboratories and can be successfully used by any consumer used MPR.

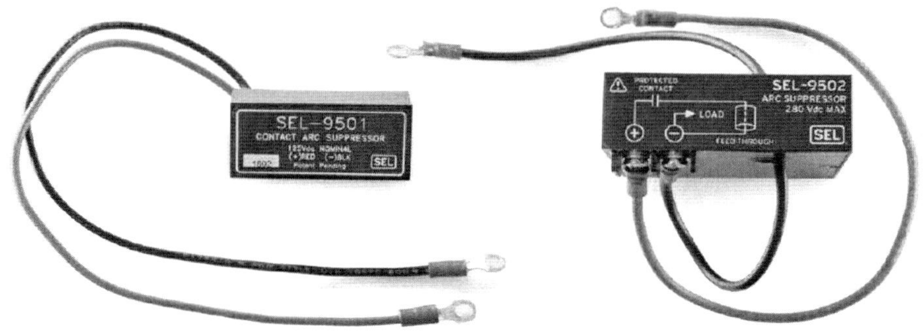

FIGURE 6.25 Smart (active type) arc-protective modules SEL-9501 and SEL-9502 types (produced by SEL).

The choice of type of protective module depends on the concrete parameters of the switching load. At "light" loads, with the time constant not exceeding 7 – 10 ms, elementary RC-modules can be used, and for heavy loads with R/L = 30 – 50 ms, active type modules are more suitable.

7

Automatic Devices for Power Engineering

7.1 ARC PROTECTION DEVICE FOR SWITCHBOARDS 6 – 24 KV

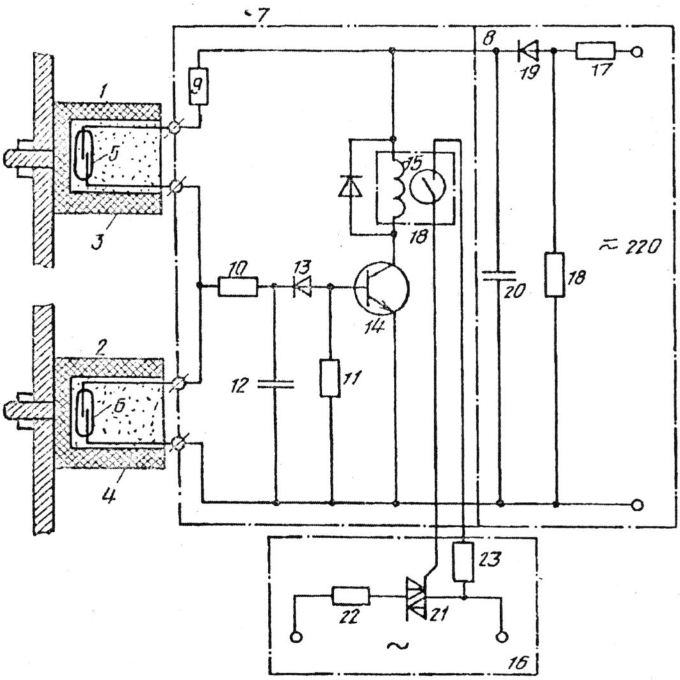

FIGURE 7.1 Basic circuit diagram for relay unit with the function of differential protection. 1, 2 – HV current sensors; 5, 6 – reed switches; 16 – output circuit.

To ensure protection against an electric arc inside factory-assembled switchboard or switchgear (FAS) cubicles, units sensitive to arc luminous radiation are usually used. However, luminous radiation sensors become contaminated during operation, and for this reason sometimes do not operate, which results in complete FAS destruction.

A new arc protection relay has been developed that uses the basic technical ideas implemented in the relay "Quasitron" series.

As shown above, the current trip level of each sensor in a "Quasitron" device can be different. Therefore, a single relay unit may operate several current circuits in complex electric equipment.

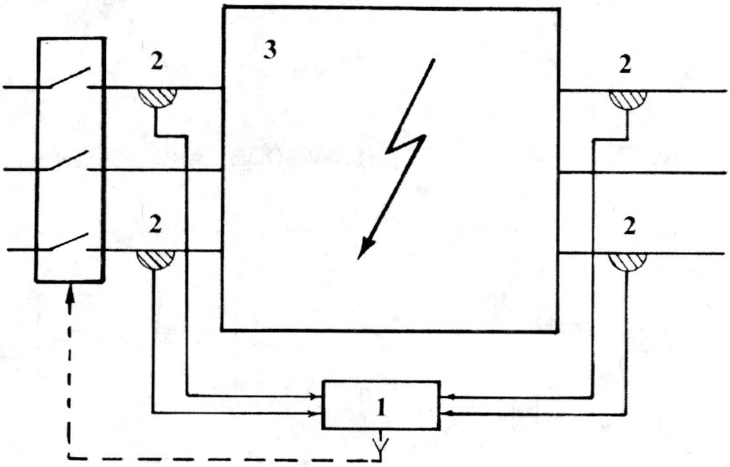

FIGURE 7.2 Arc protection device for switchboard 6 – 24 kV.
1 – relay unit; 2 – reed switch based HV current sensors; 3 – object to be protected.

Moreover, the single relay unit can be supplied with the second (differential) input (Fig. 7.1) and can be used for differential protection of a power object, for example, for arc protection of switchboard 6 – 24 kV (Fig. 7.2).

The arc protection device (APD) monitors the current at switchboard inlets and outlets. When the current of any of the inlets exceeds the over-current limit, and at the same time, the outlets are absent (low level), then the APD immediately sends a signal to actuate the solenoid-operated HV circuit breaker. The APD is a standalone device and requires no further connection to HV current transformers or other intermediary systems.

The sensors are adjusted independently of each other for a pre-set operation current value in ranges of 300 to 10,000 A.

To raise the stability of operation and tuning out of current inrush in case of powerful load connection, the APD is provided with an operation lag in the range of 0.2 s.

The device is fed from a 180 ... 250 V AC or DC network at the consumption current value about 5 mA. Safe operation of the APD is also ensured in case of deep voltage drops down to complete disappearance of voltage.

This principle can also be used to implement various applications (such as a simple differential protection).

7.2 AUTOMATIC-RESET SHORT CIRCUIT INDICATOR FOR 6 – 24 KV BUS BARS

The Short Circuit Indicator (SCI) is intended to facilitate finding the line fault in 6 – 24 kV branched cable networks without circuit breakers on each line, to reduce significantly the power supply failure duration and to improve cost effectiveness.

The importance of this problem and the promising character of the use of such devices is confirmed by the fact that they are developed and used by leading companies, such as Pacific Power & Light Co. (USA); Nortroll AS (Norway); EM Elektromechanik GmbH (Germany); East Midlands Electricity Board (England), etc.

Our device is based on the use of new technical concepts, which enables a significant increase in its reliability and reduction of its cost.

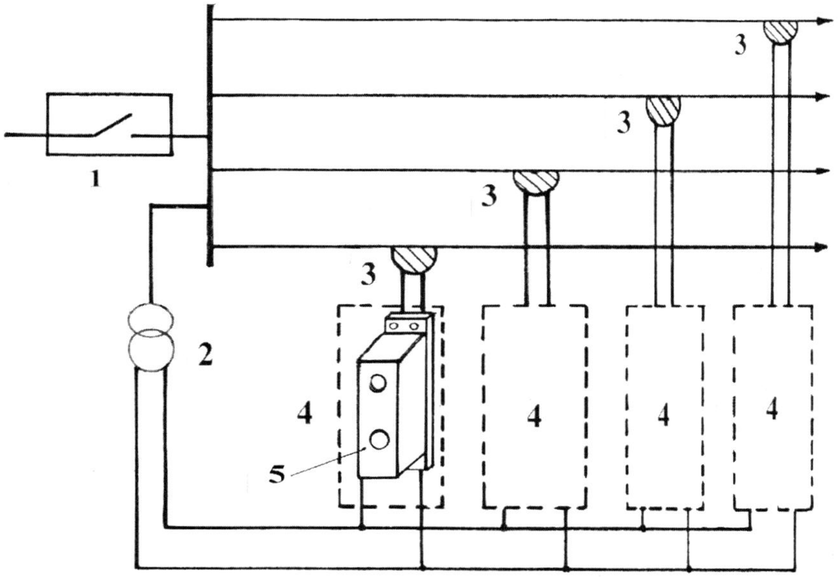

FIGURE 7.3 Short circuit indicator: connection diagram.
1 – main circuit breaker; 2 – original voltage transformer;
3 – HV current sensors; 4 – indicator units; 5 – button for manual check.

The SCI (Fig. 7.3) consists of HV current sensors based on reed switch installed directly on each current-carrying bus bar, and indicator units (Fig. 7.4).

FIGURE 7.4 Short circuit indicator: indicator unit (dimensions are 90 x 70 x 40 mm).

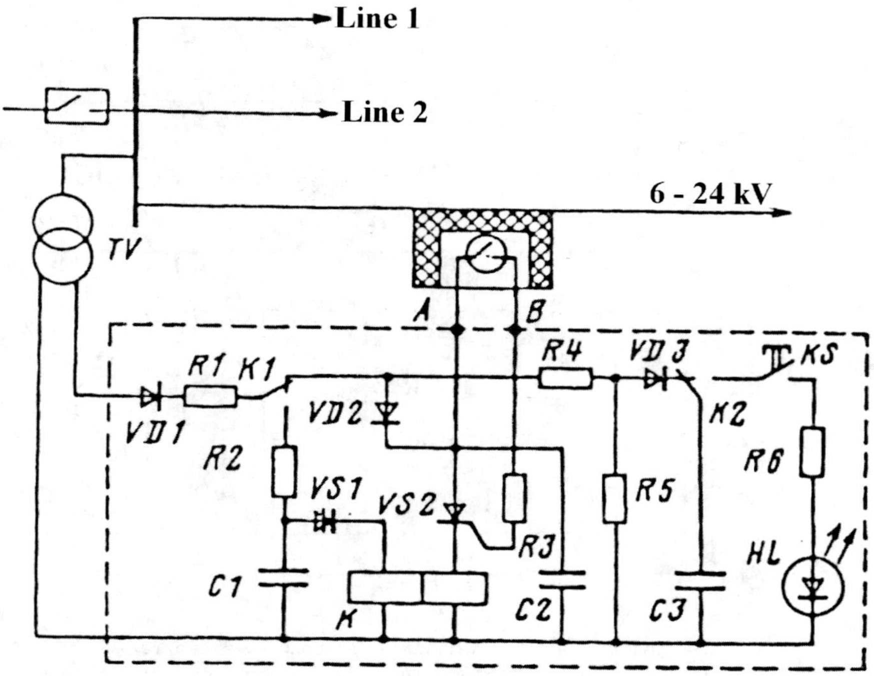

FIGURE 7.5 Circuit diagram of indicator unit.
K – two coils latching relay; HL – LED; KS – button for manual check.

The SCI automatically returns to its initial position as soon as the short circuit is eliminated, as well as under the effect of strong current inrushes after a powerful load connection.

The SCI stores phase-to-phase short circuit events even in the absence of external power supply up to 120 hours.

It is provided with the use of a special miniature latching relay K having two coils, and capacitor C3 with low leakage (Fig. 7.5).

The fault indicator can be displayed (LED) or sent to a remote system.

7.3 HIGH-CURRENT PULSE TRANSDUCER FOR METAL-OXIDE SURGE ARRESTER

The zinc-oxide arresters (varistors) used in high-voltage networks (160kV and higher), for protecting lines and high voltage equipment against over-voltages, are on the one hand rather important devices since the reliability of the power supply depends on them, and on the other hand, expensive devices demanding time-and-expense consuming maintenance. Therefore, for a fairly long time special devices facilitating diagnosis of the varistors' condition have been developed. Many manufacturers of high-voltage arresters based on zinc-oxide varistors already supply them (and have for a long time) with electromechanical surge counters.

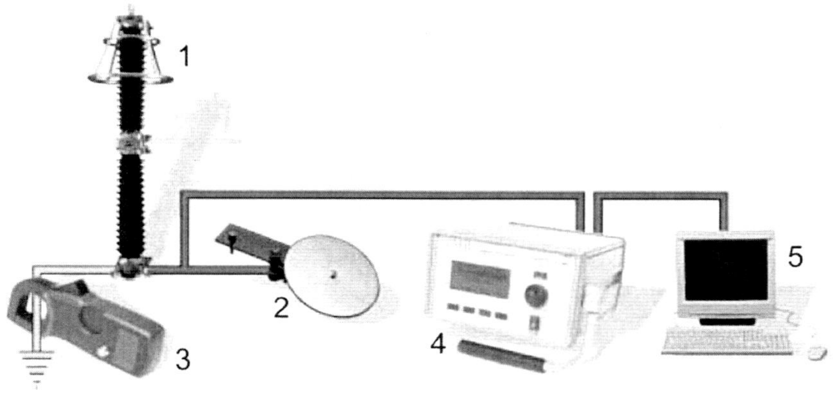

FIGURE 7.6 The Computerized diagnostic complex for monitoring of condition of high-voltage arrester.
1 – surge arrester; 2 – special field probe; 3 – clip-on current probe; 4 – leakage current monitor; 5 – computer.

One basic reason for the lack of such counters is that personnel of substations, who should periodically read information on quantity discharges and list counter indications in a log-book, seldom do. Such records are seldom kept in real life, and the information that is recorded is frequently doubtful, as it is not known what currents passed through the arrester at discharge, and when, and that is the major parameter determining the resources of an arrester.

Original detectors of arresters' operation on the basis of ferrite cores were developed in the Kharkov Polytechnic University in the 1970s. According to the degree of magnetization of cores of the multistage detector it was possible not only to fix the fact of arrester operation, but also to estimate the current level that passed through the arrester. The basic item that was lacking – the necessity of participation of personnel in reading indications of the detector, was also kept along with these developments.

FIGURE 7.7 Electronic pulses counter, records quantity of arrester discharges (produced by ABB).

In the last few years many versions of devices for diagnosing the condition of varistors built on various principles have appeared in the market. The spectrum offered is very wide: From the most complicated computerized systems (Fig. 7.6) in which it is possible to receive a broad set of parameters and full information regarding the varistor's condition, up to the most elementary electronic surge counters (Fig. 7.7).

There is no doubt regarding the efficiency of the measuring systems, allowing precisely enough to estimate the condition of the varistors and to predict their service life. However, the high cost of such equipment and necessity of participation of highly skilled personnel essentially constrain its wide application. On the other hand, regarding the transition from electromechanical counters to electronic ones, little has changed from the point of view of functionalities of the device.

It was suggested to the author to solve this problem by developing a device capable of fixing a quantity of arrester discharges, the ranges of current passing through it at each operation, automatically transferring this information using the SCADA system, this being inexpensive and accessible to manufacturing even by a power company's own staff. Dr. E. Volpov from the Israel Electric Corp. also took part in the formulation of this problem and the formation of initial requirements for this device.

The developed device will consist of one three-band transducer of pulse currents on each arrester and a low-power nine-channel send/receive device (SRD) on a group of three arresters (three phases). The reception part of everyone's SRD is included in the SCADA system, containing microprocessor-based transient and events recorders.

FIGURE 7.8 Developed pulse currents transducer

Usually all analog inputs of such recorders are occupied recording currents and voltages; however a lot of uninvolved logic inputs, which can be used for registration of operation of arresters, are always present. It is only necessary to appropriate a code number for each range of currents and to transfer this number together with a digital signal regarding the varistor's operation. That supposes that the majority of SRD types being used are low-power industrial type SRD's. With such use of transducers there is no necessity for introduction of function of internal storing of the information; therefore its design can be extremely simplified. Such a transducer (Fig. 7.8) is built as a metal-plastic case with universal fastening elements and a deep slot with an arrester grounding tie passing through. The fastening elements of the transducer are universal and assume it is directly fastened to the grounding tie or the concrete rack.

In the body of the transducer a "bookcase" is located with two assembly boards and a compartment with an unscrewing cover, in which a terminal board for external connections is located and in parallel are connected two tiny 12 V batteries from a car security system (MN21/23 type, for example). A standard 9V battery can also be used.

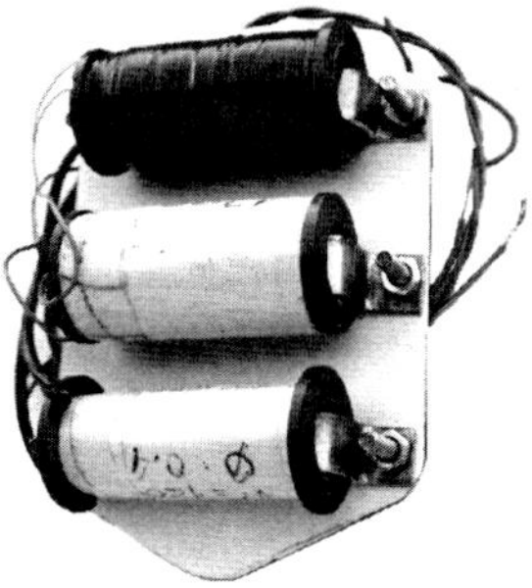

FIGURE 7.9 The unit of inductive sensors for high current pulses.

On the bottom side (closer to the slot) there are inductive current sensors (three coils without cores), wound with different diameters of wire and having different numbers of coils located (Fig. 7.9). On the top side a printed circuit board with electronic components is located (Fig. 7.10), which consists of three identical channels with inductive current sensors on the inputs.

Each channel represents a high-speed pulse expander. Receiving on input a short pulse of current of any polarity opens a thyristor VS1 and charges the capacitor C2 beginning through limit resistor R5.

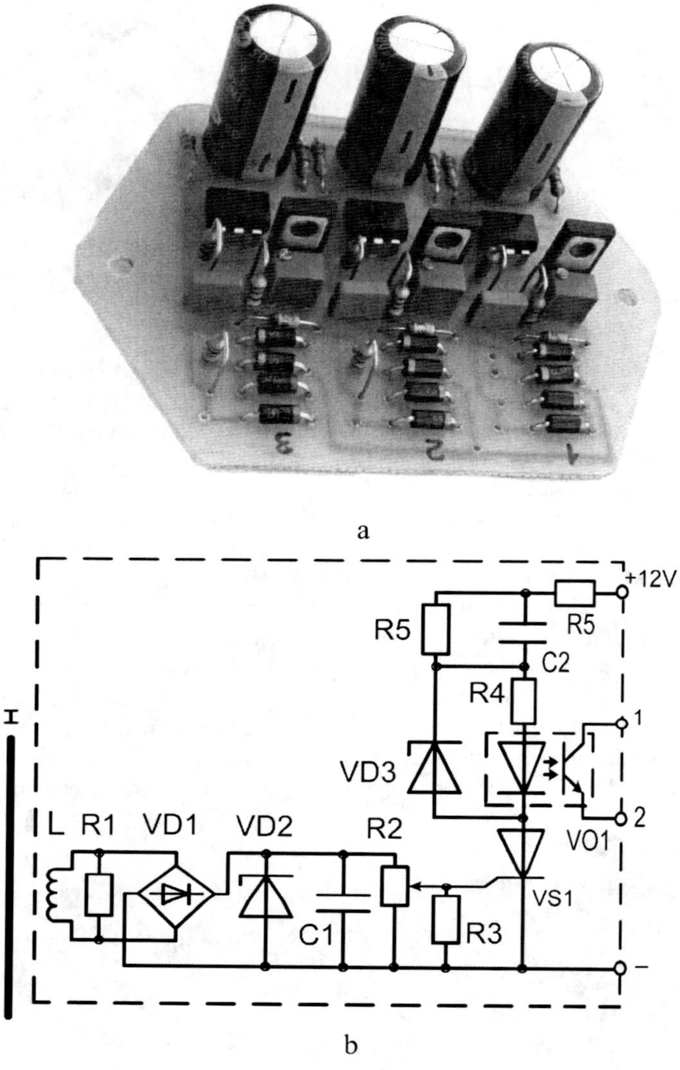

b

FIGURE 7.10 An electronic part of the device.
a – three channels printed circuit board; b – circuit diagram for one channel.

During all this time, while the value of the capacitor charge current remains higher than the thyristor holding current, the thyristor remains in a conductive condition and the output element of the VO1 optocoupler remains in a conductive condition as well. This period of time takes up to one second. Thyristor VS1 will lock and the output element of the optocoupler will also come back to a non-conducting condition after the full charge of capacitor C2. Capacitor C2 will discharge through high-resistance resistor R4 within several seconds. Then the device will again be ready to receive another current pulse. Potentiometer R2 adjusts a pickup threshold, and Zener diode VD2 protects the device from over-voltages at receiving an extremely high pulse input signal. For maintenance of a constant condition of output transistor in the optocoupler VO1 within one second at any change of current on the optocoupler's input during the charging process of the capacitor C2, this optocoupler should be chosen with a low control current (a LED's current). For this purpose the optocouplers of CS700, CH370, HCPL-2300, 6N139, SFH618A series, etc., with input currents as small as 0.5 – 1.5 мA, are recommended. For protection of such sensitive optocouplers from high current at the initial stage of capacitor C2 charge, a low-voltage and low-power Zener diode VD3 is used.

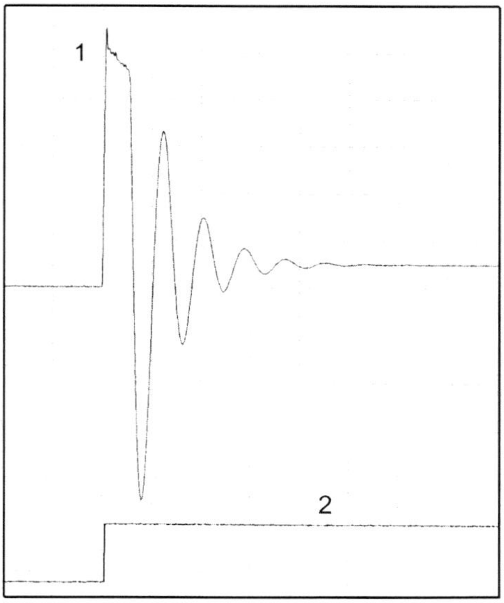

FIGURE 7.11 Oscillogram of input/output signals in device during laboratory testing with a low-voltage simulator of high currents.
1 – the pulse of a current generated by a simulator; 2 – a signal on device output.

Thus, on receiving a short current pulse on an input device with duration of a few microseconds, this scheme will generate a wide pulse with duration about one second (Fig. 7.11). The first channel is adjusted on pickups at a current of 800 A. The second - at a current of more than 5 кA, and the third – at a current of more than 10 кA in a grounding tie.

The device has passed laboratory tests on a low-voltage simulator of high currents (Fig. 7.11), and also on the Haefely high-voltage pulse generator, forming standard pulses of a current corresponding to a natural lightning pulse in all ranges of device sensitivity (Fig. 7.12).

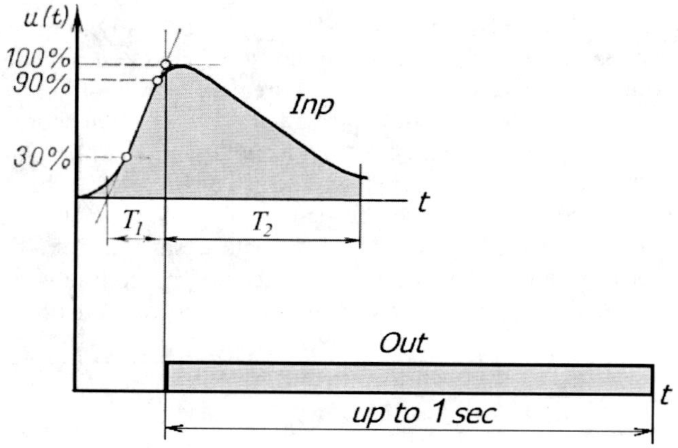

FIGURE 7.12 Pulses of current on input and output of the device during tests on a high-voltage pulse current generator, produced by Haefely.
$T_1 = 7$ μs; $T_2 = 21$ μs.

FIGURE 7.13 Some types of compact industrial radio transmitter/receiver devices, with service area up to 1 km.

The optocoupler's output elements (transistors) of each channel are connected to corresponding inputs of a low-power transmitter with a small service area. One transmitter with nine digital (on-off) channels can be used with three transducers (three-phase set).

Compact industrial radio transmitter/receivers with suitable parameters (Fig. 7.13) are offered today by many companies, including such well-known ones as Phoenix Contact, Honeywell, Sony, Ericson, Acksys, Omnex, etc.

Thus, due to integration with existing systems of transfer and registration of information available today at modern substations, the developed device allows reception of information sufficient for practical needs regarding arrester discharges (quantity of current pulses through an arrester, the range of currents of these pulses, and the exact times of operation of the arrester). Therefore it is much simpler and less expensive than existing analog devices, and its installation does not necessitate driving additional control cables in the substation territory. When desired, such a device can be constructed by the power company's own staff.

7.4 CURRENT TRANSFORMERS' PROTECTION FROM SECONDARY CIRCUIT DISCONNECTION

Current transformers (CT) should never be operated with the secondary circuit open because hazardous crest voltages may result (ANSI/IEEE C57.13-1978). Occasionally, the CT secondary circuit might become disconnected for different reasons.

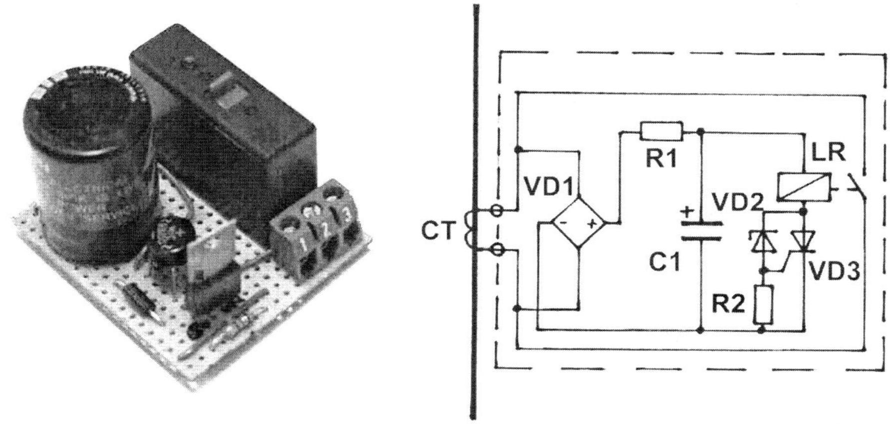

FIGURE 7.14 CT protection module for metering burdens.

In this case, high voltage (3 – 5 kV) is generated over the disconnected secondary winding resulting not only in CT's secondary winding failure but also in severe circuit damage (partial discharges in oil, fuming, and even CT detonating).

We have offered a simple automatic protection device, Fig. 7.14, which forces the CT secondary winding shortage when the voltage over it becomes higher than 100V for metering burdens (Fig. 7.14) and 600-800 V for relaying burdens (Fig. 7.15). After the fault has been cleared the device is manually returned to its initial state (reset). A remote signal indicates that the device operation can be generated.

The device circuit is very simple and cheap. The price of the most expensive element – the latching relay (LR) with switching current of up to 20A and manual reset – is $9.

For a large-size high voltage CT with a number of separate secondary windings, the corresponding number of such protection modules is mounted inside a common protective jacket fixed on the CT body. The protection device is connected to the CT outputs with a simple clamp.

For extra sensitive CT burdens such as a some types of a high impedance differential relay with very small operating current, a miniature gas discharge surge arrestor (suppressor) with very high resistance (10^8 Ohm) in normal mode can be used as a threshold element instead of the varistor (Fig. 7.15).

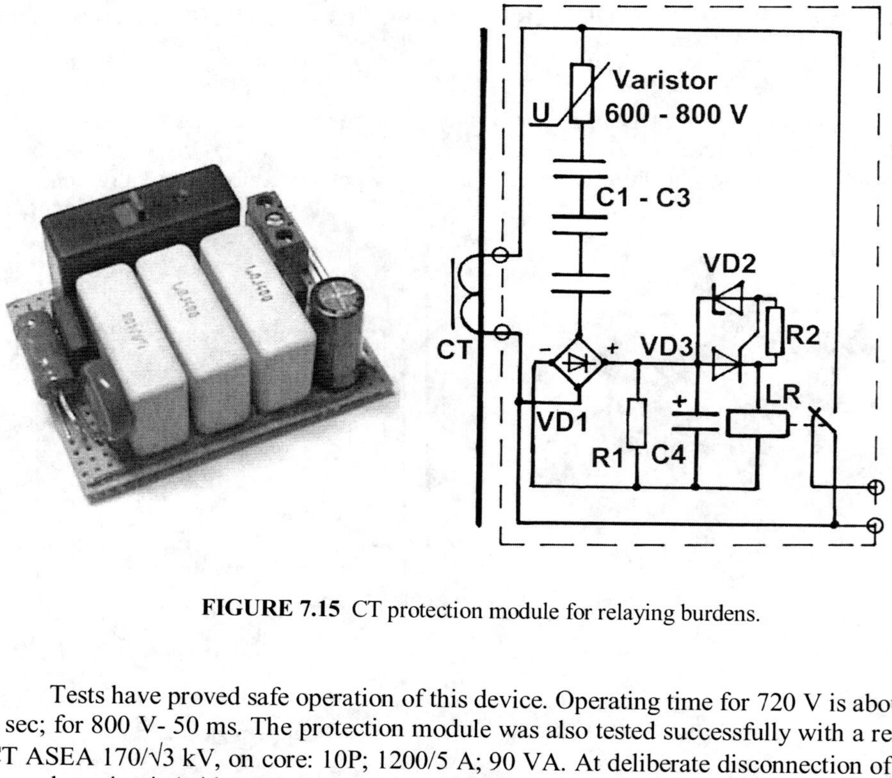

FIGURE 7.15 CT protection module for relaying burdens.

Tests have proved safe operation of this device. Operating time for 720 V is about 3 sec; for 800 V- 50 ms. The protection module was also tested successfully with a real CT ASEA 170/√3 kV, on core: 10P; 1200/5 A; 90 VA. At deliberate disconnection of a secondary circuit (with a primary current up to 1000 A) the protective module instantaneously short-circuited it. For multi-winding CT provided the multi-channels protective modules (Fig. 7.16).

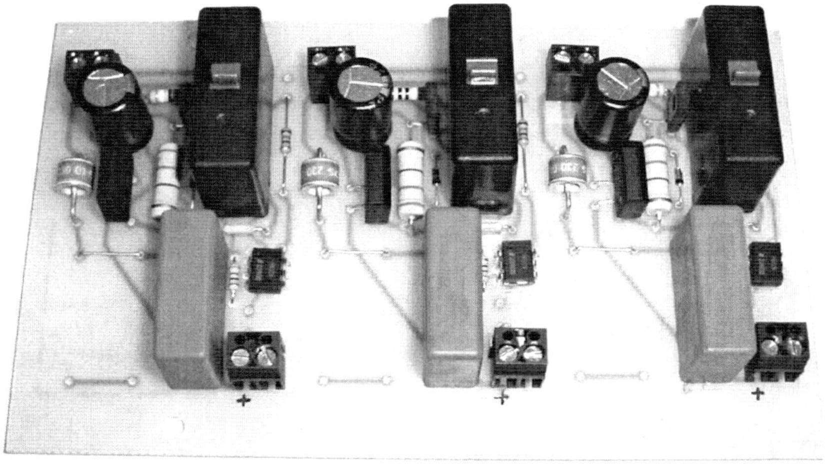

FIGURE 7.16 Three-channel CT protection module with one 1600 V capacitor and addition optocoupler as full isolated signaling element, connected in parallel to relay coil.

FIGURE 7.17 Simplified versions of patented protection devices.
a – for low power measurement CT with very low output voltage; b – for CT with relaying burden.

In simplified constructions of protective modules (Fig. 7.17) nonlinear characteristics of different semiconductor devices are used. For low-power CT with very low output voltage it is sometimes enough to shunt secondary winding with a power diode bridge for currents 20 – 30 A with short-circuited DC output circuit. If the voltage is lower than 0.8 – 1.0 V (applied to two series-connected diodes of the rectifier bridge), diodes remain closed (turned OFF) and do not influence the measuring circuit. If this level of voltage is exceeded, the diodes are opened (turned ON) and the CT secondary winding appears to be short-circuited. It is clear that the range of application of this protective module is very limited.

A second device (Fig. 7.17b) can be more widely applied. The response voltage of this device depends on barrier voltage of the SIDAC (Silicon Diode Alternating Current) and may be selected over a wide range. When the CT secondary winding circuit is open and the voltage on SIDAC increases, it opens on every voltage half-cycle and it opens the TRIAC (Triode Alternating Current Switch) short-circuiting CT winding.

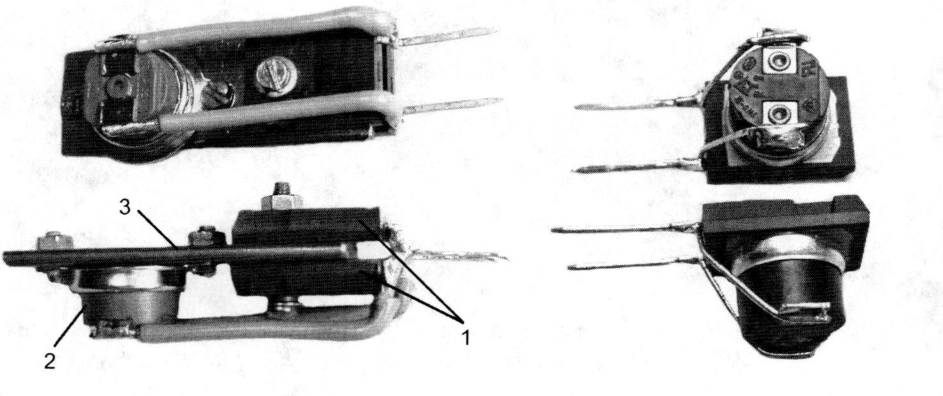

FIGURE 7.18 CT protection module based on twin diode and thermal switch. 1 – series-contrary connected diodes (or single twin diode as in right photo); 2 – thermal switch; 3 – cooper plate.

A serious disadvantage of the protective module is strong heating of the TRIAC with a current of 3 – 4 A flowing through it and the attendant risk of thermal damage. In the patent description, it is even recommended to install the TRIAC on a heat sink.

The strong heating feature of semiconductor devices with CT current (for short operating cycle only) is used for a protective module function developed by the author (Fig. 7.18). It is a simple device consisting of two series-opposite connected diodes (or one binary diode) on 20 – 30 A current and normally opened latching thermo-switch connected in parallel to diodes. The diodes are connected back-to-back between the CT outlets and do not influence their functioning in the normal mode. When the voltage runs up to 30 – 50 V for the measuring CT and 250 – 300 V for the relay protective CT, the diodes breakdown (to do this they have to be selected for the appropriate breakdown voltage) and short-circuit the CT secondary winding. In spite of the low resistance of the broken diodes when the current passes 3 – 4 A, the diodes heat up to 80 – 100 ° and this results in thermo-switch actuation and the shunting diodes with its contact. All elements cool down and the CT winding remains short-circuited by the thermo-switch contact. In the device one may use diodes of type SF301C (for measuring CT) or SF306C (for protection CT), NO thermo-switch with low temperature auto releasing (JW6-III type, for example) or latching type with manual release. The device may be provided with an additional small thermo-switch for signaling purpose, UP6 series, for example.

7.5 A SINGLE-PHASE SHORT CIRCUIT INDICATOR FOR INTERNAL HV CABLES IN MEDIUM VOLTAGE SUBSTATION

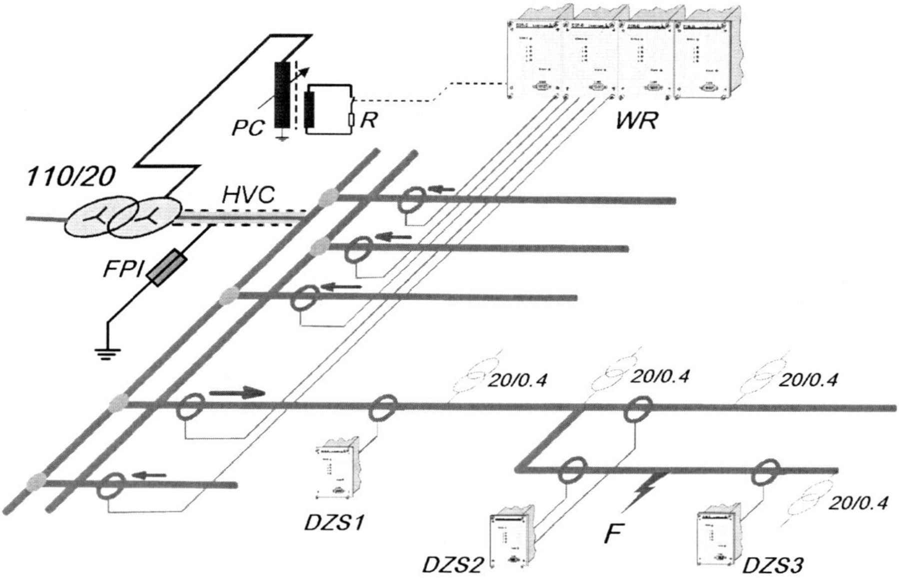

FIGURE 7.19 The block diagram of system for detecting of single-phase short circuits on the basis of Petersen's coil and the directional zero sequence relays:
PC – Petersen's coil; R – the powerful resistor 2.5 Ohm; WR – wattmetric relays; DZS – directional active zero-sequence current relays; HVC – high-voltage cable which has not been covered by detecting system; FPI – fault passage indicator; F – a place of the short circuit.

Approximately 80% of faults in medium-voltage utility overhead networks across the world stem from single phase to ground short circuits. The neutral connection in such networks is not grounded directly and the fault currents of single-phase short circuits (which as a rule are in the order of tens of amperes) are determined only by the capacitance of the conductors to earth. The detection of such short circuits in a network is not problematic and it may be performed by various known means (VT with an open delta winding connection, zero-sequence filters, etc.). Nevertheless, it is difficult to identify reliably a faulted feeder in a ramified distribution network. The simplest way to accomplish this is by serially switching-OFF (or, alternatively, serially switching-ON) each feeder (sometimes this is carried out by an automatic system) until the signal indicating the damage does not disappear (or, alternatively, does not appear).

The most popular means for identifying a faulted feeder is the Petersen coil used in conjunction with a directional zero-sequence active power relay (also called "watt-metric relays," or WR), Fig. 7.19.

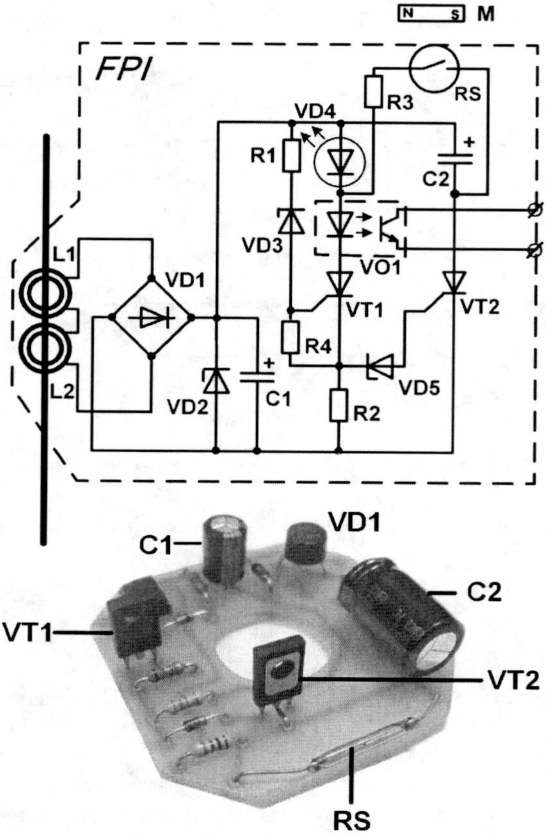

FIGURE 7.20 Circuit diagram and PCB of the fault passage indicator.

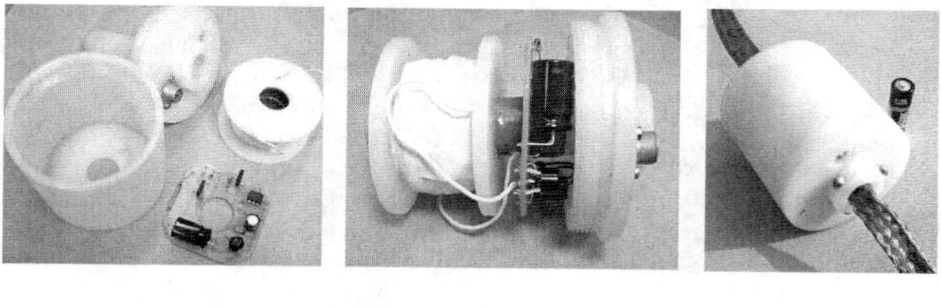

FIGURE 7.21 Construction of the fault passage indicator.

By and large, the Petersen coil (PC) in this system is not a coil at all but a transformer with several windings. One of the windings is connected between the system neutral and the ground and it has an inductance close to the capacitance of the network. Whenever a single-phase short circuit occurs, a small current develops through the PC and a low voltage appears across one of the PC auxiliary windings. This voltage is picked up by a voltage relay that connects a low resistance high-power resistor R to one of the PC windings. Thus, the short circuit current contains an appreciable active component which is registered by the watt-metric relay. Due to directional property of the watt-metric relay, only the relay installed in the faulted feeder picks up. The remaining relays are not activated due to reversed direction of the power flow at their point of installation. In order to maintain appropriate selectivity, the system can be equipped with additional directional active zero-sequence current relays (DZS), installed in the proximity of the consumer's loads.

Despite the relative complexity and efficiency of such a system, it possesses the same disadvantage as the simplest detection method that makes use of serial switching of feeders, that is, it cannot detect damage to the high-voltage cables (HVC) connecting the transformer with the feeders bus bar. A large substation contains many such cables, and not all of them are protected. Dr. A. Shkolnik from IEC described this problem.

The breakdown of cable insulation to its external shield causes capacitive currents which can be in the level of a fraction of an ampere to several amperes that would flow through the grounding bus flexible copper wire connected to the shield. In a small substation cable lengths can be from tens to hundreds of meters. The operation of a damaged cable for prolonged periods of time is dangerous for different reasons, one of which is the known occurrences of fire outbreaks in the cable channel due to severe heating at the junction points of the grounding wire. Even when the relay protection has caused the disconnection of a line, it is still necessary to search for the damaged section of the cable.

The author developed a simple electronic fault passage indicator (FPI) that reacts to such damage, Fig. 7.20. A feature of the indicator is the absence of an external power supply on the one hand, and a low power input signal on the other. These features complicate the design of the indicator's circuit.

The indicator contains two parallel nonlinear circuits that change their state during operation. The first circuit contains a light-emitting diode VD4, an optocoupler VO1, and a thyristor VT1. The second circuit contains a capacitor C2 and a thyristor VT2. In addition, there are two series connected Zener diodes VD3 and VD5 that are used as threshold elements and they are connected in series with the gates of the thyristors VT1 and VT2. The resistance values of resistors R2 and R4 are relatively high (33 and 6.2 kOhm, accordingly) and have no influence on the current through the Zener diodes. The device picks-up when the voltage levels on the secondary coils of current transformers L1 and L2 and the capacitor C1 exceeds the total rated voltage of the Zener diodes VD3 and VD5. Thus a current will be introduced into the gates of thyristors VT1 and VT2. Since the power of the input source is rather limited, the magnitudes of the currents are commensurable with the gate turn-on currents and the holding currents of the thyristors. Under such conditions, the resistances in the anode circuit influence the turning on of the thyristors. The resistance of this circuit for thyristor VT2 is close to zero (i.e, the resistance of a discharged high capacitance capacitor), and for VT1 presents of nonlinear resistance of two series connected light-emitting diodes. For this reason, thyristor VT2 is always the first to be turned on, and all the current conducted by this thyristor is consumed by capacitor C2 during charging. Capacitor C2 shunts the input signal source through its low resistance. When the capacitor is charged, and the current through it decreases down to the thyristor's (VT2) holding current, the thyristor turns off and disconnects the capacitor C2 from the input signal. At this point the appropriate conditions for turning on thyristor VT1 develop. Whenever thyristor VT1 turns on, the light-emitting diode VD4 provides a visual signal and if necessary a remote signal is provided by optocoupler VO1. If the relay protection does not disconnect this fault and the current continues to flow through the cable shield (i.e. through the indicator), this cable can be identified by means of a glowing light-emitting diode. If the protection was activated and has disconnected the cable, or if it has been disconnected by the substation's personnel, the damaged cable can be identified with the help of a permanent magnet placed in the proximity of each indicator. Thus, the magnet's magnetic field causes the internal reed switch (RS) to close and connects light-emitting diode VD4 to capacitor C2. Only the LED in the indicator with a charged capacitor C2 will glow (that is, the indicator through which the current flowed). Modern miniature electrolytic capacitors will stay charged for several days.

In view of the low power of the signal source and the absence of an external power supply, low-power high-sensitivity electronic components should be used in the device. The indicator prototype includes the following components: C106D1 thyristors, BZX85-C24 (VD2), BZX79-C18 (VD3), and BZX79-C3V0 (VD5) Zener diodes, and a H11G1 optocoupler. Structurally, the device is shaped like a small plastic cylinder, Fig. 7.21, with an axial aperture for the grounding bus flexible copper wire. On the lower face of the indicator there is a small aperture for the light-emitting diode and a miniature connector for an external circuit for a remote signal. In addition, there is a special label located opposite the reed switch which indicates where the permanent magnet should be placed in order to test the state of the indicator.

The indicator has been tested directly under field conditions in a substation equipped with a Petersen coil connected to the neutral of a 22 kV network. The test included a simulated single-phase short circuit of a cable. The indicator was shown to work reliably during the test.

7.6 GROUND CIRCUIT FAULT INDICATOR FOR UNDERGROUND HV CABLE NETWORK

The Ground Circuit Fault Indicator for underground applications (GCFI) is designed to identify the location of emergency current percolations in circuits that have underground cable screens (161 kV) connected to the ground circuit bus. GCFI also defines these currents' percolation in the circuits.

Using GCFI provides the operating staff with valuable information regarding the behavior of a cable circuit in emergency mode and enables them to make appropriate technical decisions to ensure the circuit's safety.

GCFI consists of the following units:

- Stationary leak-less unit (SU) with memory elements installed in the underground well on the metallic Cross Bounding Link Box
- Current sensor made in the form of a circular current transformer (CT) including elements of outer fixation

- Mobile manual indication unit (MU) including elements of indication and monitoring.

FIGURE 7.22 Stationary unit of GCFI system.

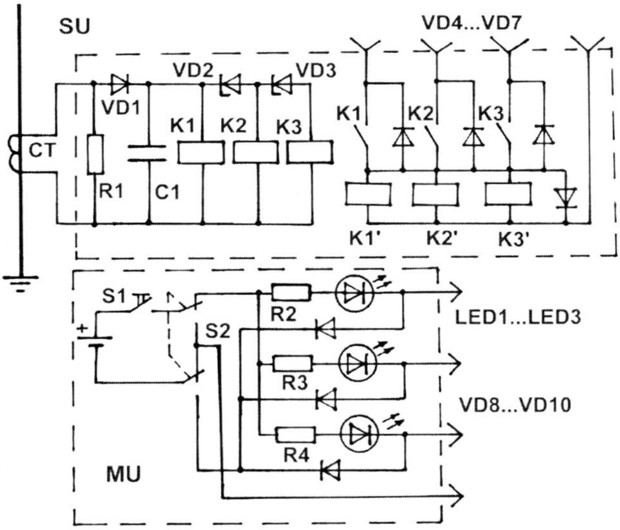

FIGURE 7.23 Circuit diagram of GCFI system.
SU – stationary unit; MU – movable unit; CT – current transformer; K1…K2 – two coil miniature latching relays; S1 – "test" button; S2 – "test – reset" toggle switch; VD2, VD3 – Zener diodes.

The Stationary Unit, Fig. 7.23, includes elements that change their state when the input current reaches the given threshold value, and remains in this new state after the input current ceases working.

Each SU has three thresholds functioning within the chosen range of emergency currents, which present approximately 30, 60, and 100 percent of the selected range of emergency currents.

The user can select one of the following ranges of emergency current: 0-10 kA, 0-20 kA, and 0-30 kA. The current range should be chosen when the device is ordered. The SU is connected to CT by a piece of twin-core isolated cable not less than 0.5 meter in length, which is provided with leak-proof input into the body of the SU. The SU is connected to the MU by means of a connector.

FIGURE 7.24 Encapsulated current transformer (2000/5A, 15P5) for GCFI system.

Threshold elements and memory elements (latching relays K1…K3, Fig. 7.23) have high interference immunity and stability, enabling them to function efficiently in a powerful electromagnetic field and difficult environmental conditions. The SU does not require using a separate power source. Its body is made of steel and is coated on the outside with two layers of waterproof varnish. The SU's interior is filled with an epoxy compound. The unit has a leak-proof and non-separable design.

FIGURE 7.25 Manual unit of GCFI system.

The Current Transformer, Fig. 7.24, is designed to be installed on the isolated cable connecting the generic point of connection of the cable screens with the grounding circuit. The CT has a non-separable circular core and is filled with an epoxy compound.

The CT is attached to the vertical wall on the cable well or on the Cross Bounding Link Box by means of bolts.

The Manual Unit, Fig. 7.25, contains:

- a deciphering unit;
- three elements of indication (LED) which determine the state of the threshold elements of the SU;
- a button and toggle switch that switches the Manual Unit into TEST and RESET modes;
- a connector to connect with the SU;
- a connector to check operating capacity of the device and battery status;
- a built-in power source (standard 9V battery);
- a control wire with two connectors.

The MU has a miniature, plastic rectangular body.

Installation of the system is shown in Fig. 7.26. The standard unit (SU) is installed in the cable well (on the inner wall of the well or on the Cross Bounding Link Box) and is affixed by two 6-mm bolts. The location of the SU fixation must be as high as possible from the surface of the well's bottom and must be maximally remote from current-conducting parts.

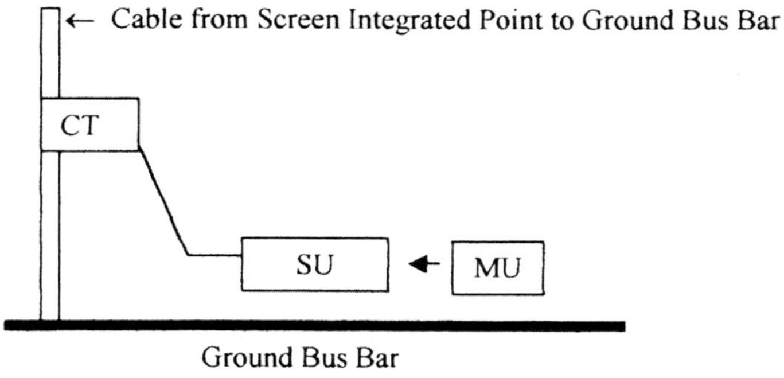

FIGURE 7.26 Installation diagram of GCFI system.

The Current Transformer (CT) is installed in the cable well (on the inner wall of the well or on the Cross Bounding Link Box, at a distance of no more than 0.5 m from the SU). It is affixed by four 6-mm bolts. The piece of cable connecting the generic point of the monitors of high-voltage cables with the grounding circuit must be passed through the CT's opening. The polarity does not matter.

The twin-core flexible cable (0.5 m in length), coming out of the SU, is connected to CT cleats. The polarity does not matter. After connection of the cable, the CT's cleat box must be coated by silicone plastering to prevent the cleats' oxidation resulting from contact with water.

Before using the device, it is necessary to check serviceability of the MU and the built-in battery. To do so, it is necessary to connect both connectors located on the MU's body, using a control wire with connectors that are part of the kit. Furthermore, it is necessary to put the toggle switch on TEST mode and press the button. In a completely serviceable MU, all three LEDs are brightly lit. Weak luminescence means that the battery is depleted. In this case, it is necessary to open the MU lid and replace the battery.

In order to define the status of SU's memory elements, it is necessary to connect it to the MU by means of the control wire with connectors.

The MU toggle switch is set on TEST mode, and the button is pressed.

If a short-circuit current of more than 10 kA surged through the location of the SU installation, LEDs must start glowing. The number of LEDs is proportional to the value of the short-circuit current:

One LED – 30% of rated operating current in damage mode.
Two LEDs – 60% of rated operating current in damage mode.
Three LEDs – 100% of rated operating current in damage mode.

After evaluation of the short-circuit current, the SU's memory must be cleared. To do so, it is necessary to set the tumbler on RESET mode and press the button.

When the clearing of the memory elements is completed, the control wire with connectors are disconnected from the SU, and the protecting plug is screwed on the connector of the SU's body.

7.7 HV INDICATORS FOR SWITCHGEARS AND SWITCHBOARDS

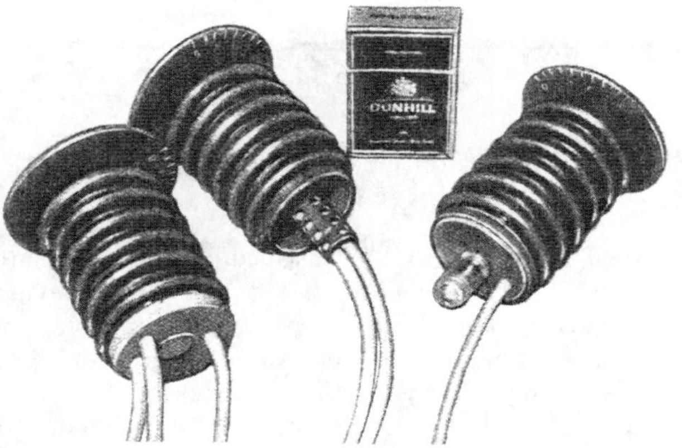

FIGURE 7.27 Portable HV indicators for direct installation on bus bars.

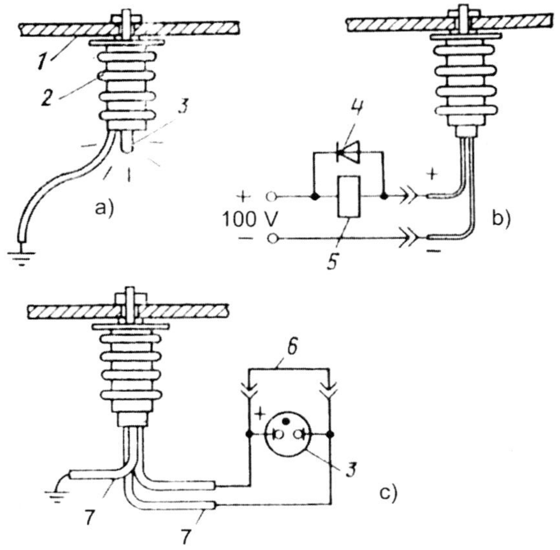

FIGURE 7.28 Connection versions for portables HV indicators.
a – with local neon lamp; **b** – with relay output; **c** – with remote indication.
1 – HV bus bar; 2 – HV plastic insulator with antenna and electronic components; 3 – neon lamp; 4 – diode; 5 – electromagnetic relay; 6 – removal jumper; 7 – HV wiring.

Many of the modern compact Switchgears and Switchboards, and in particular those filled with SF_6, are not operable with traditional portable high voltage indicators formed as long rods. In such devices built-in fixed voltage indicators with an indication unit based on neon lamps and a dedicated high-voltage sensor with a resistive and a capacitance voltage divider have been widely used.

We have proposed some novel designs of such devices (Russian Patents No. 1718129, 1821750, 2015516, 2020682).

The first group constitutes portable indicators, which can be temporarily assembled during maintenance works on high voltage bus bars of all types of high voltage equipment (Fig. 7.28, 7.30).

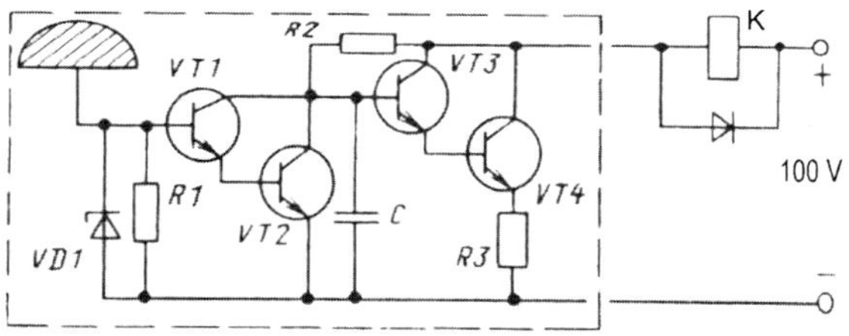

FIGURE 7.29 Circuit diagram for relay voltage sensor
K – outside electromagnetic relay.

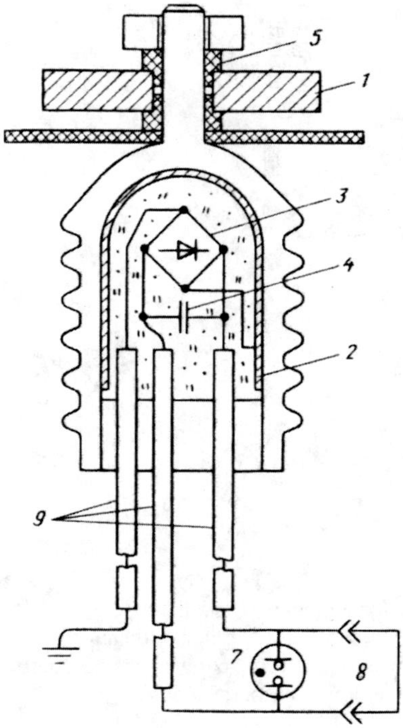

FIGURE 7.30 Construction and circuit diagram for HV indicator with blinking neon lamp.
1 – HV bus bar; 2 – antenna (conductive coating); 3 – rectifier bridge; 4 – capacitor; 5 – thermostable insertion; 7 – neon lamp; 8 – removable jumper; 9 – HV wiring.

Our HV indicator does not require any special insulators and operates with base permanent insulators already installed on a switchboard. This enables the use of HV indicators not only at switchboard manufacturing plants, but also to improve electric units of any types already in operation.

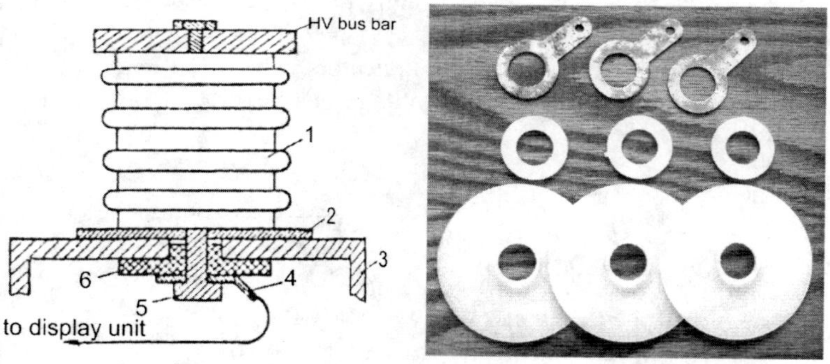

FIGURE 7.31 The stationary HV indicator: installation on bus bar and elements set.
1 – base permanent insulators; 2 – insulation washer; 3 – switchboard's console; 4 – conductive petal (output of sensor); 5 – bolt; 6 – insulation insert.

Our indicators use base permanent insulators included in switchgears and switchboards as capacitance voltage dividers (Fig. 7.31).

Such devices can be designed either in the form of a visual indicator with blinking neon lamp (Fig. 7.28, 7.30) or relay voltage sensor with an output contact (of electromagnetic relay K) used to connect indication or interlock circuit (Fig. 7.29).

Devices with output contact can also be used in addition to existing fixed indicators.

FIGURE 7.32 Display unit with neon lamps (65 x 65 x 75 mm).

Materials used to manufacture such high voltage insulators and their production procedures are same as used for HV reed switch relays described above.

The other group comprises stationary devices, which are built-in to switchgears and switchboards.

All known HV indicators usually comprise a specially designed insulator installed on a switchboard's bus bar and connected to a display unit.

The display unit case (Fig. 7.32), has a hinged self-closing cover with an additional window, providing for viewing the neon lamps even with the cover closed. A permanent magnet mounted on the cover provides for connecting reed switch 13 (Fig. 7.33) with the cover closed, and its disconnecting with the cover open.

When the display unit cover is open the switch side to which high voltage is applied can be validated. The device can also be used to verify the correctness of the line phasing.

After repair work in power system (6 – 24 kV) it is necessary to check the phase tangling of the electric cables to avoid unpredicted accidents. Our indicator has been developed to solve this problem.

With phased-in voltage at both inputs of the display unit (from the side of assembled buses and from the side of the cable input) the indication lamps will not blink when the cover is closed. In cases of phase tangling, the indicator is capable of identifying the wrongly connected phases (the lamps corresponding to the wrong phases will blink indicating to the maintenance staff which switchgear or switchboard phases should be changed).

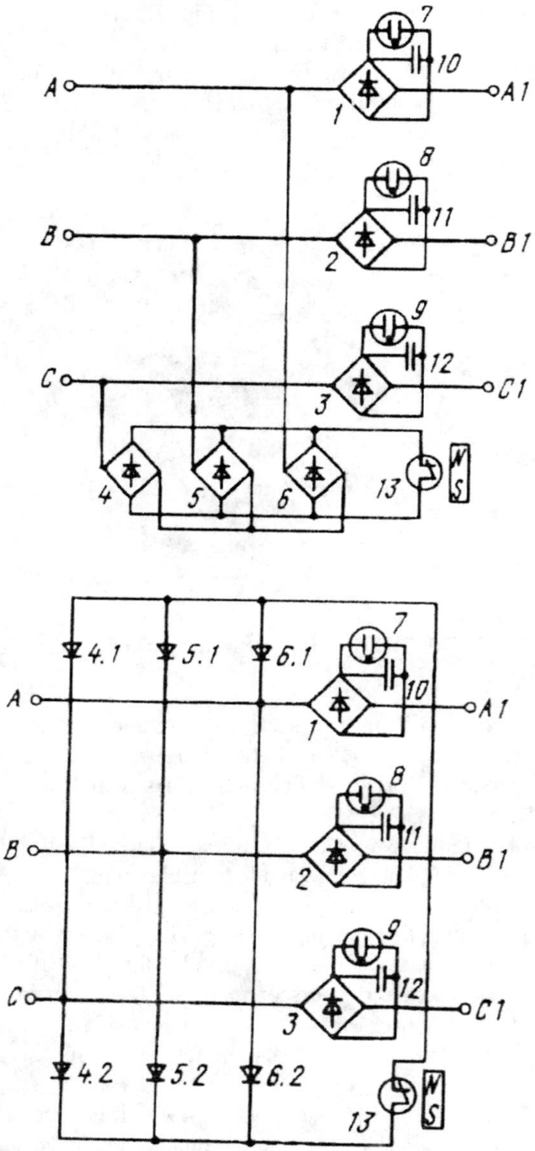

FIGURE 7.33 Variants of display unit circuit diagrams.

This device can also be used to determine some other switchgears or switchboards operation modes (Fig. 7.34).

The display unit was also subjected to fundamental upgrading and it can be connected via remote control or by a signal from the doors block-contact (Fig. 7.35).

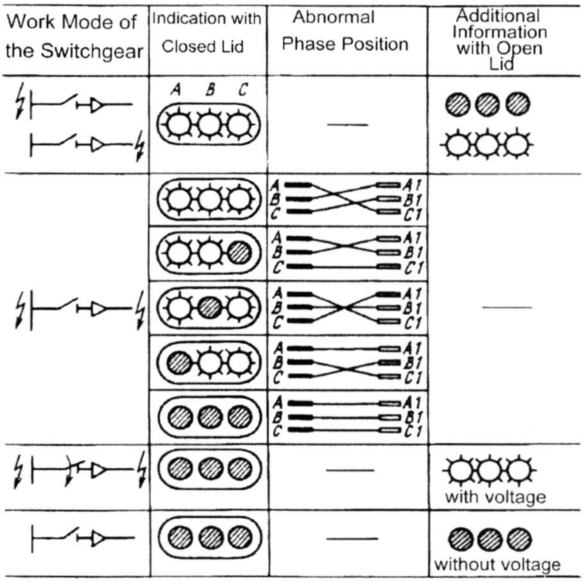

FIGURE 7.34 Mnemonic diagram for abnormal modes identification.

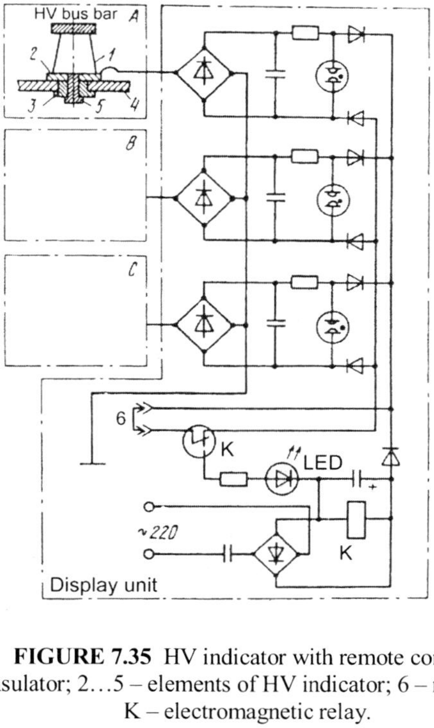

FIGURE 7.35 HV indicator with remote control.
1 – HV insulator; 2…5 – elements of HV indicator; 6 – removal jumper;
K – electromagnetic relay.

Appendix A1: High-Speed Miniature Reed Switches

Reed switch type	**HSR-V100R**
Manufacturer	**Hermetic Switch, Inc.**

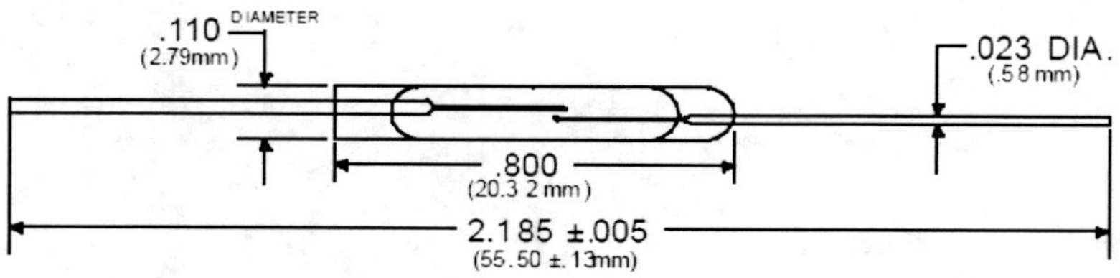

Contact form	NO (form A)
Contact material	Rhodium
Max. contact rating (resistive), W	25
Max. switching voltage, V DC or peak AC	500
Max. switching current, A DC or peak AC	1.0
Max. carry current, A DC or peak AC	2.0
Dielectric strength, V DC	1750
Max. initial contact resistance, Ω	0.15
Pull in value (AT range)	15 – 40
Max. contact capacitance, pF	0.6
Min. insulation resistance, Ω	10^{11}
Typ. resonance frequency, kHz	–
Operate time, ms	1.0
Release time, ms	0.9
Operating temperature, °C	-60 +125

Reed switch type	KSK-1A75
Manufacturer	Meder Electronics

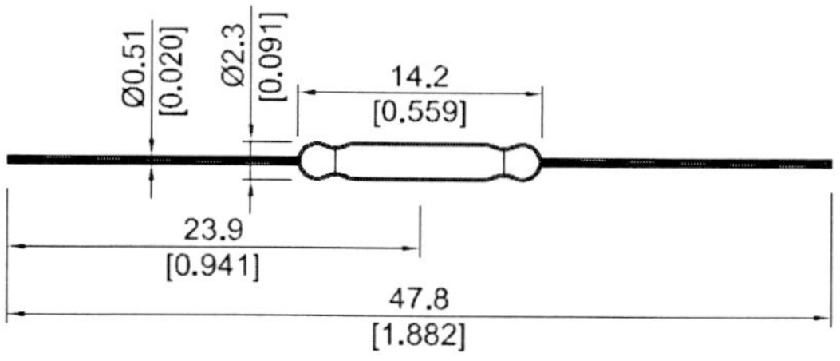

Dimensions in mm (inches)

Contact form	NO (form A)
Contact material	–
Max. contact rating (resistive), W	10
Max. switching voltage, V DC or peak AC	1000
Max. switching current, A DC or peak AC	0.5
Max. carry current, A DC or peak AC	1.0
Dielectric strength, V DC	1500
Max. initial contact resistance, Ω	0.2
Pull in value (AT range)	15 – 40
Max. contact capacitance, pF	0.4
Min. insulation resistance, Ω	10^{10}
Typ. resonance frequency, kHz	-
Operate time, ms	0.5
Release time, ms	0.1
Operating temperature, °C	-20 +130

Reed switch type	**KSK-1A85**
Manufacturer	Meder Electronics

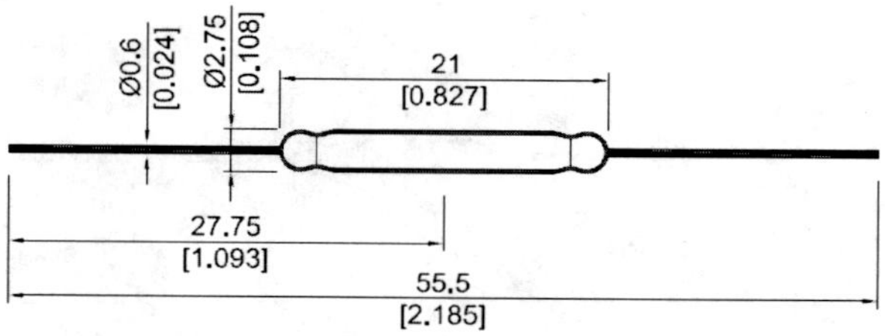

Dimensions in mm (inches)

Contact form	NO (form A)
Contact material	-
Max. contact rating (resistive load), W	100
Max. switching voltage, V DC or peak AC	1000
Max. switching current, A DC or peak AC	1.0
Max. carry current, A DC or peak AC	2.5
Dielectric strength, V DC	4000
Max. initial contact resistance, Ω	0.2
Pull in value (AT range)	20 – 60
Max. contact capacitance, pF	0.15
Min. insulation resistance, Ω	10^{10}
Typ. resonance frequency, kHz	-
Operate time, ms	1.0
Release time, ms	0.1
Operating temperature, °C	-20 +130

Reed switch type	**MARR-5**
Manufacturer	**Hamlin**

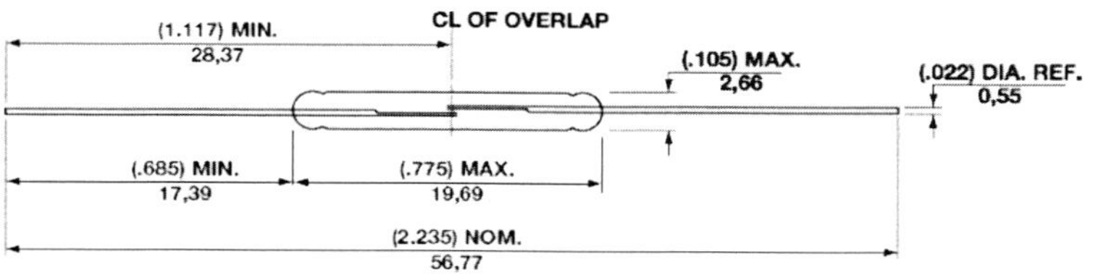

Dimensions in mm (inches)

Contact form	NO (form A)
Contact material	-
Max. contact rating (resistive), W	10
Max. switching voltage, V DC or peak AC	1000
Max. switching current, A DC or peak AC	0.5
Max. carry current, A DC or peak AC	1.3
Dielectric strength, V DC	2000
Max. initial contact resistance, Ω	0.1
Pull in value (AT range)	17 – 38
Max. contact capacitance, pF	0.2
Min. insulation resistance, Ω	10^{12}
Typ. resonance frequency, kHz	3.2
Operate time, ms	0.75
Release time, ms	0.3
Operating temperature, °C	-75 +125

Reed switch type	**MRA560G**
Manufacturer	**Crydom**

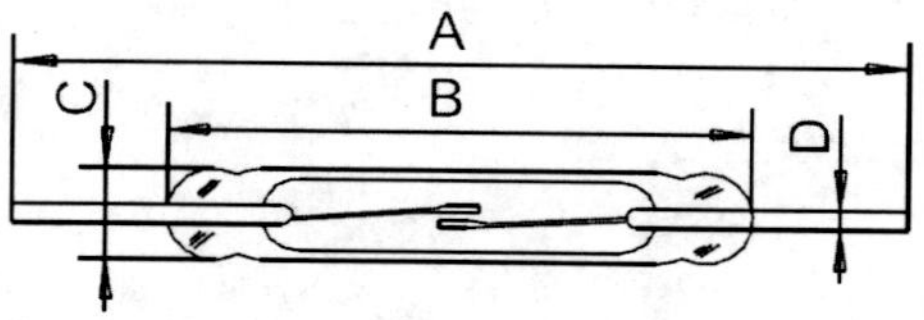

Dimensions in mm

A	56
B	21
C	2.8
D	0.60

Contact form	NO (form A)
Contact material	Ruthenium
Max. contact rating (resistive), W	100
Max. switching voltage, V DC or peak AC	1000
Max. switching current, A DC or peak AC	1.0
Max. carry current, A DC or peak AC	2.5
Dielectric strength, V DC	1500
Max. initial contact resistance, Ω	0.1
Pull in value (AT range)	20 – 40
Max. contact capacitance, pF	0.5
Min. insulation resistance, Ω	10^{10}
Typ. resonance frequency, kHz	2500
Operate time, ms	1.0
Release time, ms	0.05
Operating temperature, °C	-40 + 150

Reed switch type	**R1-48C**
Manufacturer	**Coto Technology**

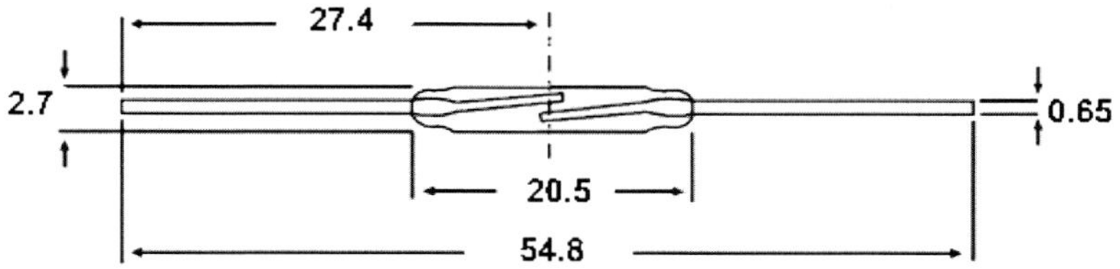

Dimensions in mm

Contact form	NO (form A)
Max. contact rating (resistive), W	70
Max. switching voltage, V DC or peak AC	250
Max. switching current, A DC or peak AC	1.0
Max. carry current, A DC or peak AC	2.25
Dielectric strength, V DC	780
Max. initial contact resistance, Ω	0.6
Pull in value (AT range)	46 – 70
Max. contact capacitance, pF	0.2
Min. insulation resistance, Ω	10^6
Typ. resonance frequency, kHz	3.2
Operate time, ms	0.35
Release time, ms	0.03
Operating temperature, °C	-55 +125

Appendix A2: High-Voltage Vacuum Reed Switches

Reed switch type	**AV10**
Manufacturer	**Celduc relais**

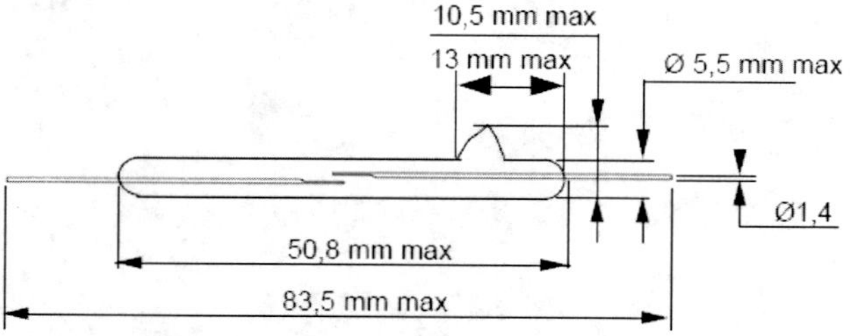

Contact form	NO (form A)
Contact material	-
Max. contact rating (resistive), W	50
Max. switching voltage, V DC or peak AC	7500
Max. switching current, A DC or peak AC	0.3
Max. carry current, A DC or peak AC	3.0
Dielectric strength, V DC	10,000
Max. initial contact resistance, Ω	0.1
Pull in value (AT range)	100 – 120
Max. contact capacitance, pF	-
Min. insulation resistance, Ω	10^{10}
Typ. resonance frequency, kHz	-
Operate time, ms	3.0
Release time, ms	-
Operating temperature, °C	-

Reed switch type	**GC 6520**
Manufacturer	**Comus Group**

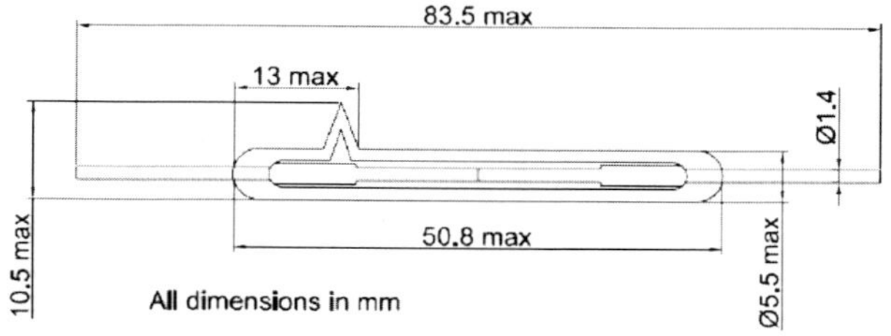

Vacuumized

Contact form	NO (form A)
Contact material	Tungsten
Max. contact rating (resistive), W	10
Max. switching voltage, V DC or peak AC	12,000
Max. switching current, A DC or peak AC	1.0
Max. carry current, A DC or peak AC	3.0
Dielectric strength, V DC	19,000
Max. initial contact resistance, Ω	0.1
Pull in value (AT range)	160 – 200
Max. contact capacitance, pF	0.5
Min. insulation resistance, Ω	10^{10}
Typ. resonance frequency, kHz	-
Operate time, ms	3.6
Release time, ms	0.5
Operating temperature, °C	-55 +125

Reed switch type	**GC 6522**
Manufacturer	**Comus Group**

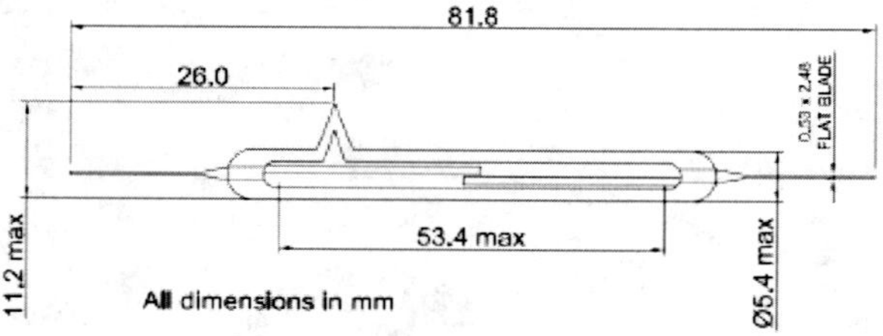

Vacuumized

Contact form	NO (form A)
Contact material	Tungsten
Max. contact rating (resistive), W	50
Max. switching voltage, V DC or peak AC	7000
Max. switching current, A DC or peak AC	3.0
Max. carry current, A DC or peak AC	5.0
Dielectric strength, V DC	10,000
Max. initial contact resistance, Ω	0.1
Pull in value (AT range)	100 – 150
Max. contact capacitance, pF	1.0
Min. insulation resistance, Ω	10^{10}
Typ. resonance frequency, kHz	-
Operate time, ms	3.0
Release time, ms	1.0
Operating temperature, °C	-55 +125

Reed switch type	**GC 6523**
Manufacturer	**Comus Group**

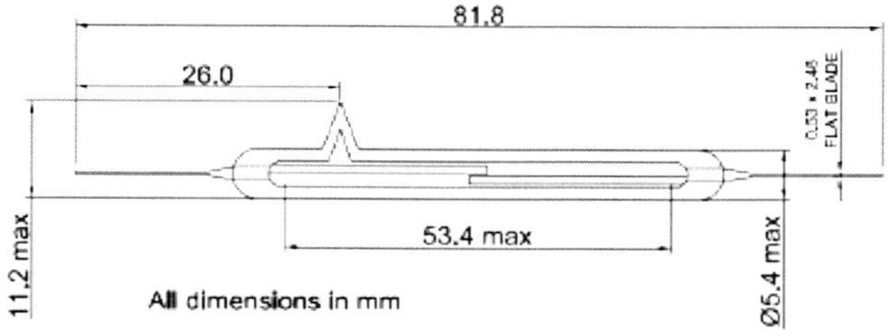

Vacuumized

Contact form	NO (form A)
Contact material	Tungsten
Max. contact rating (resistive), W	50
Max. switching voltage, V DC or peak AC	10,000
Max. switching current, A DC or peak AC	3.0
Max. carry current, A DC or peak AC	5.0
Dielectric strength, V DC	15,000
Max. initial contact resistance, Ω	0.1
Pull in value (AT range)	130 – 170
Max. contact capacitance, pF	1.0
Min. insulation resistance, Ω	10^{10}
Operate time, ms	3.0
Release time, ms	1.0
Operating temperature, °C	-55 +125

Reed switch type	**HBS-10KVDC**
Manufacturer	**Comus Group**

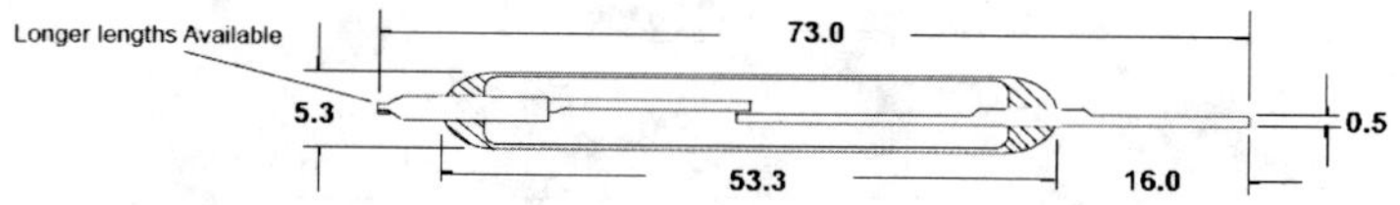

Dimensions in mm

Contact form	NO (form A)
Contact material	Tungsten
Max. contact rating (resistive), W	50
Max. switching voltage, V DC or peak AC	7500
Max. switching current, A DC or peak AC	3.0
Max. carry current, A DC or peak AC	5.0
Dielectric strength, V DC	10,000
Max. initial contact resistance, Ω	0.1
Pull in value (AT range)	90 – 200
Max. contact capacitance, pF	0.5
Min. insulation resistance, Ω	10^{10}
Typ. resonance frequency, kHz	-
Operate time, ms	3.6
Release time, ms	0.5
Operating temperature, °C	-55 +125

Reed switch type	**HSR-V10K**
Manufacturer	**Hermetic Switch, Inc.**

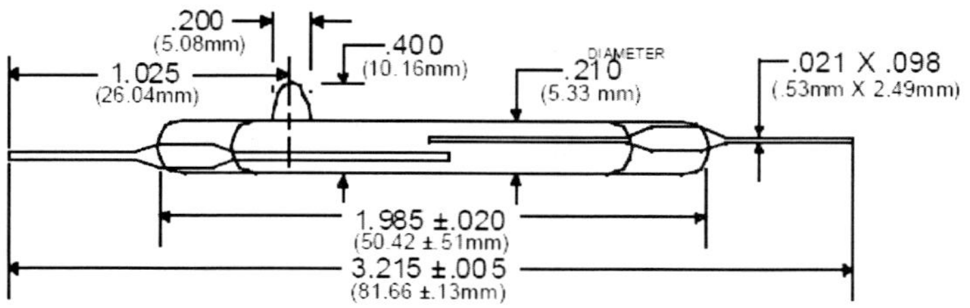

Contact form	NO (form A)
Contact material	Tungsten in vacuum
Max. contact rating (resistive), W	50
Max. switching voltage, V DC or peak AC	7500
Max. switching current, A DC or peak AC	3.0
Max. carry current, A DC or peak AC	4.0
Dielectric strength, V DC	10,000
Max. initial contact resistance, Ω	0.1
Pull in value (AT range)	100 – 150
Max. contact capacitance, pF	1
Min. insulation resistance, Ω	10^{11}
Typ. resonance frequency, kHz	-
Operate time, ms	3.2
Release time, ms	1.5
Operating temperature, °C	-60 +125

Reed switch type	HSR-V15K
Manufacturer	Hermetic Switch, Inc.

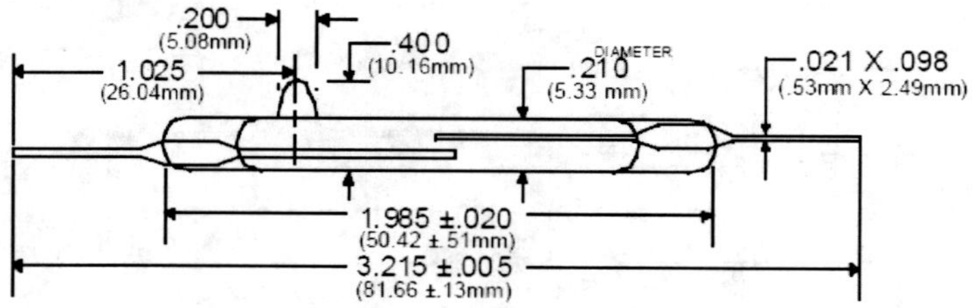

Contact form	NO (form A)
Contact material	Tungsten in vacuum
Max. contact rating (resistive), W	50
Max. switching voltage, V DC or peak AC	10,000
Max. switching current, A DC or peak AC	3.0
Max. carry current, A DC or peak AC	4.0
Dielectric strength, V DC	15,000
Max. initial contact resistance, Ω	0.1
Pull in value (AT range)	140 – 170
Max. contact capacitance, pF	1
Min. insulation resistance, Ω	10^{11}
Typ. resonance frequency, kHz	-
Operate time, ms	3.2
Release time, ms	1.5
Operating temperature, °C	-60 +125

Reed switch type	**HSR-V207**
Manufacturer	**Hermetic Switch, Inc.**

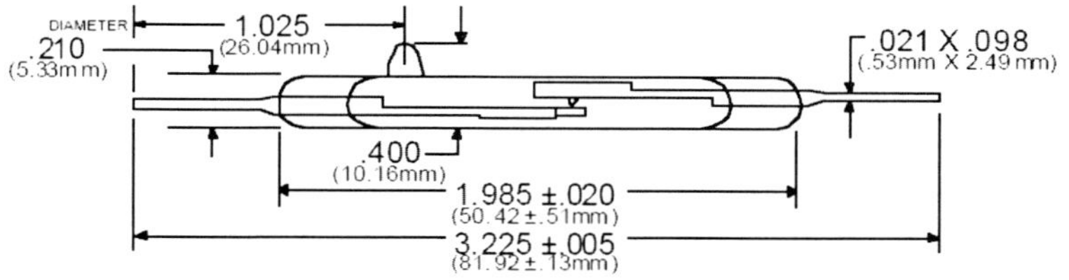

Contact form	NO (form A)
Contact material	Tungsten in vacuum
Max. contact rating (resistive), W	200
Max. switching voltage, V DC or peak AC	2500
Max. switching current, A DC or peak AC	3.0
Max. carry current, A DC or peak AC	-
Dielectric strength, V DC	10,000
Max. initial contact resistance, Ω	0.5
Pull in value (AT range)	100 – 150
Max. contact capacitance, pF	0.8
Min. insulation resistance, Ω	10^{11}
Typ. resonance frequency, kHz	-
Operate time, ms	4.6
Release time, ms	0.8
Operating temperature, °C	-60 +125

Reed switch type	**HYR5008**
Manufacturer	**Aleph International**

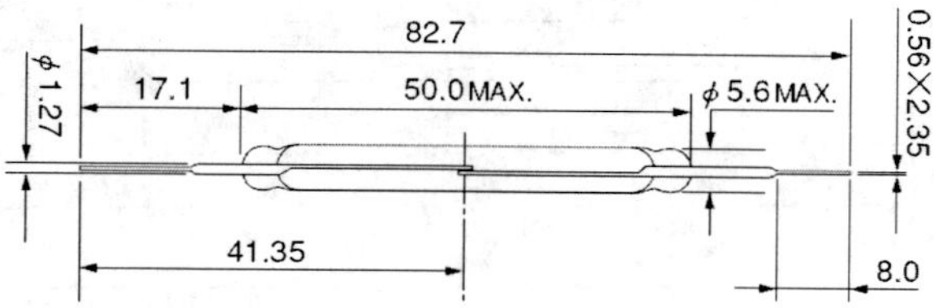

Dimensions in mm (inches)

Contact form	NO (form A)
Contact material	Rhodium
Max. contact rating (resistive), W	25
Max. switching voltage, V DC or peak AC	3500
Max. switching current, A DC or peak AC	0.5
Max. carry current, A DC or peak AC	-
Dielectric strength, V DC	15,000
Max. initial contact resistance, Ω	0.6
Pull in value (AT range)	200 – 250
Max. contact capacitance, pF	0.8
Min. insulation resistance, Ω	10^{10}
Typ. resonance frequency, kHz	0.8
Operate time, ms	-
Release time, ms	-

Reed switch type	**KSK-1A77**
Manufacturer	Meder Electronics

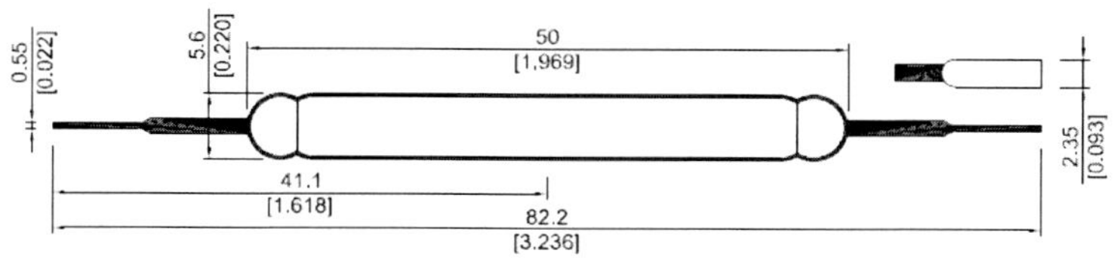

Dimensions in mm (inches)

Contact form	NO (form A)
Contact material	-
Max. contact rating (resistive), W	25
Max. switching voltage, V DC or peak AC	3500
Max. switching current, A DC or peak AC	1.0
Max. carry current, A DC or peak AC	6.0
Dielectric strength, V DC	15,000
Max. initial contact resistance, Ω	0.15
Pull in value (AT range)	200 – 250
Max. contact capacitance, pF	0.8
Min. insulation resistance, Ω	10^{10}
Typ. resonance frequency, kHz	-
Operate time, ms	3.0
Release time, ms	3.0
Operating temperature, °C	-20 +130

Appendix A3: Mercury Wetted Reed Switches

Reed switch type	**HGZ**
Manufacturer	**Comus Group**

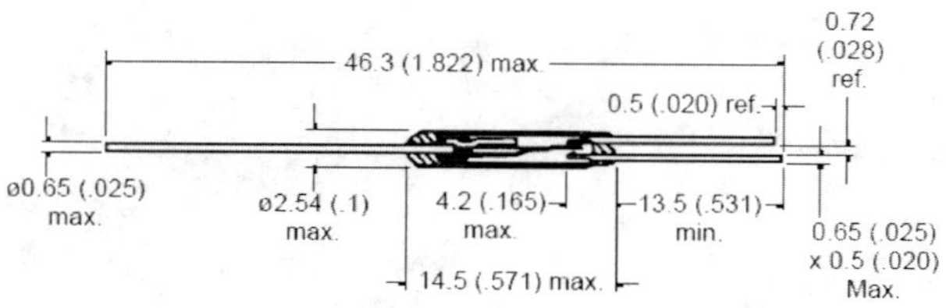

Vertical mounted required (± 30º)

Dimensions in mm (inches)

Contact form	CO (form C)
Contact material	Hg, mercury wetted
Max. contact rating (resistive), W	50
Max. switching voltage, V DC or peak AC	500
Max. switching current, A DC or peak AC	2.0
Max. carry current, A DC or peak AC	3.0
Dielectric strength, V DC	1000
Max. initial contact resistance, Ω	0.025
Pull in value (AT range)	40 – 70
Max. contact capacitance, pF	0.5
Min. insulation resistance, Ω	10^{10}
Operate time, ms	3.0
Release time, ms	2.0
Operating temperature, ºC	-38 +125

Reed switch type	**HYR-9001-1**
Manufacturer	**Aleph International**

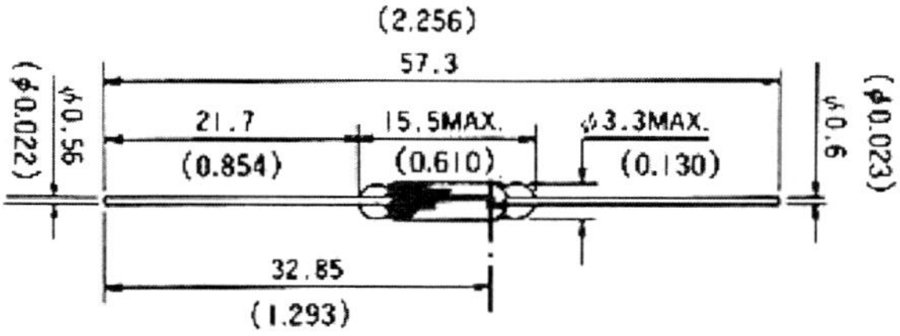

Dimensions in mm (inches)

Vertical mounting required

Contact form	NO (form A)
Contact material	Mercury wetted
Max. contact rating (resistive), W	50
Max. switching voltage, V DC or peak AC	500
Max. switching current, A DC or peak AC	2.0
Dielectric strength, V DC	1000
Max. initial contact resistance, Ω	0.05
Pull in value (AT range)	40 – 70
Max. contact capacitance, pF	0.3
Min. insulation resistance, Ω	10^{10}
Typ. resonance frequency, kHz	0.3
Operate time, ms	-
Release time, ms	-
Operating temperature, °C	-

Reed switch type	**HYR-9001-9**
Manufacturer	**Aleph International**

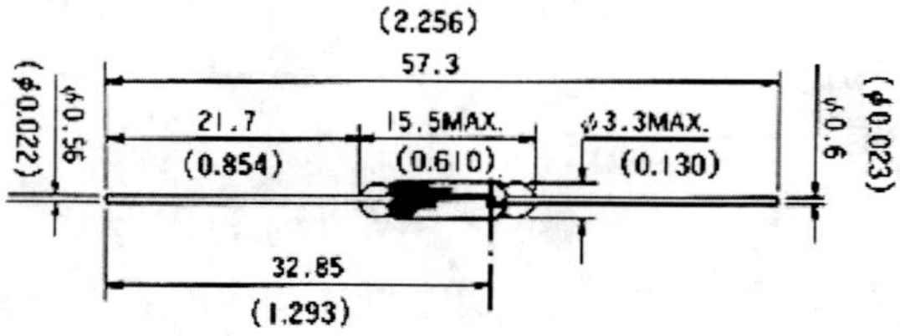

Dimensions in mm (inches)

Vertical mounting required

Contact form	NO (form A)
Contact material	Mercury wetted
Max. contact rating (resistive), W	50
Max. switching voltage, V DC or peak AC	1500
Max. switching current, A DC or peak AC	2.0
Max. carry current, A DC or peak AC	-
Dielectric strength, V DC	2000
Max. initial contact resistance, Ω	0.05
Pull in value (AT range)	40 – 70
Max. contact capacitance, pF	0.3
Min. insulation resistance, Ω	10^{13}
Typ. resonance frequency, kHz	0.3
Operate time, ms	-
Release time, ms	-

Reed switch type	**HYR-9004-9**
Manufacturer	**Aleph International**

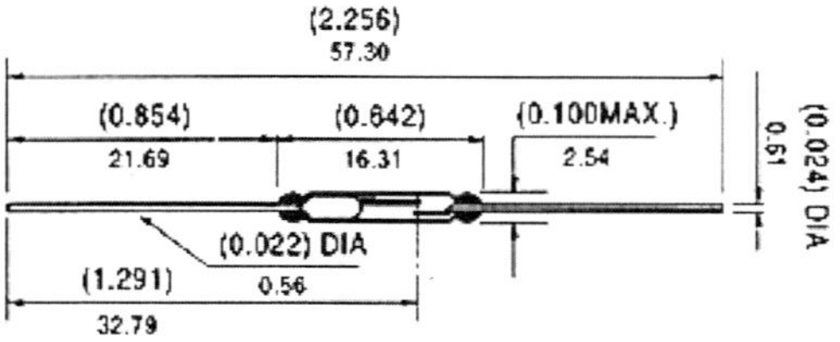

Dimensions in mm (inches)

Vertical mounting required

Contact form	NO (form A)
Contact material	Mercury wetted
Max. contact rating (resistive), W	35
Max. switching voltage, V DC or peak AC	1000
Max. switching current, A DC or peak AC	1.0
Max. carry current, A DC or peak AC	-
Dielectric strength, V DC	1500
Max. initial contact resistance, Ω	0.05
Pull in value (AT range)	35 – 70
Max. contact capacitance, pF	0.3
Min. insulation resistance, Ω	10^{13}
Typ. resonance frequency, kHz	0.3
Operate time, ms	-
Release time, ms	-
Operating temperature, °C	-

Reed switch type	**HYR-9005-9**
Manufacturer	**Aleph International**

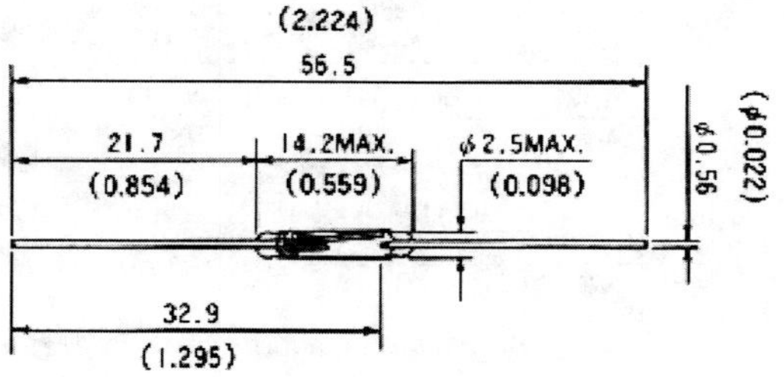

Dimensions in mm (inches), Vertical mounting required

Contact form	NO (form A)
Contact material	Mercury wetted
Max. contact rating (resistive), W	50
Max. switching voltage, V DC or peak AC	1000
Max. switching current, A DC or peak AC	1.0
Dielectric strength, V DC	1500
Max. initial contact resistance, Ω	0.6
Pull in value (AT range)	40 – 60
Max. contact capacitance, pF	0.4
Min. insulation resistance, Ω	10^{13}
Typ. resonance frequency, kHz	0.3
Operate time, ms	-
Release time, ms	-
Operating temperature, °C	-

Appendix A4: Industrial Dry Reed Switches

Reed switch type	**AL44**
Manufacturer	**Celduc relais**

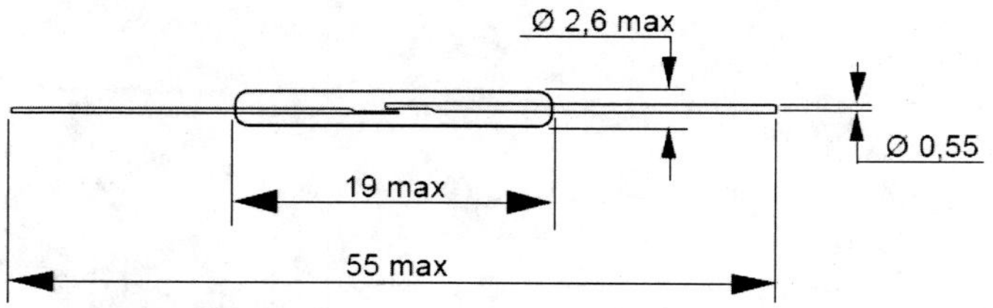

Contact form	NO (form A)
Contact material	Rhodium
Max. contact rating (resistive), W	10
Max. switching voltage, V DC or peak AC	500
Max. switching current, A DC or peak AC	0.5
Max. carry current, A DC or peak AC	1.0
Dielectric strength, V DC	1300
Max. initial contact resistance, Ω	0.1
Pull in value (AT range)	15 – 35
Max. contact capacitance, pF	0.5
Min. insulation resistance, Ω	10^{11}
Typ. resonance frequency, kHz	2.9
Operate time, ms	2.0
Release time, ms	0.1
Operating temperature, °C	-40 +85

Reed switch type	**1965**
Manufacturer	**Bare Reed**

Lead Length	Lead Diameter	Glass Length	Glass Diameter	Overall Length
1.417	0.028	0.748	0.102	2.165

Contact form	NO (form A)
Contact material	-
Max. contact rating (resistive), W	40
Max. switching voltage, V DC or peak AC	400
Max. switching current, A DC or peak AC	2.0
Max. carry current, A DC or peak AC	2.0
Dielectric strength, V DC	1000
Max. initial contact resistance, Ω	-
Pull in value (AT range)	20 – 25
Max. contact capacitance, pF	-
Min. insulation resistance, Ω	-
Typ. resonance frequency, kHz	-
Operate time, ms	2.0
Release time, ms	-
Operating temperature, °C	-

Reed switch type	**CD30**
Manufacturer	**Celduc relais**

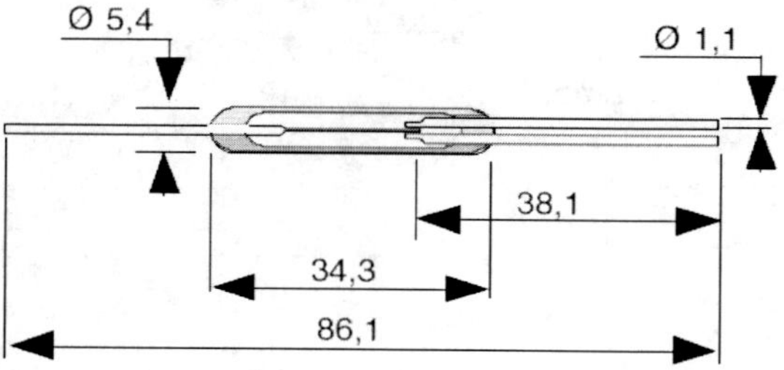

Contact form	CO (form C)
Contact material	Tungsten
Max. contact rating (resistive), W	100
Max. switching voltage, V DC or peak AC	500
Max. switching current, A DC or peak AC	3.0
Max. carry current, A DC or peak AC	-
Dielectric strength, V DC	1000
Max. initial contact resistance, Ω	0.5
Pull in value (AT range)	60 – 100
Max. contact capacitance, pF	1
Min. insulation resistance, Ω	10^8
Typ. resonance frequency, kHz	-
Operate time, ms	3.0
Release time, ms	-
Operating temperature, °C	-20 +125

Reed switch type	**CRC500H**		
Manufacturer	**Crydom**		
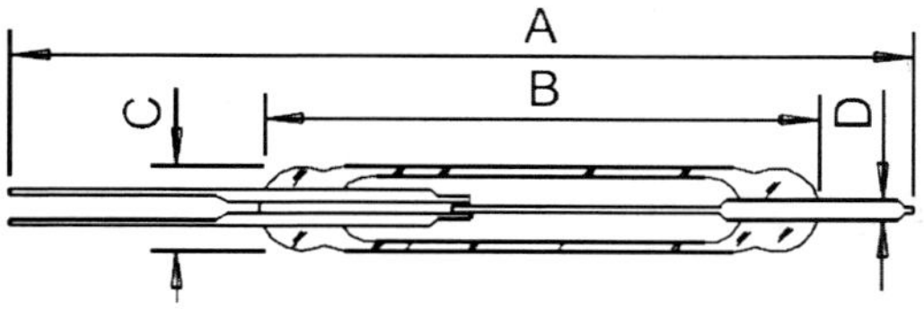 Dimensions in mm		A	87
		B	35
		C	5.4
		D	1.0

Contact form	CO (form C)
Contact material	Ruthenium
Max. contact rating (resistive), W	25
Max. switching voltage, V DC or peak AC	250
Max. switching current, A DC or peak AC	1.0
Max. carry current, A DC or peak AC	-
Dielectric strength, V DC	1000
Max. initial contact resistance, Ω	0.1
Pull in value (AT range)	50 – 90
Max. contact capacitance, pF	2.0
Min. insulation resistance, Ω	10^8
Typ. resonance frequency, kHz	-
Operate time, ms	3.0
Release time, ms	1.0
Operating temperature, °C	0 + 125

Reed switch type	**CTC500H**
Manufacturer	**Crydom**

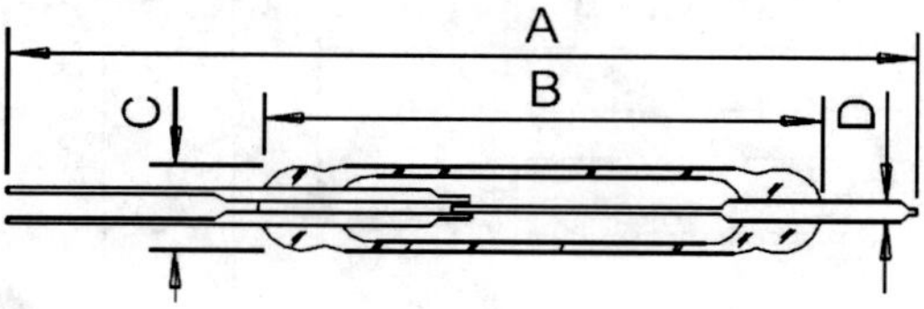

Dimensions in mm

A	87
B	35
C	5.4
D	1.0

Contact form	CO (form C)
Contact material	Tungsten
Max. contact rating (resistive), W	100
Max. switching voltage, V DC or peak AC	500
Max. switching current, A DC or peak AC	3.0
Max. carry current, A DC or peak AC	-
Dielectric strength, V DC	1000
Max. initial contact resistance, Ω	0.5
Pull in value (AT range)	60 – 100
Max. contact capacitance, pF	2.0
Min. insulation resistance, Ω	10^8
Typ. resonance frequency, kHz	-
Operate time, ms	3.5
Release time, ms	1.0
Operating temperature, °C	0 + 125

Reed switch type	DRA200G
Manufacturer	Crydom

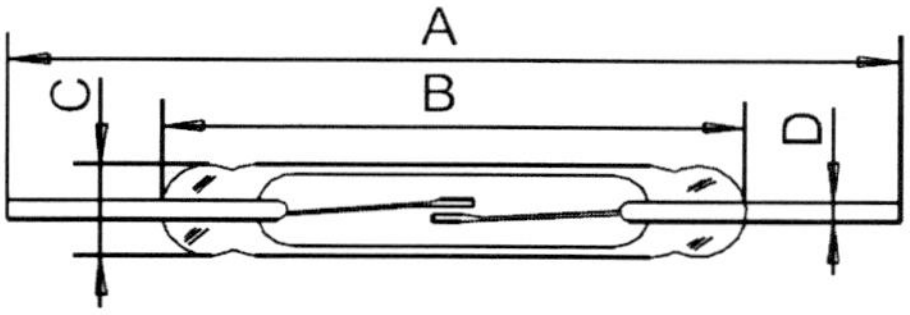

Dimensions in mm

A	79
B	52
C	5.4
D	2.5 x 0.5

Contact form	NO (form A)
Contact material	Ruthenium
Max. contact rating (resistive), W	80
Max. switching voltage, V DC or peak AC	250
Max. switching current, A DC or peak AC	1.3
Max. carry current, A DC or peak AC	2.0
Dielectric strength, V DC	800
Max. initial contact resistance, Ω	0.08
Pull in value (AT range)	79 – 95
Max. contact capacitance, pF	0.8
Min. insulation resistance, Ω	10^{11}
Typ. resonance frequency, kHz	900
Operate time, ms	4.0
Release time, ms	0.2
Operating temperature, °C	-40 + 150

Reed switch type	DRA282G	
Manufacturer	**Crydom**	
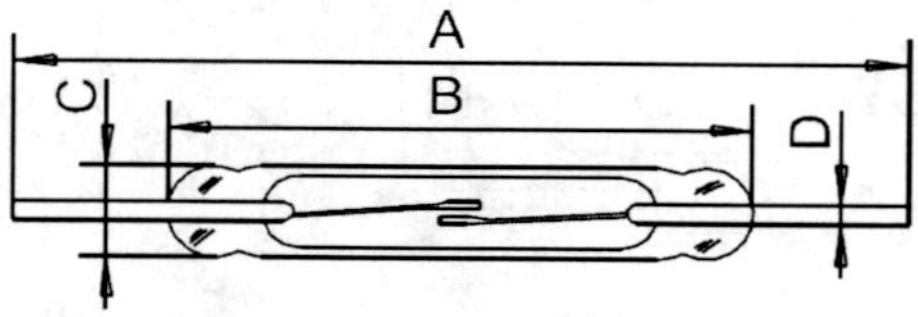 Dimensions in mm	A	79
	B	52
	C	5.4
	D	2.5 x 0.5
Contact form	NO (form A)	
Contact material	Ruthenium	
Max. contact rating (resistive), W	120	
Max. switching voltage, V DC or peak AC	250	
Max. switching current, A DC or peak AC	3.0	
Max. carry current, A DC or peak AC	5.0	
Dielectric strength, V DC	800	
Max. initial contact resistance, Ω	0.08	
Pull in value (AT range)	79 – 95	
Max. contact capacitance, pF	0.8	
Min. insulation resistance, Ω	10^{11}	
Typ. resonance frequency, kHz	900	
Operate time, ms	3.5	
Release time, ms	0.2	
Operating temperature, °C	-40 + 150	

Reed switch type	**DRA283**		
Manufacturer	**Crydom**		
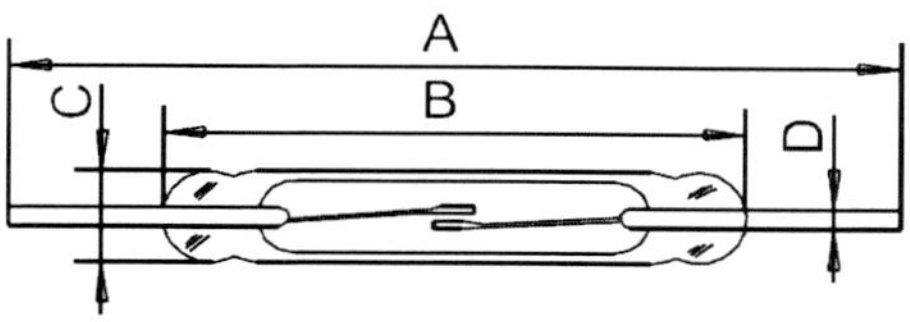 Dimensions in mm	A	84	
	B	51	
	C	5.4	
	D	2.5 x 0.5	

Contact form	NO (form A)
Contact material	Ruthenium
Max. contact rating (resistive), W	250
Max. switching voltage, V DC or peak AC	250
Max. switching current, A DC or peak AC	5.0
Max. carry current, A DC or peak AC	5.0
Dielectric strength, V DC	575
Max. initial contact resistance, Ω	0.1
Pull in value (AT range)	60 – 120
Max. contact capacitance, pF	0.6
Min. insulation resistance, Ω	10^{10}
Typ. resonance frequency, kHz	900
Operate time, ms	5.0
Release time, ms	0.2
Operating temperature, °C	-40 + 150

Reed switch type	DRA500H	
Manufacturer	**Crydom**	

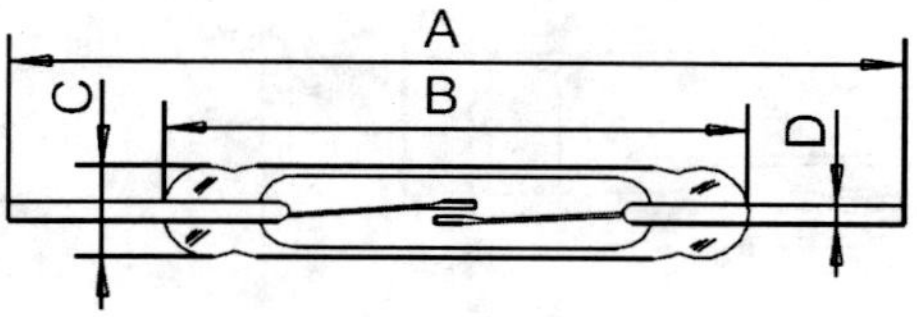

Dimensions in mm

A	82
B	51
C	5.5
D	2.5 x 0.5

Contact form	NO (form A)
Contact material	Ruthenium
Max. contact rating (resistive), W	25
Max. switching voltage, V DC or peak AC	500
Max. switching current, A DC or peak AC	1.5
Max. carry current, A DC or peak AC	-
Dielectric strength, V DC	2500
Max. initial contact resistance, Ω	0.1
Pull in value (AT range)	60 – 100
Max. contact capacitance, pF	0.8
Min. insulation resistance, Ω	10^8
Typ. resonance frequency, kHz	-
Operate time, ms	3.0
Release time, ms	1.5
Operating temperature, °C	0 + 125

Reed switch type	**DRR-129**
Manufacturer	**Hamlin**

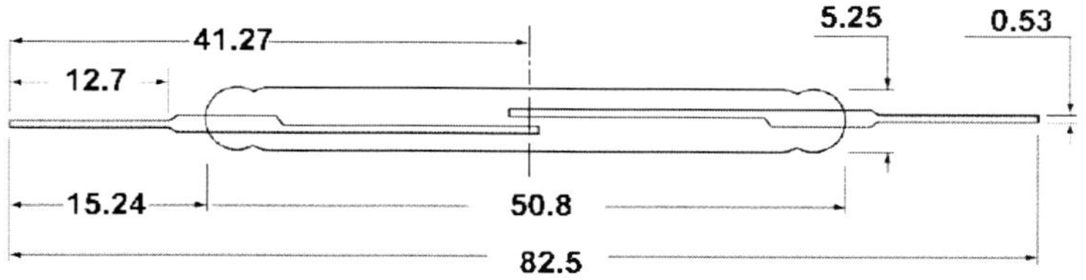

Contact form	NO (form A)
Contact material	-
Max. contact rating (resistive), W	100
Max. switching voltage, V DC or peak AC	400
Max. switching current, A DC or peak AC	3.0
Max. carry current, A DC or peak AC	6.0
Dielectric strength, V DC	600
Max. initial contact resistance, Ω	0.1
Pull in value (AT range)	42 – 83
Max. contact capacitance, pF	0.6
Min. insulation resistance, Ω	10^{10}
Typ. resonance frequency, kHz	3.2
Operate time, ms	4.5
Release time, ms	2.5
Operating temperature, °C	-40 +125

Reed switch type	DRR-DTH
Manufacturer	Hamlin

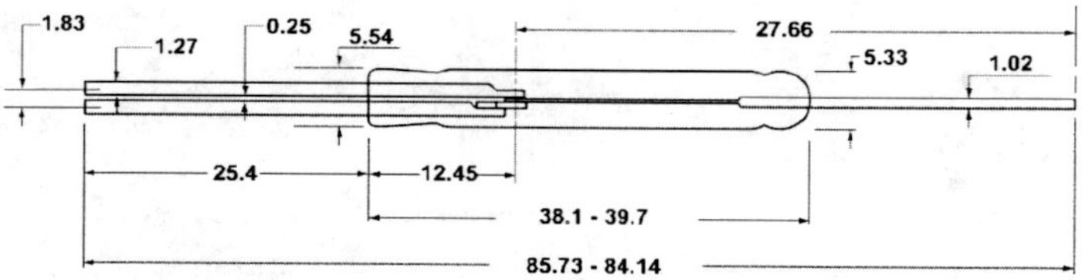

Dimensions in mm

Contact form	CO (form C)
Contact material	-
Max. contact rating (resistive), W	30
Max. switching voltage, V DC or peak AC	500
Max. switching current, A DC or peak AC	0.5
Max. carry current, A DC or peak AC	3.0
Dielectric strength, V DC	1200
Max. initial contact resistance, Ω	0.12
Pull in value (AT range)	50 – 80
Max. contact capacitance, pF	2.0
Min. insulation resistance, Ω	10^9
Typ. resonance frequency, kHz	2.75
Operate time, ms	4.5
Release time, ms	7.0
Operating temperature, °C	-20 +125

Reed switch type	DRT-DTH
Manufacturer	**Hamlin**

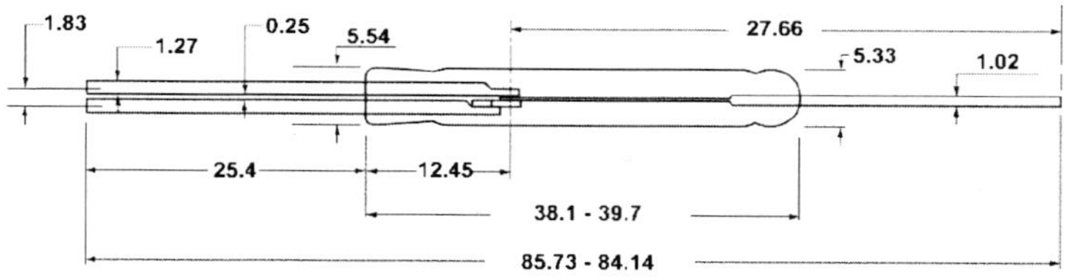

Dimensions in mm

Contact form	CO (form C)
Contact material	-
Max. contact rating (resistive), W	50
Max. switching voltage, V DC or peak AC	500
Max. switching current, A DC or peak AC	1.5
Max. carry current, A DC or peak AC	2.0
Dielectric strength, V DC	1000
Max. initial contact resistance, Ω	0.5
Pull in value (AT range)	50 – 80
Max. contact capacitance, pF	2.0
Min. insulation resistance, Ω	10^9
Typ. resonance frequency, kHz	2.75
Operate time, ms	5.5
Release time, ms	8.0
Operating temperature, °C	-20 +125

Reed switch type	**DTA500H**	
Manufacturer	**Crydom**	
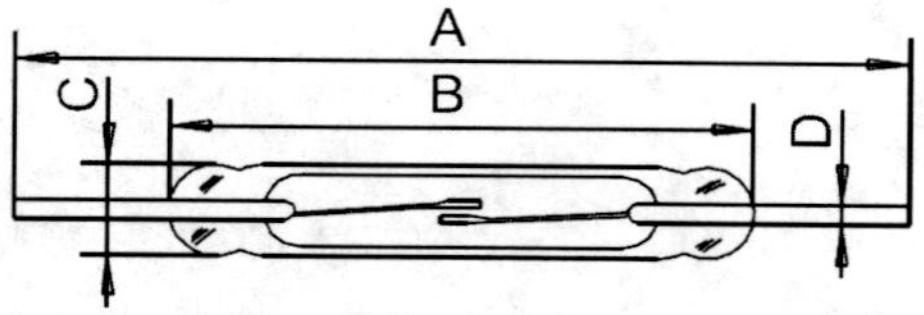 Dimensions in mm	A	82
	B	51
	C	5.5
	D	2.5 x 0.5
Contact form	NO (form A)	
Contact material	Tungsten	
Max. contact rating (resistive), W	50	
Max. switching voltage, V DC or peak AC	1000	
Max. switching current, A DC or peak AC	2.5	
Max. carry current, A DC or peak AC	-	
Dielectric strength, V DC	2500	
Max. initial contact resistance, Ω	0.1	
Pull in value (AT range)	60 – 100	
Max. contact capacitance, pF	1.5	
Min. insulation resistance, Ω	10^8	
Typ. resonance frequency, kHz	-	
Operate time, ms	3.0	
Release time, ms	1.5	
Operating temperature, °C	0 + 125	

Reed switch type	DTA810H	
Manufacturer	Crydom	
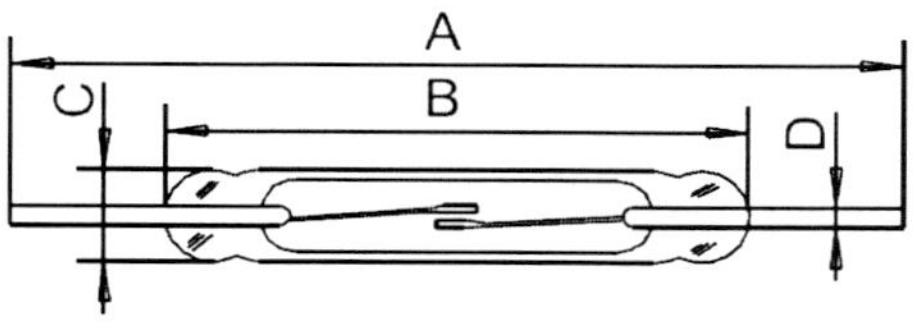 Dimensions in mm	A	82
	B	54
	C	5.5
	D	0.50
Contact form	NO (form A)	
Contact material	Tungsten	
Max. contact rating (resistive), W	50	
Max. switching voltage, V DC or peak AC	7500	
Max. switching current, A DC or peak AC	3.0	
Max. carry current, A DC or peak AC	-	
Dielectric strength, V DC	10,000	
Max. initial contact resistance, Ω	0.1	
Pull in value (AT range)	100 – 150	
Max. contact capacitance, pF	1.0	
Min. insulation resistance, Ω	10^{10}	
Typ. resonance frequency, kHz	-	
Operate time, ms	3.0	
Release time, ms	1.0	
Operating temperature, °C	0 + 125	

Reed switch type	**GC3817**
Manufacturer	**Comus Group**

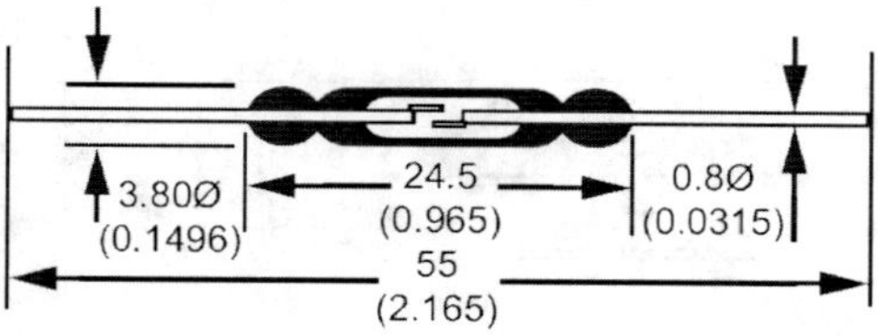

Dimensions in mm (inches)

Contact form	NO (form A)
Contact material	Rhodium
Max. contact rating (resistive), W	60
Max. switching voltage, V DC or peak AC	400
Max. switching current, A DC or peak AC	3.0
Max. carry current, A DC or peak AC	4.0
Dielectric strength, V DC	1000
Max. initial contact resistance, Ω	0.08
Pull in value (AT range)	30 – 70
Max. contact capacitance, pF	0.5
Min. insulation resistance, Ω	10^{11}
Typ. resonance frequency, kHz	2.4
Operate time, ms	2.5
Release time, ms	0.1
Operating temperature, °C	-40 +125

Reed switch type	**GC 1513**
Manufacturer	**Comus Group**

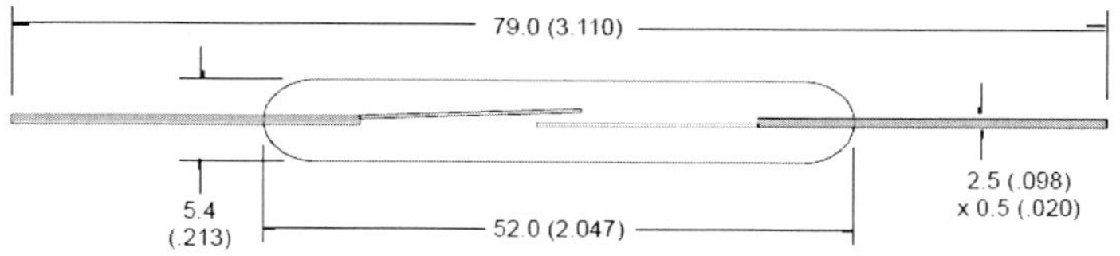

Dimensions in mm (inches)

Contact form	NO (form A)
Contact material	Rhodium
Max. contact rating (resistive), W	120
Max. switching voltage, V DC or peak AC	1000
Max. switching current, A DC or peak AC	3.0
Max. carry current, A DC or peak AC	5.0
Dielectric strength, V DC	3000
Max. initial contact resistance, Ω	0.08
Pull in value (AT range)	75 – 95
Max. contact capacitance, pF	0.8
Min. insulation resistance, Ω	10^{11}
Typ. resonance frequency, kHz	900
Operate time, ms	3.5
Release time, ms	0.2
Operating temperature, °C	-40 +125

Reed switch type	**GC 1917**
Manufacturer	**Comus Group**

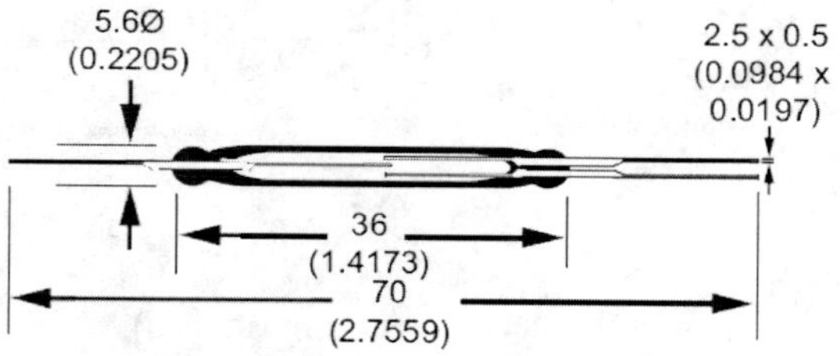

Dimensions in mm (inches)

Contact form	CO (form C)
Contact material	Rhodium
Max. contact rating (resistive), W	60
Max. switching voltage, V DC or peak AC	400
Max. switching current, A DC or peak AC	1.0
Max. carry current, A DC or peak AC	2.0
Dielectric strength, V DC	1000
Max. initial contact resistance, Ω	0.1
Pull in value (AT range)	50 – 80
Max. contact capacitance, pF	1.0
Min. insulation resistance, Ω	10^9
Operate time, ms	4.0
Release time, ms	0.15
Operating temperature, °C	-40 +125

Reed switch type	**GC 2314**
Manufacturer	**Comus Group**

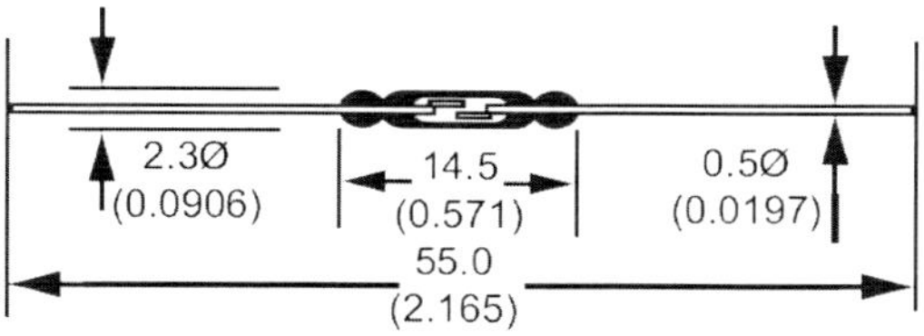

Dimensions in mm (inches)

Contact form	NO (form A)
Contact material	Rhodium
Max. contact rating (resistive), W	10
Max. switching voltage, V DC or peak AC	400
Max. switching current, A DC or peak AC	0.5
Max. carry current, A DC or peak AC	1.0
Dielectric strength, V DC	600
Max. initial contact resistance, Ω	0.15
Pull in value (AT range)	10 – 35
Max. contact capacitance, pF	0.7
Min. insulation resistance, Ω	10^{11}
Typ. resonance frequency, kHz	5.0
Operate time, ms	2.0
Release time, ms	0.05
Operating temperature, °C	-40 +125

Reed switch type	HSR-780R
Manufacturer	**Hermetic Switch, Inc.**

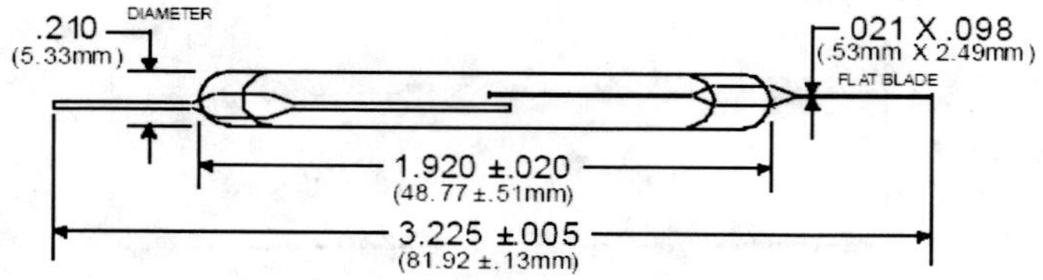

Contact form	NO (form A)
Contact material	Rhodium
Max. contact rating (resistive), W	25
Max. switching voltage, V DC or peak AC	500
Max. switching current, A DC or peak AC	1.5
Max. carry current, A DC or peak AC	8.8
Dielectric strength, V DC	2500
Max. initial contact resistance, Ω	0.1
Pull in value (AT range)	60 – 100
Max. contact capacitance, pF	0.9
Min. insulation resistance, Ω	10^7
Typ. resonance frequency, kHz	-
Operate time, ms	2.5
Release time, ms	2.1
Operating temperature, °C	-40 +200

Reed switch type	**HSR-790W**
Manufacturer	**Hermetic Switch, Inc.**

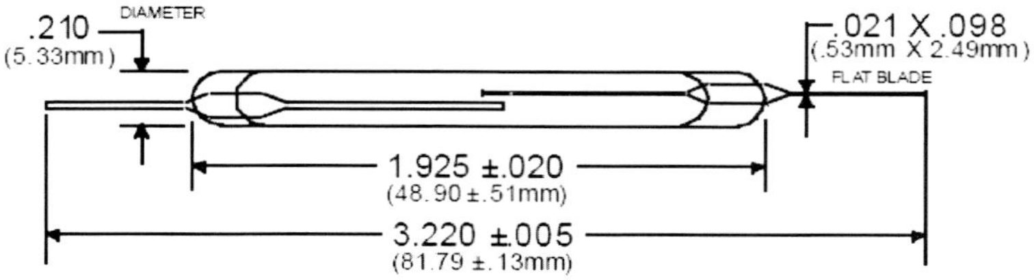

Contact form	NO (form A)
Contact material	Tungsten
Max. contact rating (resistive), W	50
Max. switching voltage, V DC or peak AC	1000
Max. switching current, A DC or peak AC	2.5
Max. carry current, A DC or peak AC	5.5
Dielectric strength, V DC	2500
Max. initial contact resistance, Ω	0.1
Pull in value (AT range)	60 – 100
Max. contact capacitance, pF	1.1
Min. insulation resistance, Ω	10^8
Typ. resonance frequency, kHz	-
Operate time, ms	3.0
Release time, ms	1.6
Operating temperature, °C	0 +125

Reed switch type	HSR-830R
Manufacturer	**Hermetic Switch, Inc.**

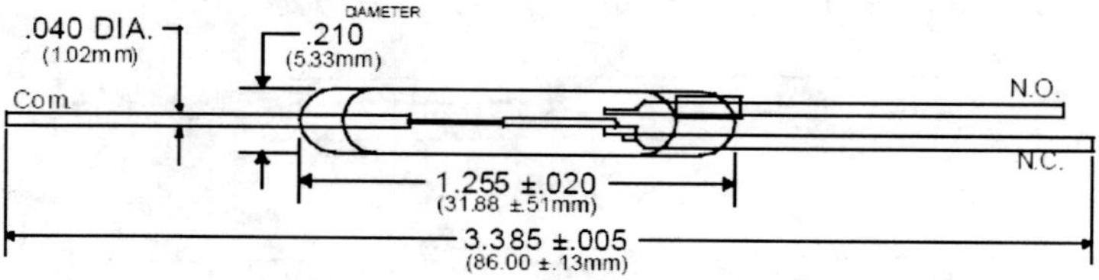

Contact form	CO (form C)
Contact material	Rhodium
Max. contact rating (resistive), W	25
Max. switching voltage, V DC or peak AC	250
Max. switching current, A DC or peak AC	1.0
Max. carry current, A DC or peak AC	-
Dielectric strength, V DC	1000
Max. initial contact resistance, Ω	0.1
Pull in value (AT range)	50 – 100
Max. contact capacitance, pF	2.5
Min. insulation resistance, Ω	10^7
Typ. resonance frequency, kHz	-
Operate time, ms	3.6
Release time, ms	4.2
Operating temperature, °C	-25 +125

Reed switch type	**HSR-834**
Manufacturer	**Hermetic Switch, Inc.**

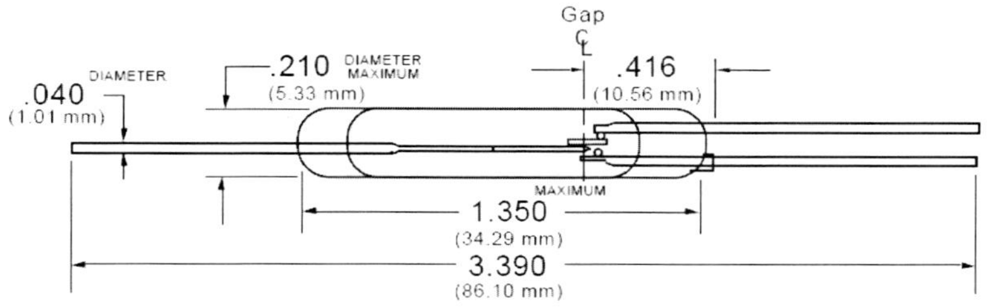

Contact form	CO (form C)
Contact material	Tungsten
Max. contact rating (resistive), W	100
Max. switching voltage, V DC or peak AC	500
Max. switching current, A DC or peak AC	3.0
Max. carry current, A DC or peak AC	-
Dielectric strength, V DC	1000
Max. initial contact resistance, Ω	0.5
Pull in value (AT range)	60 – 100
Max. contact capacitance, pF	2.0
Min. insulation resistance, Ω	10^8
Typ. resonance frequency, kHz	-
Operate time, ms	2.0
Release time, ms	1.0
Operating temperature, °C	0 - 125

Reed switch type	HSR-V933W
Manufacturer	**Hermetic Switch, Inc.**

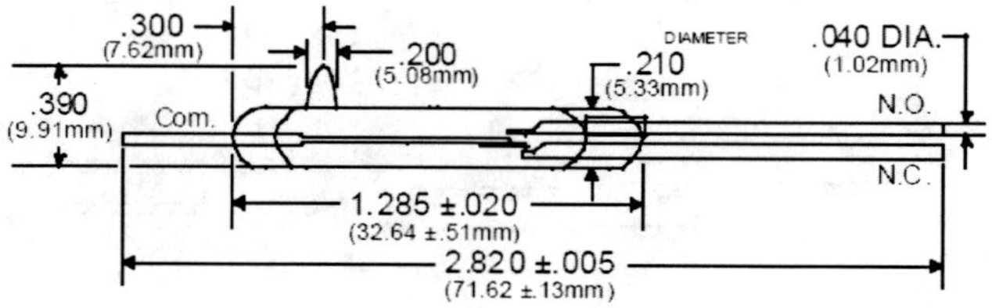

Contact form	CO (form C)
Contact material	Tungsten
Max. contact rating (resistive), W	100
Max. switching voltage, V DC or peak AC	500
Max. switching current, A DC or peak AC	3.0
Max. carry current, A DC or peak AC	-
Dielectric strength, V DC	1500
Max. initial contact resistance, Ω	0.5
Pull in value (AT range)	40 – 100
Max. contact capacitance, pF	3.0
Min. insulation resistance, Ω	10^{10}
Typ. resonance frequency, kHz	-
Operate time, ms	4.2
Release time, ms	3.7
Operating temperature, °C	-60 +125

Reed switch type	**HYR1559**
Manufacturer	**Aleph International**

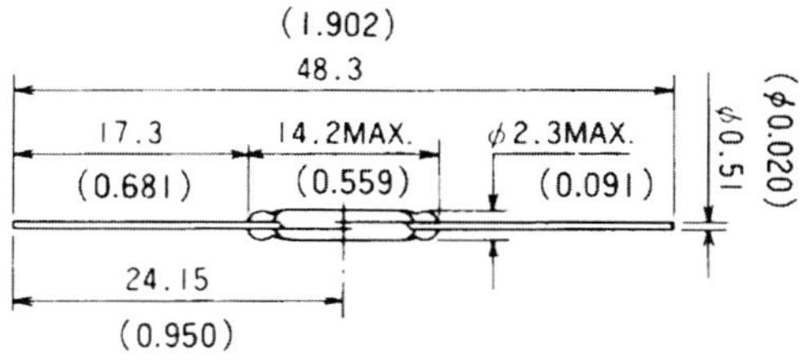

Dimensions in mm (inches)

Contact form	NO (form A)
Contact material	Ruthenium
Max. contact rating (resistive), W	10
Max. switching voltage, V DC or peak AC	500
Max. switching current, A DC or peak AC	0.5
Max. carry current, A DC or peak AC	-
Dielectric strength, V DC	1500
Max. initial contact resistance, Ω	0.2
Pull in value (AT range)	15 – 50
Max. contact capacitance, pF	0.4
Min. insulation resistance, Ω	10^{10}
Typ. resonance frequency, kHz	4.8
Operate time, ms	-
Operating temperature, °C	-

Reed switch type	HYR2016
Manufacturer	**Aleph International**

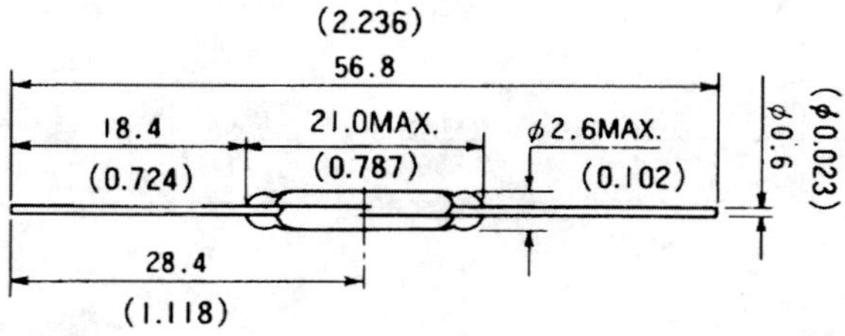

Dimensions in mm (inches)

Contact form	NO (form A)
Contact material	Rhodium
Max. contact rating (resistive), W	25
Max. switching voltage, V DC or peak AC	1000
Max. switching current, A DC or peak AC	1.0
Max. carry current, A DC or peak AC	-
Dielectric strength, V DC	2500
Max. initial contact resistance, Ω	0.1
Pull in value (AT range)	15 – 70
Max. contact capacitance, pF	0.4
Min. insulation resistance, Ω	10^{11}
Typ. resonance frequency, kHz	2.2
Operate time, ms	-
Release time, ms	-

Reed switch type	KSK-1A84
Manufacturer	Meder Electronics

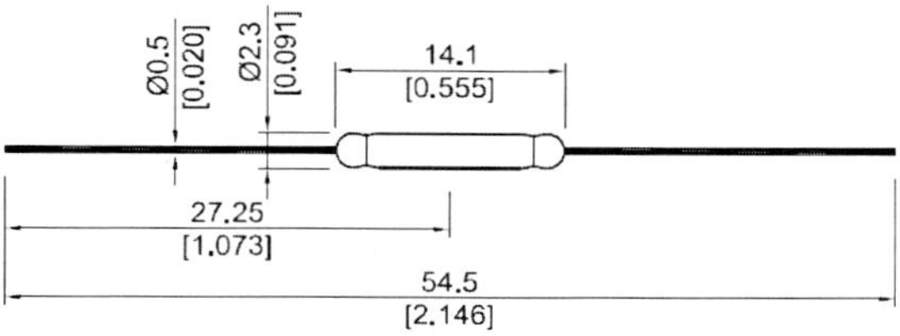

Dimensions in mm (inches)

Contact form	NO (form A)
Contact material	-
Max. contact rating (resistive), W	10
Max. switching voltage, V DC or peak AC	400
Max. switching current, A DC or peak AC	0.5
Max. carry current, A DC or peak AC	1.0
Dielectric strength, V DC	700
Max. initial contact resistance, Ω	0.15
Pull in value (AT range)	15 – 50
Max. contact capacitance, pF	0.7
Min. insulation resistance, Ω	10^{11}
Operate time, ms	2.0
Release time, ms	0.1
Operating temperature, °C	-20 +130

Reed switch type	MH5
Manufacturer	Comus Group

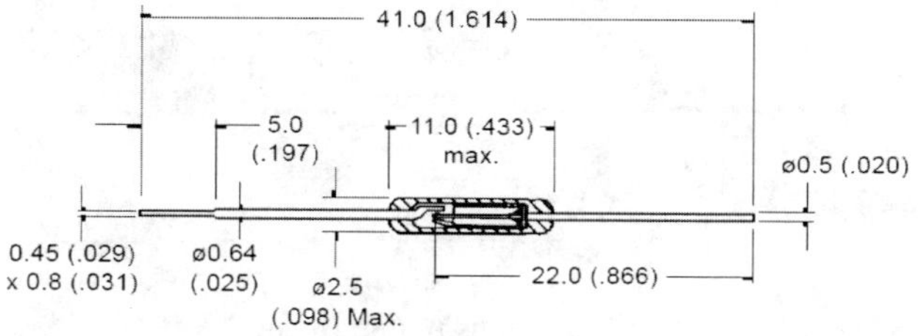

Dimensions in mm (inches)

Contact form	NO (form A)
Contact material	Hg
Max. contact rating (resistive), W	50
Max. switching voltage, V DC or peak AC	500
Max. switching current, A DC or peak AC	2.0
Max. carry current, A DC or peak AC	2.0
Dielectric strength, V DC	1500
Max. initial contact resistance, Ω	0.03
Pull in value (AT range)	30 – 65
Max. contact capacitance, pF	0.3
Min. insulation resistance, Ω	10^{10}
Operate time, ms	1.2
Release time, ms	1.0
Operating temperature, °C	-38 + 125

Reed switch type	**R14U, R15U (Bestact™)**
Manufacturer	Yaskawa Electric America, Ltd

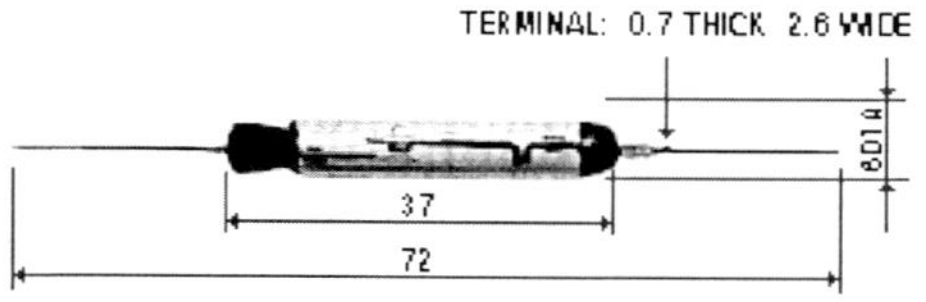

Dimensions in mm

Contact form	NO (form A)
Contact material	two stage commutation
Max. contact rating	360VA, 69W
Max. switching voltage, V DC and AC	240
Max. making current, A AC	30
Max. breaking current, A (at 115V DC)	0.6
Max. carry current, A DC or peak AC	5.0
Dielectric strength, V AC, rms	800
Max. initial contact resistance, Ω	0.1 – 0.5
Pull in value (AT range)	180 – 230
Max. contact capacitance, pF	0.5
Min. insulation resistance, Ω	10^9
Operate time, ms	4.0
Release time, ms	2.0
Operating temperature, °C	-50 +150

Reed switch type	**R24U, R25U (Bestact™)**
Manufacturer	Yaskawa Electric America, Ltd

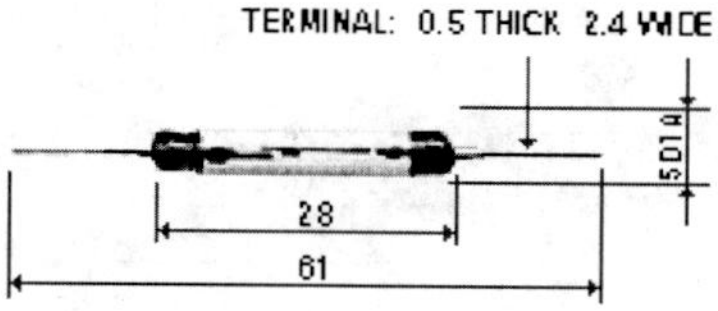

Dimensions in mm

Contact form	NO (form A)
Contact material	two stage commutation
Max. contact rating (resistive)	180VA, 69W
Max. switching voltage, V DC and AC	240
Max. making current, A AC	15
Max. breaking current, A (at 115V DC)	0.5
Max. carry current, A DC or peak AC	3.0
Dielectric strength, V AC	500
Max. initial contact resistance, Ω	0.2
Pull in value (AT range)	100 – 130
Max. contact capacitance, pF	0.5
Min. insulation resistance, Ω	10^9
Operate time, ms	4.0
Release time, ms	2.0
Operating temperature, °C	-50 +150

Appendix B1: High-Voltage Bipolar Transistors

Transistor type	**2SC4635LS**	
Manufacturer	**Sanyo**	

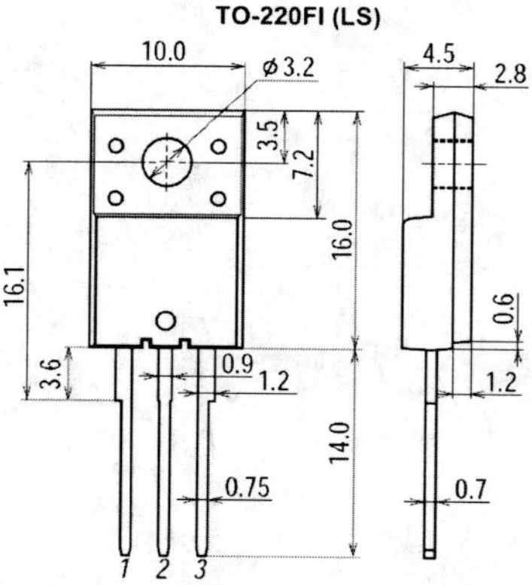

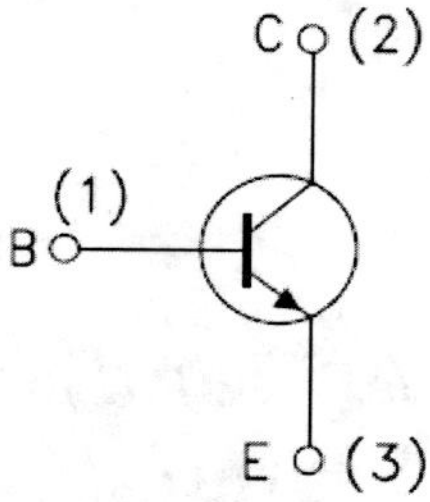

Transistor kind	N-P-N, Bipolar	
Collector-base voltage ($I_E = 0$), V	V_{CBO}	1500
Collector-emitter voltage ($I_B = 0$), V	V_{CEO}	1500
Emitter-base voltage ($I_C = 0$), V	V_{EBO}	5
Collector current: continuous/peak, A	I_C	0.02/0.06
Base current, A	I_B	-
Total dissipation (T_C = 25°C), W	P_{tot}	2
Collector-emitter saturation voltage, V	$V_{CE\,(SAT)}$*	5
DC current gain	h_{FE}*	10 – 60
Collector cut-off current ($V_{BE} = 0$), μA	I_{CES}	1
Max. operating junction temperature, °C	T_j	150
Thermal resistance: junction-case, °C/W	$R_{TH\ j\text{-}case}$	8.3

Transistor type	2SC4637LS
Manufacturer	**Sanyo**

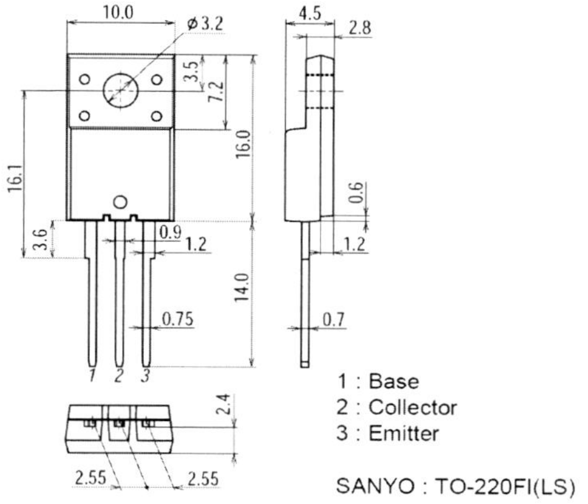

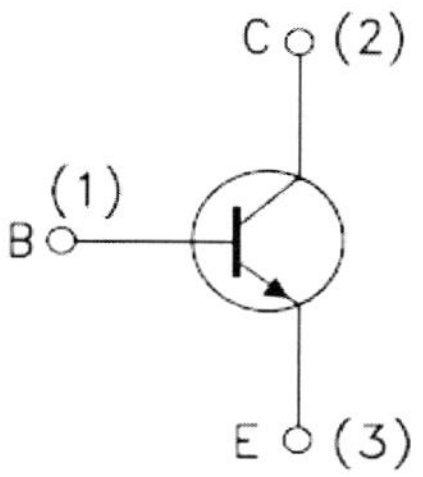

Transistor kind	N-P-N, Bipolar	
Collector-base voltage ($I_E = 0$), V	V_{CBO}	2000
Collector-emitter voltage ($I_B = 0$), V	V_{CEO}	1800
Emitter-base voltage ($I_C = 0$), V	V_{EBO}	5
Collector current: continuous/pulse, A	I_C	0.015/0.0.05
Base current, A	I_B	-
Total dissipation ($T_C = 25°C$), W	P_{tot}	2
Collector-emitter saturation voltage, V	$V_{CE\,(SAT)}$*	5
DC current gain	h_{FE}*	10 – 60
Collector cut-off current ($V_{BE} = 0$), μA	I_{CES}	1
Max. operating junction temperature, °C	T_j	150
Thermal resistance: junction-case, °C/W	$R_{TH\,j\text{-}case}$	-

Transistor type	**2SC5300, TS7992**
Manufacturer	**Sanyo**

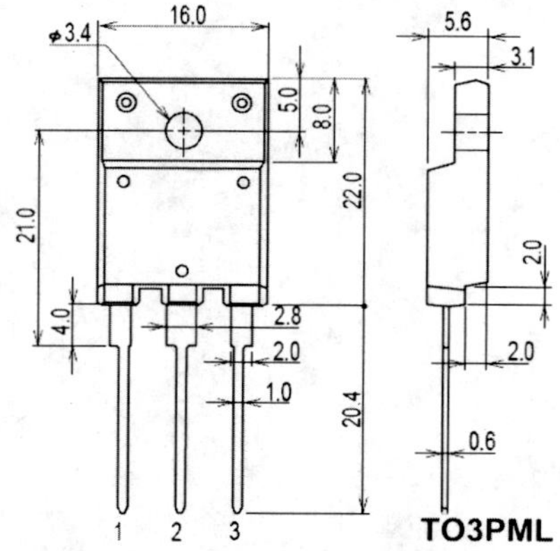

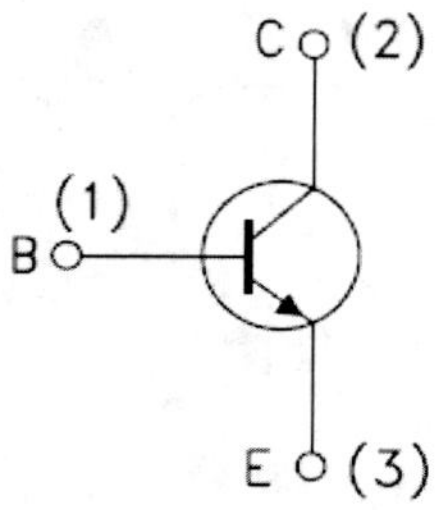

Transistor kind		N-P-N, Bipolar
Collector-base voltage ($I_E = 0$), V	V_{CBO}	1500 (2SC5300) 1600 (TS7992)
Collector-emitter voltage ($I_B = 0$), V	V_{CEO}	800
Emitter-base voltage ($I_C = 0$), V	V_{EBO}	6
Collector current: continuous/peak, A	I_C	20/40
Total dissipation ($T_C = 25$ºC), W	P_{tot}	75
Collector-emitter saturation voltage, V	$V_{CE\,(SAT)}$*	5
DC current gain	h_{FE}*	4 – 30
Collector cut-off current ($V_{BE} = 0$), μA	I_{CES}	1000
Max. operating junction temperature, ºC	T_j	150
Thermal resistance: junction-case, ºC/W	$R_{TH\,j\text{-}case}$	-

Transistor type	2SC5612	
Manufacturer	Toshiba	
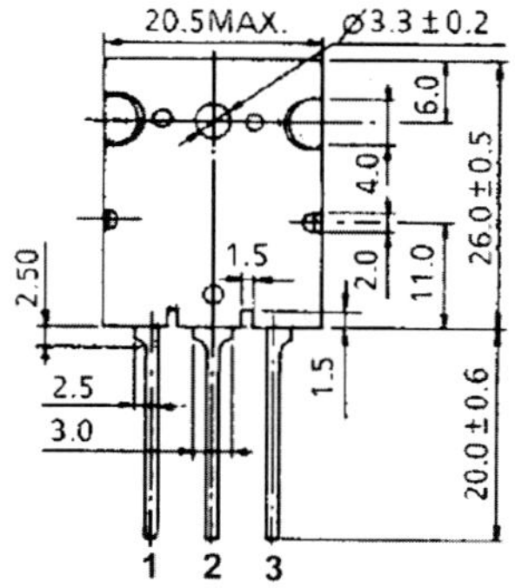	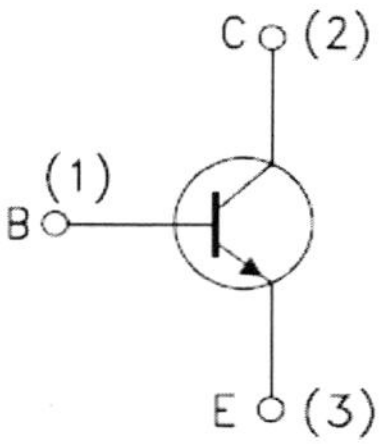	
Transistor kind	N-P-N, Bipolar	
Collector-base voltage (I_E = 0), V	V_{CBO}	2000
Collector-emitter voltage (I_B = 0), V	V_{CEO}	900
Emitter-base voltage (I_C = 0), V	V_{EBO}	5
Collector current: continuous/peak, A	I_C	22/44
Base current, A	I_B	11
Total dissipation (T_C = 25°C), W	P_{tot}	220
Collector-emitter saturation voltage, V	$V_{CE\,(SAT)}$*	3
DC current gain	h_{FE}*	5 – 50
Collector cut-off current (V_{BE} = 0), μA	I_{CES}	1000
Max. operating junction temperature, °C	T_j	150
Thermal resistance: junction-case, °C/W	$R_{TH\ j\text{-}case}$	-

Transistor type	2SC5748	
Manufacturer	**Toshiba**	
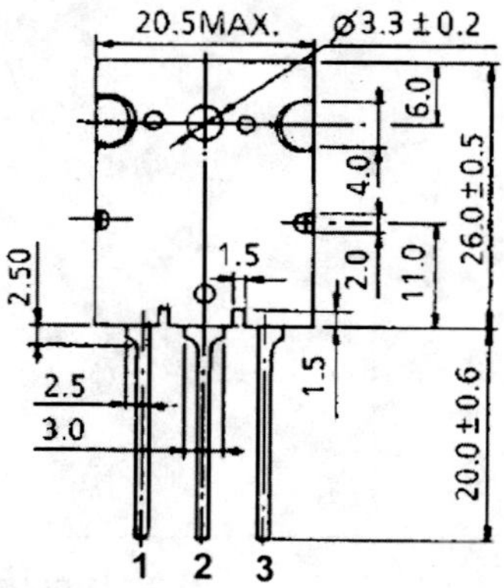	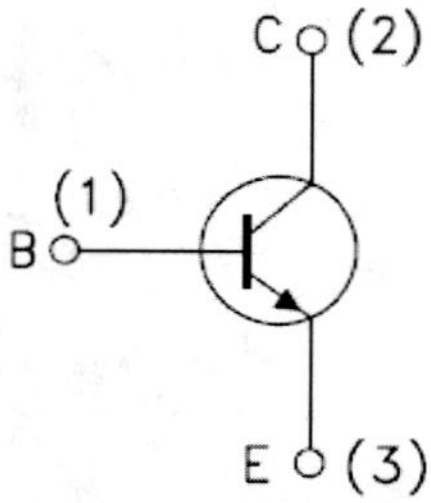	
Transistor kind	N-P-N, Bipolar	
Collector-base voltage ($I_E = 0$), V	V_{CBO}	2000
Collector-emitter voltage ($I_B = 0$), V	V_{CEO}	900
Emitter-base voltage ($I_C = 0$), V	V_{EBO}	5
Collector current: continuous/peak, A	I_C	16/32
Base current, A	I_B	8
Total dissipation ($T_C = 25°C$), W	P_{tot}	210
Collector-emitter saturation voltage, V	$V_{CE\ (SAT)}$*	3
DC current gain	h_{FE}*	5 – 55
Collector cut-off current ($V_{BE} = 0$), μA	I_{CES}	1000
Max. operating junction temperature, °C	T_j	150
Thermal resistance: junction-case, °C/W	$R_{TH\ j\text{-}case}$	-

Transistor type	**2SD2580**	
Manufacturer	**Sanyo**	
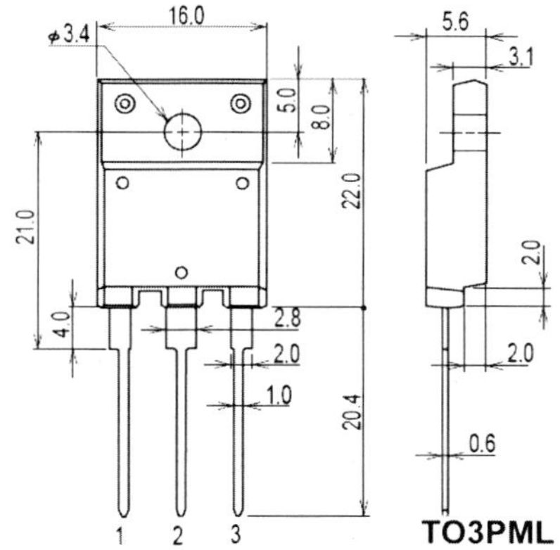	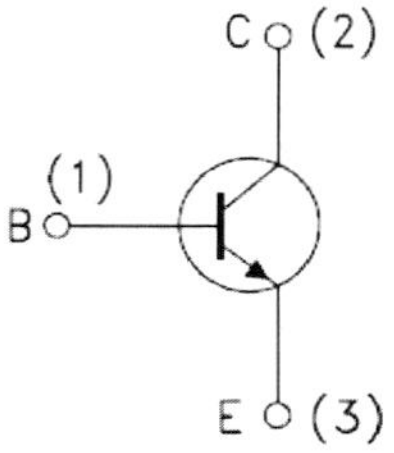	
Transistor kind	N-P-N, Bipolar	
Collector-base voltage (I_E = 0), V	V_{CBO}	1500
Collector-emitter voltage (I_B = 0), V	V_{CEO}	800
Emitter-base voltage (I_C = 0), V	V_{EBO}	6
Collector current: continuous/peak, A	I_C	10/30
Base current, A	I_B	-
Total dissipation (T_C = 25ºC), W	P_{tot}	70
Collector-emitter saturation voltage, V	$V_{CE\,(SAT)}$*	5
DC current gain	h_{FE}*	5 – 30
Dielectric strength (leads to case), V RMS	V_{ISOL}	-
Collector cut-off current (V_{BE} = 0), µA	I_{CES}	1000
Max. operating junction temperature, ºC	T_j	150

Transistor type	2T8277A	
Manufacturer	**OKB “Iskra” (Ulianovsk, Russia)**	
 KT-9M	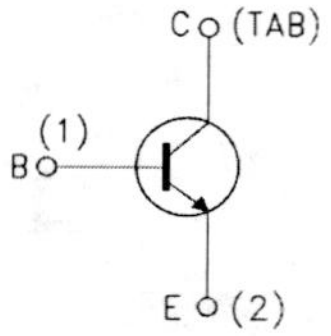	
Transistor kind	N-P-N, Bipolar	
Collector-base voltage ($I_E = 0$), V	V_{CBO}	1500
Collector-emitter voltage ($I_B = 0$), V	V_{CEO}	700
Collector-emitter voltage ($U_{BE} = 0$)	V_{CES}	-
Emitter-base voltage ($I_C = 0$), V	V_{EBO}	10
Collector current: continuous/peak, A	I_C	16/22
Base current, A	I_B	9
Total dissipation ($T_C = 25$ºC), W	P_{tot}	200
Collector-emitter saturation voltage, V	$V_{CE(SAT)}$*	1.2 – 1.7
DC current gain	h_{FE}*	>5
Collector cut-off current ($V_{BE} = 0$), μA	I_{CES}	1000
Max. operating junction temperature, ºC	T_j	150
Thermal resistance: junction-case, ºC/W	$R_{TH\ j\text{-}case}$	0.63

Transistor type	2T8292A	
Manufacturer	**OKB "Iskra" (Ulianovsk, Russia)**	
 KT-9M	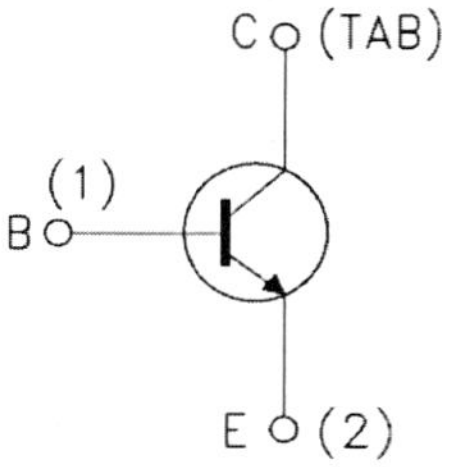	
Transistor kind	N-P-N, Bipolar	
Collector-base voltage ($I_E = 0$), V	V_{CBO}	850
Collector-emitter voltage ($I_B = 0$), V	V_{CEO}	450
Emitter-base voltage ($I_C = 0$), V	V_{EBO}	7
Collector current: continuous/peak, A	I_C	60/90
Base current, A	I_B	15
Total dissipation ($T_C = 25°C$), W	P_{tot}	300
Collector-emitter saturation voltage, V	$V_{CE\,(SAT)}$*	0.9
DC current gain	h_{FE}*	> (2 – 10)
Dielectric strength (leads to case), V RMS	V_{ISOL}	-
Collector cut-off current ($V_{BE} = 0$), μA	I_{CES}	1000
Max. operating junction temperature, °C	T_j	150
Thermal resistance: junction-case, °C/W	$R_{TH\,j\text{-}case}$	-

Transistor type	**2N5014**	
Manufacturer	**Solid State Devices, Inc.**	
 **TO-205AD** **TO39** PIN CONFIGURATION 1. EMITTER 2. BASE 3. COLLECTOR	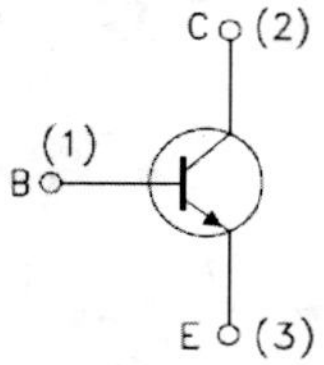	
Transistor kind	N-P-N, Bipolar	
Collector-base voltage ($I_E = 0$), V	V_{CBO}	900
Collector-emitter voltage ($I_B = 0$), V	V_{CEO}	-
Collector-emitter voltage ($R_{BE} = 1\ k\Omega$)	V_{CER}	900
Emitter-base voltage ($I_C = 0$), V	V_{EBO}	5
Collector current: continuous, A	I_C	0.5
Base current, A	I_B	0.05
Total dissipation ($T_C = 25°C$), W	P_{tot}	2
Collector-emitter saturation voltage, V	$V_{CE\ (SAT)}$*	1.5
DC current gain	h_{FE}*	30 – 180
Dielectric strength (leads to case), V RMS	V_{ISOL}	-
Collector cut-off current ($V_{BE} = 0$), μA	I_{CES}	100
Max. operating junction temperature, °C	T_j	200
Thermal resistance: junction-case, °C/W	$R_{TH\ j\text{-}case}$	50

Transistor type	2N5015	
Manufacturer	**Solid State Devices, Inc.**	
 TO-205AD TO39 PIN CONFIGURATION 1. EMITTER 2. BASE 3. COLLECTOR	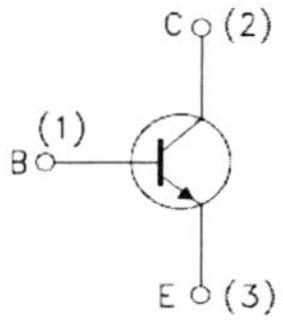	
Transistor kind	N-P-N, Bipolar	
Collector-base voltage (I_E = 0), V	V_{CBO}	1000
Collector-emitter voltage (I_B = 0), V	V_{CEO}	-
Collector-emitter voltage (R_{BE} = 1 kΩ)	V_{CER}	1000
Emitter-base voltage (I_C = 0), V	V_{EBO}	5
Collector current: continuous, A	I_C	0.5
Base current, A	I_B	0.05
Total dissipation (T_C = 25°C), W	P_{tot}	2
Collector-emitter saturation voltage, V	$V_{CE(SAT)}$*	1.5
DC current gain	h_{FE}*	30 – 180
Dielectric strength (leads to case), V RMS	V_{ISOL}	-
Collector cut-off current (V_{BE} = 0), µA	I_{CES}	100
Max. operating junction temperature, °C	T_j	200
Thermal resistance: junction-case, °C/W	$R_{TH\ j\text{-}case}$	50

Transistor type	2SC4030
Manufacturer	Sanyo

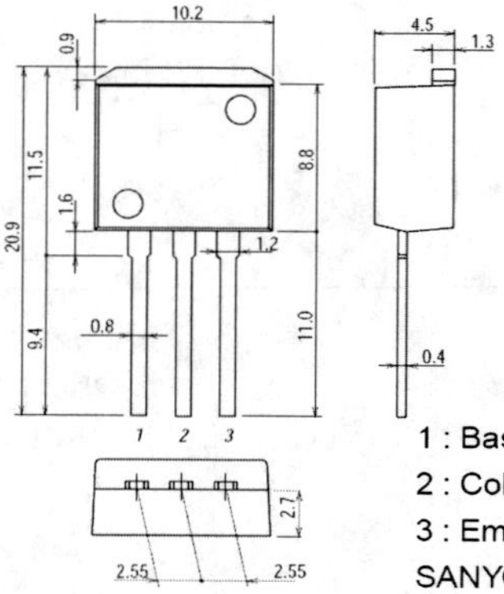

1 : Base
2 : Collector
3 : Emitter
SANYO : TO-220MF

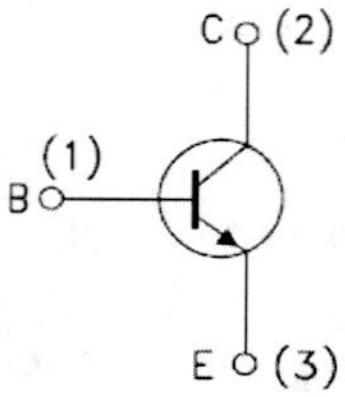

Parameter	Symbol	Value
Transistor kind	N-P-N, Bipolar	
Collector-base voltage ($I_E = 0$), V	V_{CBO}	1700
Collector-emitter voltage ($I_B = 0$), V	V_{CEO}	900
Emitter-base voltage ($I_C = 0$), V	V_{EBO}	5
Collector current: continuous/pulse, A	I_C	0.05/0.15
Base current, A	I_B	-
Total dissipation ($T_C = 25°C$), W	P_{tot}	1.65
Collector-emitter saturation voltage, V	$V_{CE\,(SAT)}$*	5
DC current gain	h_{FE}*	20 – 120
Collector cut-off current ($V_{BE} = 0$), μA	I_{CES}	1
Max. operating junction temperature, °C	T_j	150
Thermal resistance: junction-case, °C/W	$R_{TH\ j\text{-}case}$	-

Transistor type	**2SC4633LS**
Manufacturer	**Sanyo**

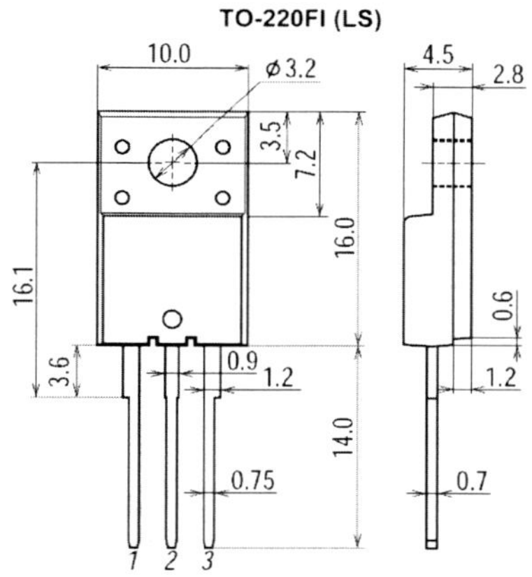

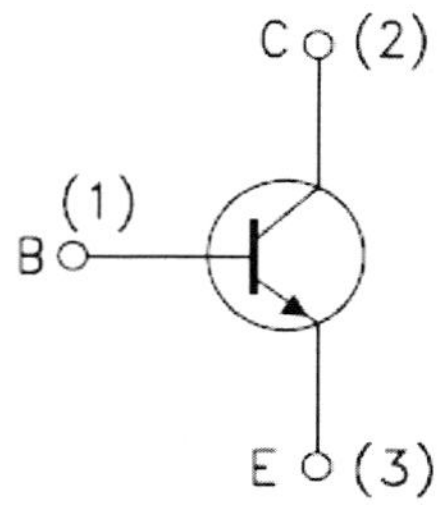

Transistor kind	N-P-N, Bipolar	
Collector-base voltage (I_E = 0), V	V_{CBO}	1500
Collector-emitter voltage (I_B = 0), V	V_{CEO}	1200
Emitter-base voltage (I_C = 0), V	V_{EBO}	5
Collector current: continuous/peak, A	I_C	0.03/0.1
Base current, A	I_B	-
Total dissipation (T_C = 25°C), W	P_{tot}	2
Collector-emitter saturation voltage, V	$V_{CE\ (SAT)}$*	5
DC current gain	h_{FE}*	10 – 60
Collector cut-off current (V_{BE} = 0), μA	I_{CES}	1
Max. operating junction temperature, °C	T_j	150
Thermal resistance: junction-case, °C/W	$R_{TH\ j\text{-}case}$	8.3

Transistor type	**BU208A (TO-3), BU508A (TO-218), BU508AF1 (ISOWATT)**
Manufacturer	**STMicroelectronics, SGS Thomson Microelectronics**

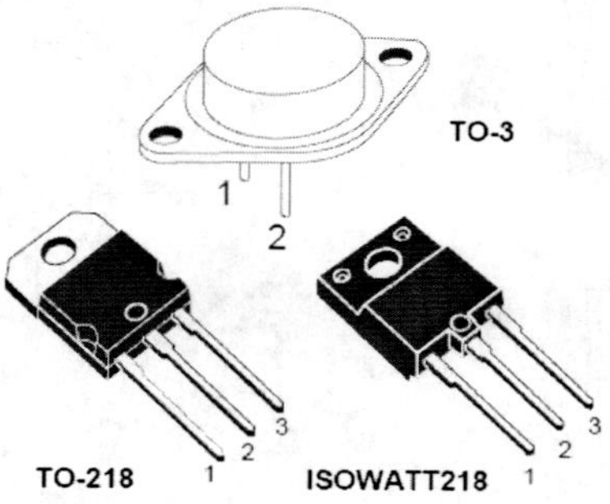

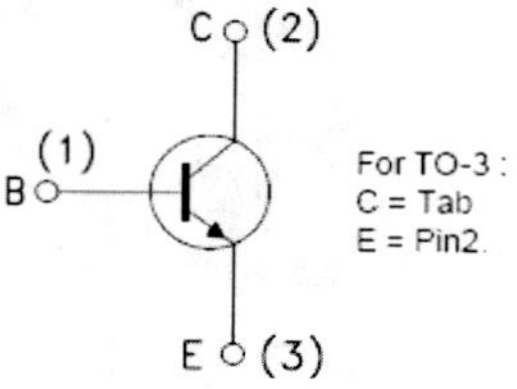

Transistor kind	N-P-N, Bipolar	
Collector-emitter voltage (I_B = 0), V	V_{CEO}	700
Collector-emitter voltage (U_{BE} = 0)	V_{CES}	1500
Emitter-base voltage (I_C = 0), V	V_{EBO}	10
Collector current: continuous/peak, A	I_C	8/15
Total dissipation (T_C = 25ºC), W	P_{tot}	150 (TO-3), 125 (TO-218), 50 (ISOWATT)
Collector-emitter saturation voltage, V	V_{CE}	1
Dielectric strength (leads to case), V RMS	V_{ISOL}	2500 (ISOWATT)
Collector cut-off current (V_{BE} = 0), μA	I_{CES}	100
Max. operating junction temperature, ºC	T_j	150

Transistor type	**BU2508A**
Manufacturer	**Phillips Semiconductors**

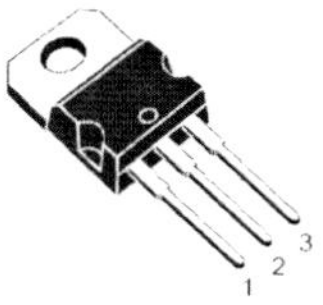

TO-220

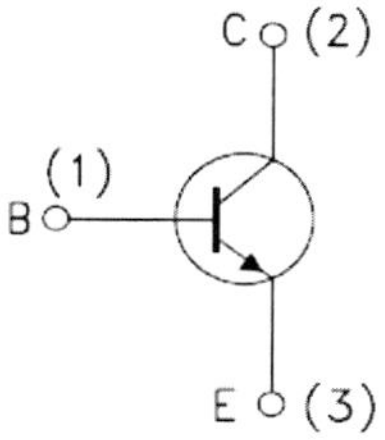

Transistor kind	N-P-N, Bipolar	
Collector-base voltage (I_E = 0), V	V_{CBO}	-
Collector-emitter voltage (I_B = 0), V	V_{CEO}	700
Collector-emitter voltage (U_{BE} = 0)	V_{CES}	1500
Collector current: continuous/peak, A	I_C	8/15
Base current, A	I_B	4
Total dissipation (T_C = 25°C), W	P_{tot}	125
Collector-emitter saturation voltage, V	$V_{CE (SAT)}$*	5
DC current gain	h_{FE}*	4 – 26
Collector cut-off current (V_{BE} = 0), μA	I_{CES}	1000
Max. operating junction temperature, °C	T_j	150
Thermal resistance: junction-case, °C/W	$R_{TH\ j\text{-}case}$	-

Transistor type	**BU2520AF**
Manufacturer	**Phillips Semiconductors**

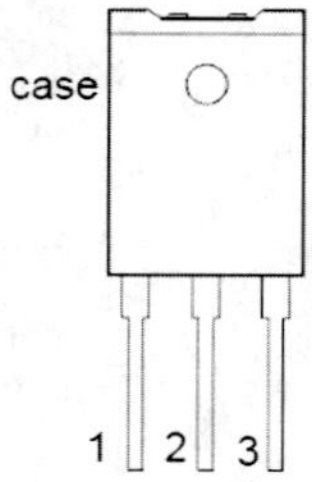

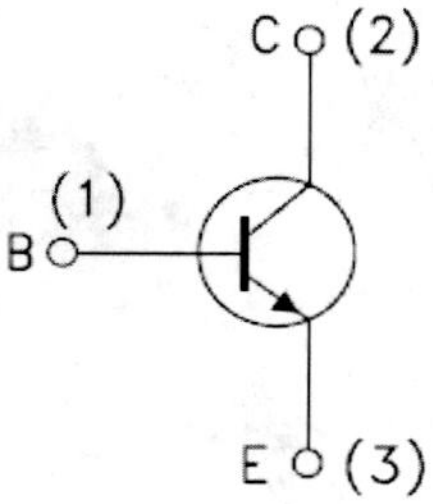

Transistor kind	N-P-N, Bipolar	
Collector-emitter voltage ($I_B = 0$), V	V_{CEO}	800
Collector-emitter voltage ($U_{BE} = 0$)	V_{CES}	1500
Collector current: continuous/peak, A	I_C	10/25
Base current, A	I_B	6
Total dissipation ($T_C = 25$ºC), W	P_{tot}	45
Collector-emitter saturation voltage, V	$V_{CE\,(SAT)}$*	5
DC current gain	h_{FE}*	5 – 9.5
Dielectric strength (leads to case), V RMS	V_{ISOL}	2500
Collector cut-off current ($V_{BE} = 0$), μA	I_{CES}	1000
Max. operating junction temperature, ºC	T_j	150
Thermal resistance: junction-case, ºC/W	$R_{TH\ j\text{-}case}$	2.8 – 3.7

Transistor type	**BU2522AF**	
Manufacturer	**Phillips Semiconductors**	
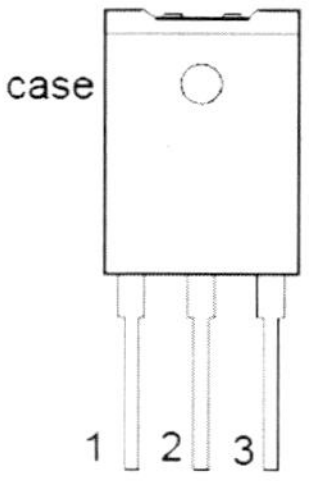	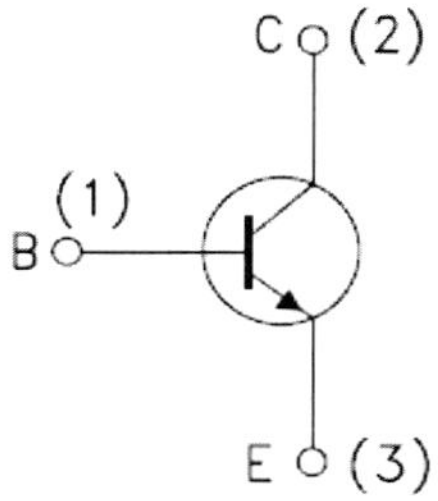	
Transistor kind	N-P-N, Bipolar	
Collector-emitter voltage (I_B = 0), V	V_{CEO}	800
Collector-emitter voltage (U_{BE} = 0)	V_{CES}	1500
Collector current: continuous/peak, A	I_C	10/25
Base current, A	I_B	6
Total dissipation (T_C = 25ºC), W	P_{tot}	45
Collector-emitter saturation voltage, V	$V_{CE\,(SAT)}$*	5
DC current gain	h_{FE}*	5 – 8
Dielectric strength (leads to case), V RMS	V_{ISOL}	2500
Collector cut-off current (V_{BE} = 0), μA	I_{CES}	250
Max. operating junction temperature, ºC	T_j	150
Thermal resistance: junction-case, ºC/W	$R_{TH\ j\text{-}case}$	2.8 – 3.7

Transistor type	**BU2527A**	
Manufacturer	**Phillips Semiconductors**	
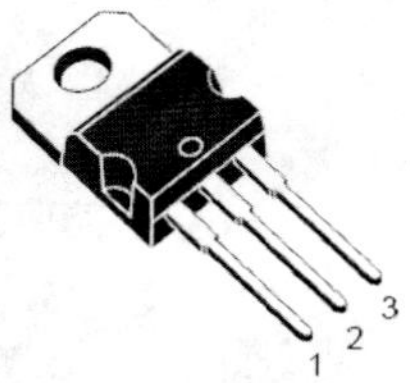 TO-220	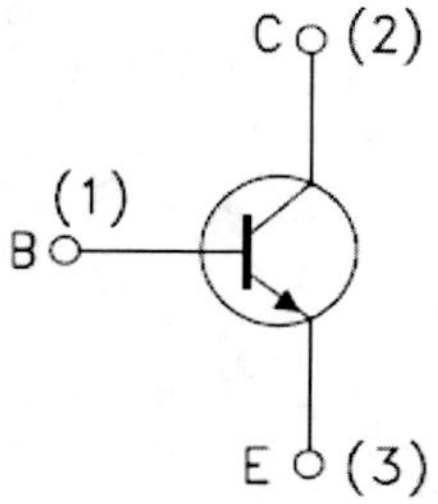	
Transistor kind	N-P-N, Bipolar	
Collector-emitter voltage (I_B = 0), V	V_{CEO}	800
Collector-emitter voltage (U_{BE} = 0)	V_{CES}	1500
Collector current: continuous/peak, A	I_C	12/30
Base current, A	I_B	8
Total dissipation (T_C = 25ºC), W	P_{tot}	125
Collector-emitter saturation voltage, V	$V_{CE\ (SAT)}$*	5
DC current gain	h_{FE}*	5 – 9
Dielectric strength (leads to case), V RMS	V_{ISOL}	-
Collector cut-off current (V_{BE} = 0), μA	I_{CES}	250
Max. operating junction temperature, ºC	T_j	150
Thermal resistance: junction-case, ºC/W	$R_{TH\ j\text{-}case}$	-

Transistor type	**BU2727A**
Manufacturer	**Phillips Semiconductors**

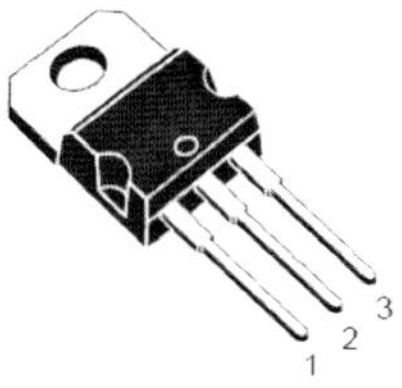

TO-220

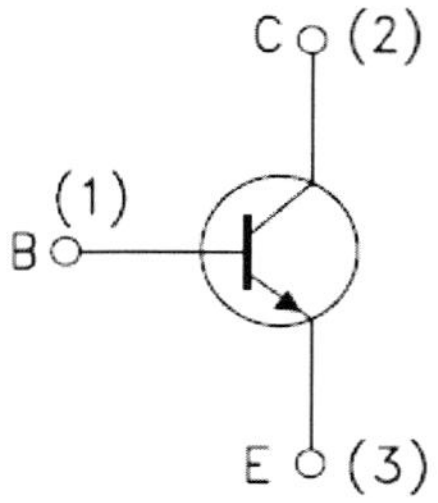

Transistor kind	N-P-N, Bipolar	
Collector-base voltage (I_E = 0), V	V_{CBO}	-
Collector-emitter voltage (I_B = 0), V	V_{CEO}	825
Collector-emitter voltage (U_{BE} = 0)	V_{CES}	1700
Collector current: continuous/peak, A	I_C	12/30
Base current, A	I_B	12
Total dissipation (T_C = 25°C), W	P_{tot}	125
Collector-emitter saturation voltage, V	$V_{CE\,(SAT)}$*	1
DC current gain	h_{FE}*	5 – 11
Dielectric strength (leads to case), V RMS	V_{ISOL}	-
Collector cut-off current (V_{BE} = 0), μA	I_{CES}	1000
Max. operating junction temperature, °C	T_j	150
Thermal resistance: junction-case, °C/W	$R_{TH\ j\text{-}case}$	-

Transistor type	BUF410A
Manufacturer	**STMicroelectronics**

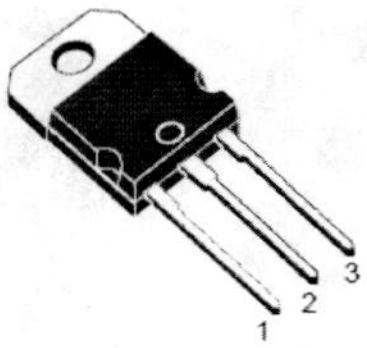

TO-218

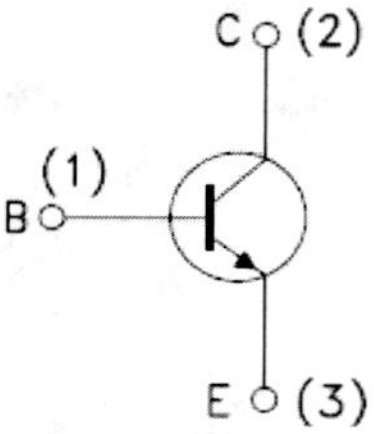

Transistor kind	N-P-N, Bipolar	
Collector-base voltage (I_E = 0), V	V_{CBO}	-
Collector-emitter voltage (I_B = 0), V	V_{CEO}	450
Collector-emitter voltage (I_B = - 1.5 V)	V_{CEV}	1000
Emitter-base voltage (I_C = 0), V	V_{EBO}	7
Collector current: continuous/peak, A	I_C	15/30
Base current, A	I_B	3
Total dissipation (T_C = 25ºC), W	P_{tot}	125
Collector-emitter saturation voltage, V	$V_{CE\,(SAT)}$*	0.8
DC current gain	h_{FE}*	-
Dielectric strength (leads to case), V RMS	V_{ISOL}	-
Collector cut-off current (V_{BE} = 0), μA	I_{CES}	200
Max. operating junction temperature, ºC	T_j	150
Thermal resistance: junction-case, ºC/W	$R_{TH\ j\text{-}case}$	1

Transistor type	BUF420AW	
Manufacturer	**STMicroelectronics**	
 TO-247	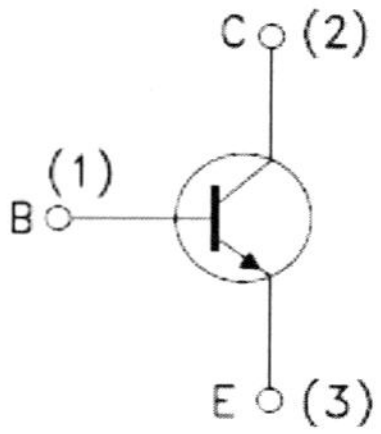	
Transistor kind	N-P-N, Bipolar	
Collector-base voltage (I_E = 0), V	V_{CBO}	-
Collector-emitter voltage (I_B = 0), V	V_{CEO}	450
Collector-emitter voltage (U_{BE} = - 1.5 V)	V_{CEV}	1000
Emitter-base voltage (I_C = 0), V	V_{EBO}	7
Collector current: continuous/peak, A	I_C	30/60
Base current, A	I_B	6
Total dissipation (T_C = 25°C), W	P_{tot}	200
Collector-emitter saturation voltage, V	$V_{CE\,(SAT)}$*	0.8
DC current gain	h_{FE}*	-
Dielectric strength (leads to case), V RMS	V_{ISOL}	-
Collector cut-off current (V_{BE} = 0), μA	I_{CES}	200
Max. operating junction temperature, °C	T_j	150
Thermal resistance: junction-case, °C/W	$R_{TH\,j\text{-}case}$	0.63

Transistor type	**BUH315DFH**
Manufacturer	**STMicroelectronics**

TO-220FH

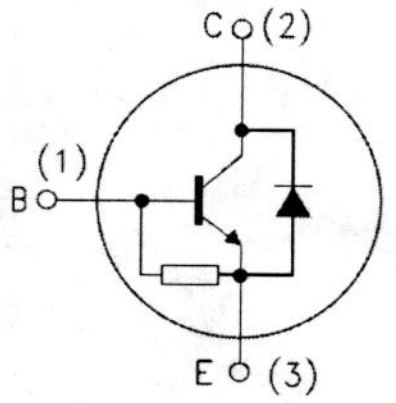

Transistor kind	N-P-N, Bipolar	
Collector-base voltage ($I_E = 0$), V	V_{CBO}	1500
Collector-emitter voltage ($I_B = 0$), V	V_{CEO}	700
Collector-emitter voltage ($U_{BE} = 0$)	V_{CES}	1200
Emitter-base voltage ($I_C = 0$), V	V_{EBO}	10
Collector current: continuous/peak, A	I_C	6/12
Base current, A	I_B	3
Total dissipation ($T_C = 25°C$), W	P_{tot}	40
Collector-emitter saturation voltage, V	$V_{CE\,(SAT)}$*	1.5
DC current gain	h_{FE}*	4 – 9
Dielectric strength (leads to case), V RMS	V_{ISOL}	2500
Collector cut-off current ($V_{BE} = 0$), μA	I_{CES}	200
Max. operating junction temperature, °C	T_j	150
Thermal resistance: junction-case, °C/W	$R_{TH\ j\text{-}case}$	3.12

Transistor type	BUH415	
Manufacturer	STMicroelectronics	
 ISOWATT218	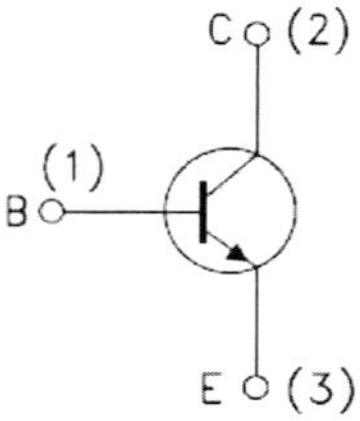	
Transistor kind	N-P-N, Bipolar	
Collector-base voltage ($I_E = 0$), V	V_{CBO}	1500
Collector-emitter voltage ($I_B = 0$), V	V_{CEO}	700
Collector-emitter voltage ($U_{BE} = 0$)	V_{CES}	-
Emitter-base voltage ($I_C = 0$), V	V_{EBO}	10
Collector current: continuous/peak, A	I_C	8/12
Base current, A	I_B	5
Total dissipation ($T_C = 25$°C), W	P_{tot}	50
Collector-emitter saturation voltage, V	$V_{CE\,(SAT)}$*	1.5
DC current gain	h_{FE}*	6 – 12
Dielectric strength (leads to case), V RMS	V_{ISOL}	2500
Collector cut-off current ($V_{BE} = 0$), μA	I_{CES}	200
Max. operating junction temperature, °C	T_j	150
Thermal resistance: junction-case, °C/W	$R_{TH\ j\text{-}case}$	2.5

Transistor type	**BUH417**	
Manufacturer	**STMicroelectronics**	
 ISOWATT218	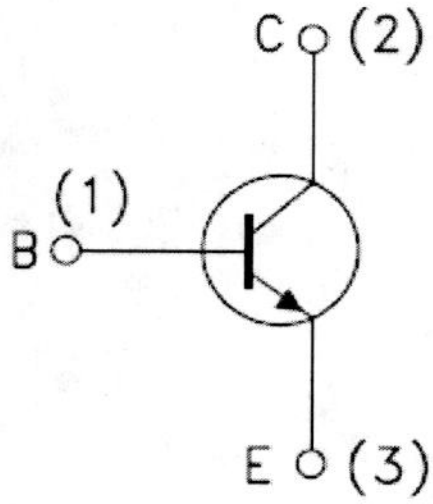	
Transistor kind	N-P-N, Bipolar	
Collector-base voltage ($I_E = 0$), V	V_{CBO}	1700
Collector-emitter voltage ($I_B = 0$), V	V_{CEO}	700
Collector-emitter voltage ($U_{BE} = 0$)	V_{CES}	-
Emitter-base voltage ($I_C = 0$), V	V_{EBO}	10
Collector current: continuous/peak, A	I_C	7/12
Base current, A	I_B	4
Total dissipation ($T_C = 25$°C), W	P_{tot}	55
Collector-emitter saturation voltage, V	$V_{CE\,(SAT)}$*	1.5
DC current gain	h_{FE}*	>6
Dielectric strength (leads to case), V RMS	V_{ISOL}	-
Collector cut-off current ($V_{BE} = 0$), μA	I_{CES}	100
Max. operating junction temperature, °C	T_j	150
Thermal resistance: junction-case, °C/W	$R_{TH\ j\text{-}case}$	2.27

Transistor type	**BUH715**	
Manufacturer	**STMicroelectronics**	
 ISOWATT218	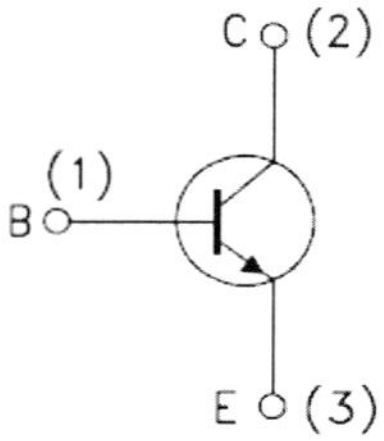	
Transistor kind	N-P-N, Bipolar	
Collector-base voltage (I_E = 0), V	V_{CBO}	1500
Collector-emitter voltage (I_B = 0), V	V_{CEO}	700
Collector-emitter voltage (U_{BE} = 0)	V_{CES}	-
Emitter-base voltage (I_C = 0), V	V_{EBO}	10
Collector current: continuous/peak, A	I_C	10/20
Base current, A	I_B	5
Total dissipation (T_C = 25°C), W	P_{tot}	57
Collector-emitter saturation voltage, V	$V_{CE\,(SAT)}$*	1.5
DC current gain	h_{FE}*	8 – 16
Dielectric strength (leads to case), V r.m.s.	V_{ISOL}	-
Collector cut-off current (V_{BE} = 0), μA	I_{CES}	1000
Max. operating junction temperature, °C	T_j	150
Thermal resistance: junction-case, °C/W	$R_{TH\ j\text{-}case}$	2.2

Transistor type	**BUH1215**	
Manufacturer	**STMicroelectronics**	

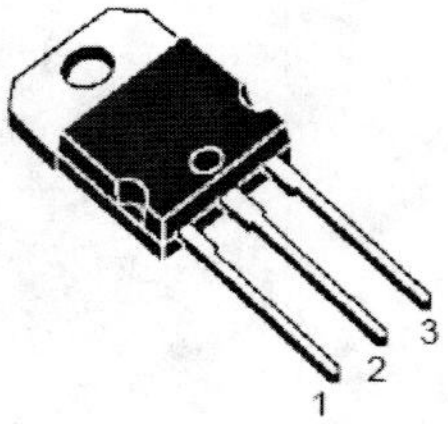

TO-218

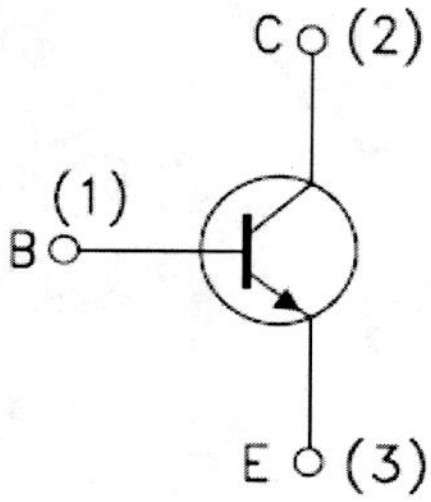

Transistor kind	N-P-N, Bipolar	
Collector-base voltage ($I_E = 0$), V	V_{CBO}	1500
Collector-emitter voltage ($I_B = 0$), V	V_{CEO}	700
Collector-emitter voltage ($U_{BE} = 0$)	V_{CES}	-
Emitter-base voltage ($I_C = 0$), V	V_{EBO}	10
Collector current: continuous/peak, A	I_C	16/22
Base current, A	I_B	9
Total dissipation ($T_C = 25$ºC), W	P_{tot}	200
Collector-emitter saturation voltage, V	$V_{CE\ (SAT)}$*	1.5
DC current gain	h_{FE}*	5 – 14
Dielectric strength (leads to case), V RMS	V_{ISOL}	-
Collector cut-off current ($V_{BE} = 0$), µA	I_{CES}	100
Max. operating junction temperature, ºC	T_j	150
Thermal resistance: junction-case, ºC/W	$R_{TH\ j\text{-}case}$	0.63

Transistor type	**BUL213**	
Manufacturer	**STMicroelectronics**	
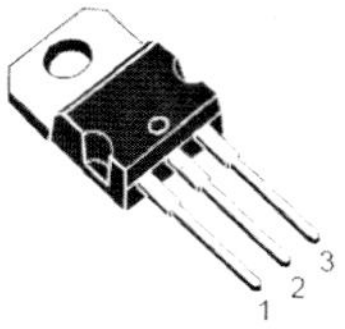 **TO-220**	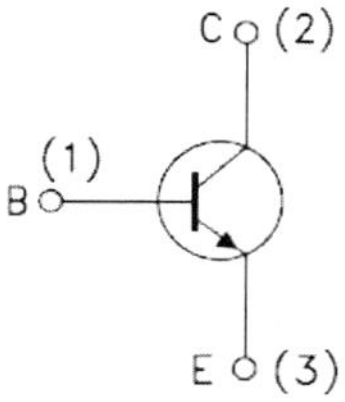	
Transistor kind	N-P-N, Bipolar	
Collector-base voltage (I_E = 0), V	V_{CBO}	-
Collector-emitter voltage (I_B = 0), V	V_{CEO}	600
Collector-emitter voltage (U_{BE} = 0)	V_{CES}	1300
Emitter-base voltage (I_C = 0), V	V_{EBO}	9
Collector current: continuous/peak, A	I_C	3/6
Base current, A	I_B	2
Total dissipation (T_C = 25°C), W	P_{tot}	60
Collector-emitter saturation voltage, V	$V_{CE\,(SAT)}$*	0.9
DC current gain	h_{FE}*	16 – 36
Dielectric strength, V, r.m.s	V_{ISOL}	-
Collector cut-off current (V_{BE} = 0), μA	I_{CES}	100
Max. operating temperature, °C	T_j	150
Therm. resistance: junction-case, °C/W	$R_{TH\ j\text{-}case}$	2.08

Transistor type	**BUL216**	
Manufacturer	**STMicroelectronics**	

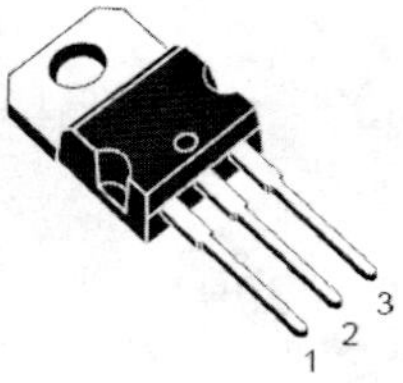

TO-220

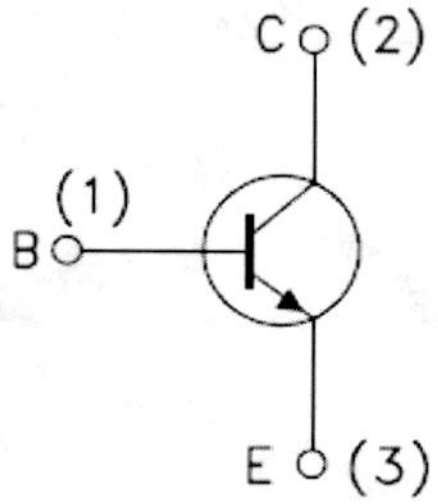

Transistor kind	N-P-N, Bipolar	
Collector-base voltage (I_E = 0), V	V_{CBO}	-
Collector-emitter voltage (I_B = 0), V	V_{CEO}	800
Collector-emitter voltage (U_{BE} = 0)	V_{CES}	1600
Emitter-base voltage (I_C = 0), V	V_{EBO}	9
Collector current: continuous/peak, A	I_C	4/6
Base current, A	I_B	2
Total dissipation (T_C = 25ºC), W	P_{tot}	90
Collector-emitter saturation voltage, V	$V_{CE\ (SAT)}$*	1.2
DC current gain	h_{FE}*	10 – 40
Dielectric strength (leads to case), V RMS	V_{ISOL}	-
Collector cut-off current (V_{BE} = 0), μA	I_{CES}	100
Max. operating junction temperature, ºC	T_j	150
Thermal resistance: junction-case, ºC/W	$R_{TH\ j\text{-}case}$	1.39

Transistor type	**BUL310**	
Manufacturer	**STMicroelectronics**	

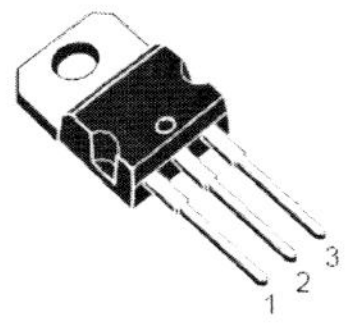

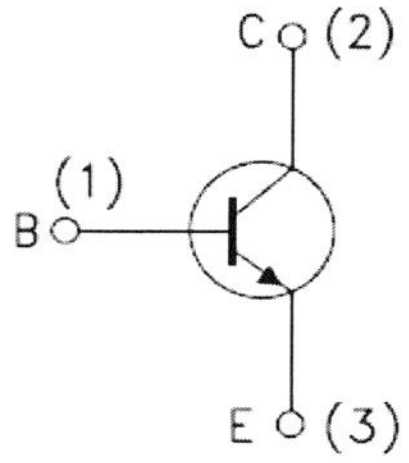

Transistor kind	N-P-N, Bipolar	
Collector-base voltage ($I_E = 0$), V	V_{CBO}	-
Collector-emitter voltage ($I_B = 0$), V	V_{CEO}	500
Collector-emitter voltage ($U_{BE} = 0$)	V_{CES}	1000
Emitter-base voltage ($I_C = 0$), V	V_{EBO}	9
Collector current: continuous/peak, A	I_C	5/10
Base current, A	I_B	3
Total dissipation ($T_C = 25$°C), W	P_{tot}	75
Collector-emitter saturation voltage, V	$V_{CE(SAT)}$*	0.9
DC current gain	h_{FE}*	6 – 14
Collector cut-off current ($V_{BE} = 0$), μA	I_{CES}	100
Max. operating junction temperature, °C	T_j	150
Thermal resistance: junction-case, °C/W	$R_{TH\ j\text{-}case}$	1.65

Transistor type	BUL312FP
Manufacturer	STMicroelectronics

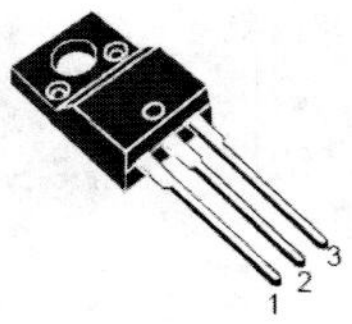

TO-220FP

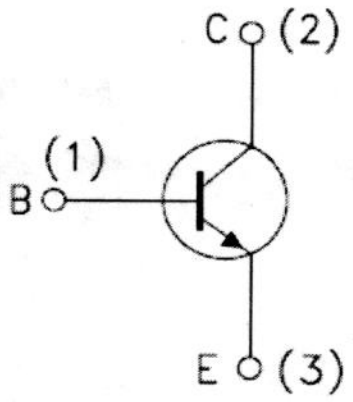

Transistor kind	N-P-N, Bipolar	
Collector-base voltage (I_E = 0), V	V_{CBO}	-
Collector-emitter voltage (I_B = 0), V	V_{CEO}	500
Collector-emitter voltage (U_{BE} = 0)	V_{CES}	1150
Emitter-base voltage (I_C = 0), V	V_{EBO}	9
Collector current: continuous/peak, A	I_C	5/10
Base current, A	I_B	3
Total dissipation (T_C = 25ºC), W	P_{tot}	36
Collector-emitter saturation voltage, V	$V_{CE\,(SAT)}$*	1.2
DC current gain	h_{FE}*	8 – 13.5
Dielectric strength (leads to case), V RMS	V_{ISOL}	1500
Collector cut-off current (V_{BE} = 0), µA	I_{CES}	1000
Max. operating junction temperature, ºC	T_j	150
Thermal resistance: junction-case, ºC/W	$R_{TH\ j\text{-}case}$	3.5

Transistor type	BUL416	
Manufacturer	STMicroelectronics	
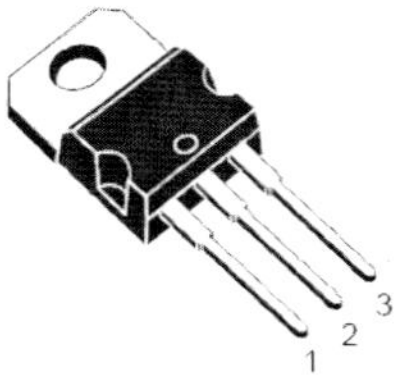 TO-220	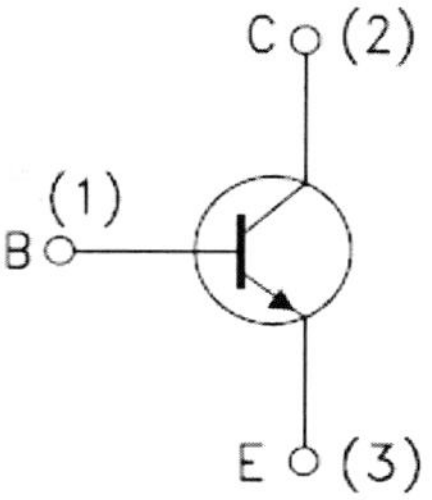	
Transistor kind	N-P-N, Bipolar	
Collector-base voltage ($I_E = 0$), V	V_{CBO}	-
Collector-emitter voltage ($I_B = 0$), V	V_{CEO}	800
Collector-emitter voltage ($U_{BE} = 0$)	V_{CES}	1600
Emitter-base voltage ($I_C = 0$), V	V_{EBO}	9
Collector current: continuous/peak, A	I_C	6/9
Base current, A	I_B	5
Total dissipation ($T_C = 25°C$), W	P_{tot}	110
Collector-emitter saturation voltage, V	$V_{CE\,(SAT)}$*	1.5 – 3.0
DC current gain	h_{FE}*	10 – 40
Dielectric strength (leads to case), V RMS	V_{ISOL}	-
Collector cut-off current ($V_{BE} = 0$), μA	I_{CES}	100
Max. operating junction temperature, °C	T_j	150
Thermal resistance: junction-case, °C/W	$R_{TH\ j\text{-}case}$	1.14

Transistor type	**BUL510**
Manufacturer	**STMicroelectronics**

TO-220

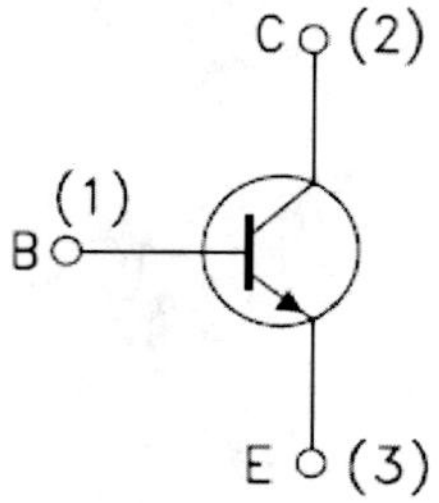

Transistor kind	N-P-N, Bipolar	
Collector-base voltage (I_E = 0), V	V_{CBO}	-
Collector-emitter voltage (I_B = 0), V	V_{CEO}	450
Collector-emitter voltage (U_{BE} = 0)	V_{CES}	1000
Emitter-base voltage (I_C = 0), V	V_{EBO}	9
Collector current: continuous/peak, A	I_C	10/18
Base current, A	I_B	3.5
Total dissipation (T_C = 25°C), W	P_{tot}	100
Collector-emitter saturation voltage, V	$V_{CE\,(SAT)}$*	1.5
DC current gain	h_{FE}*	15 – 45
Dielectric strength (leads to case), V RMS	V_{ISOL}	-
Collector cut-off current (V_{BE} = 0), μA	I_{CES}	100
Max. operating junction temperature, °C	T_j	150
Thermal resistance: junction-case, °C/W	$R_{TH\,j\text{-}case}$	1.25

Transistor type	BUL742C		
Manufacturer	STMicroelectronics		
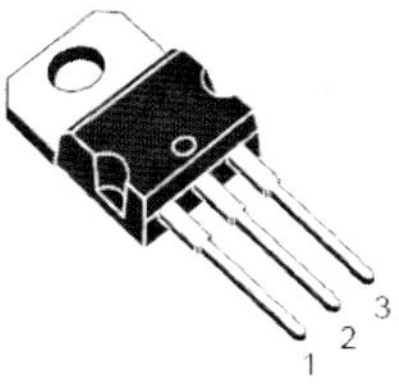 TO-220	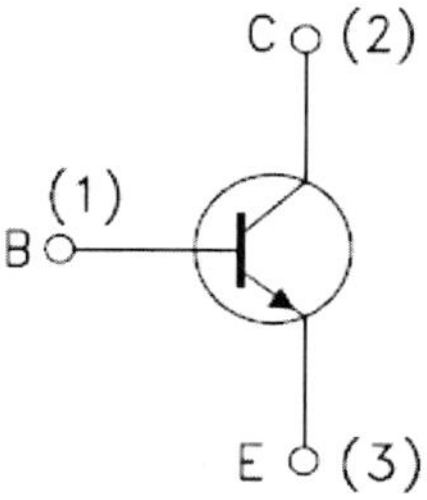		
Transistor kind	N-P-N, Bipolar		
Collector-base voltage (I_E = 0), V	V_{CBO}	-	
Collector-emitter voltage (I_B = 0), V	V_{CEO}	400	
Collector-emitter voltage (U_{BE} = 0)	V_{CES}	1050	
Emitter-base voltage (I_C = 0), V	V_{EBO}	-	
Collector current: continuous/peak, A	I_C	4/8	
Base current, A	I_B	2	
Total dissipation (T_C = 25°C), W	P_{tot}	70	
Collector-emitter saturation voltage, V	$V_{CE\,(SAT)}$*	1.5	
DC current gain	h_{FE}*	25 – 50	
Dielectric strength (leads to case), V RMS	V_{ISOL}	-	
Collector cut-off current (V_{BE} = 0), μA	I_{CES}	100	
Max. operating junction temperature, °C	T_j	150	
Thermal resistance: junction-case, °C/W	$R_{TH\ j\text{-}case}$	1.79	

Transistor type	**BUL810**	
Manufacturer	**STMicroelectronics**	
 TO-218	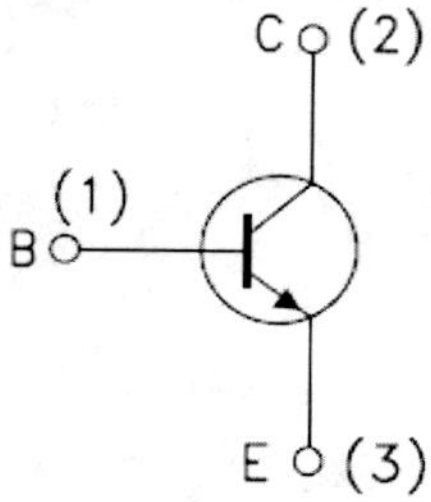	
Transistor kind	N-P-N, Bipolar	
Collector-base voltage ($I_E = 0$), V	V_{CBO}	-
Collector-emitter voltage ($I_B = 0$), V	V_{CEO}	450
Collector-emitter voltage ($U_{BE} = 0$)	V_{CES}	1000
Emitter-base voltage ($I_C = 0$), V	V_{EBO}	9
Collector current: continuous/peak, A	I_C	15/22
Base current, A	I_B	5
Total dissipation ($T_C = 25°C$), W	P_{tot}	125
Collector-emitter saturation voltage, V	$V_{CE\,(SAT)}$*	1.5
DC current gain	h_{FE}*	10 – 40
Dielectric strength (leads to case), V RMS	V_{ISOL}	-
Collector cut-off current ($V_{BE} = 0$), μA	I_{CES}	100
Max. operating junction temperature, °C	T_j	150
Thermal resistance: junction-case, °C/W	$R_{TH\ j\text{-}case}$	1

Transistor type	**BUL1102E**
Manufacturer	**STMicroelectronics**

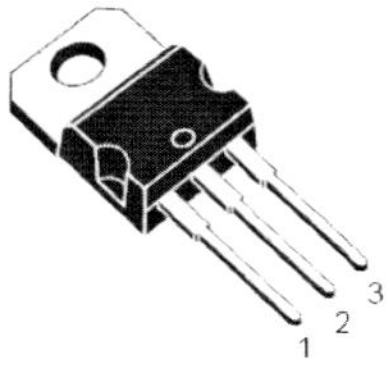

TO-220

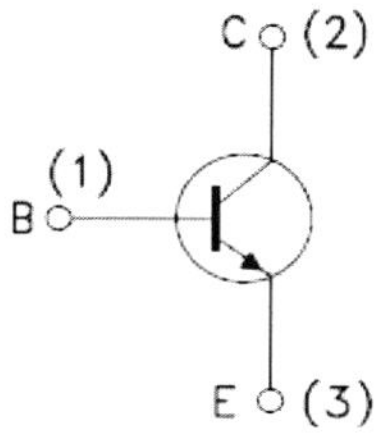

Transistor kind	N-P-N, Bipolar	
Collector-base voltage ($I_E = 0$), V	V_{CBO}	-
Collector-emitter voltage ($I_B = 0$), V	V_{CEO}	450
Collector-emitter voltage ($U_{BE} = 0$)	V_{CES}	1100
Emitter-base voltage ($I_C = 0$), V	V_{EBO}	12
Collector current: continuous/peak, A	I_C	4/8
Base current, A	I_B	2
Total dissipation ($T_C = 25°C$), W	P_{tot}	70
Collector-emitter saturation voltage, V	$V_{CE\,(SAT)}$*	0.8
DC current gain	h_{FE}*	12 – 20
Dielectric strength (leads to case), V RMS	V_{ISOL}	-
Collector cut-off current ($V_{BE} = 0$), μA	I_{CES}	100
Max. operating junction temperature, °C	T_j	150
Thermal resistance: junction-case, °C/W	$R_{TH\ j\text{-}case}$	1.78

Transistor type	**BUL1603ED**	
Manufacturer	**STMicroelectronics**	
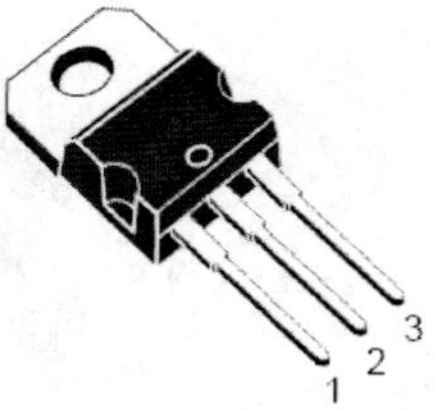 TO-220	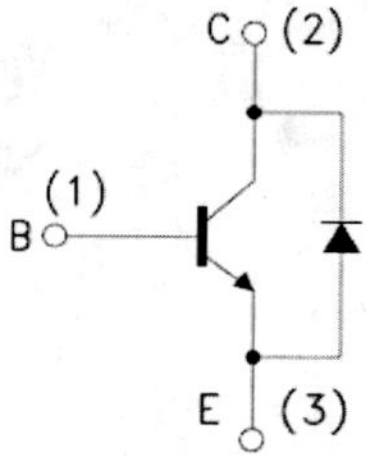	
Transistor kind	N-P-N, Bipolar	
Collector-base voltage ($I_E = 0$), V	V_{CBO}	-
Collector-emitter voltage ($I_B = 0$), V	V_{CEO}	650
Collector-emitter voltage ($U_{BE} = 0$)	V_{CES}	1600
Emitter-base voltage ($I_C = 0$), V	V_{EBO}	11
Collector current: continuous/peak, A	I_C	3/6
Base current, A	I_B	2
Total dissipation ($T_C = 25$°C), W	P_{tot}	80
Collector-emitter saturation voltage, V	$V_{CE\ (SAT)}$*	1.5
DC current gain	h_{FE}*	4 – 40
Dielectric strength (leads to case), V RMS	V_{ISOL}	-
Collector cut-off current ($V_{BE} = 0$), μA	I_{CES}	100
Max. operating junction temperature, °C	T_j	150
Thermal resistance: junction-case, °C/W	$R_{TH\ j\text{-}case}$	1.56

Transistor type	**BUL7216**	
Manufacturer	**STMicroelectronics**	
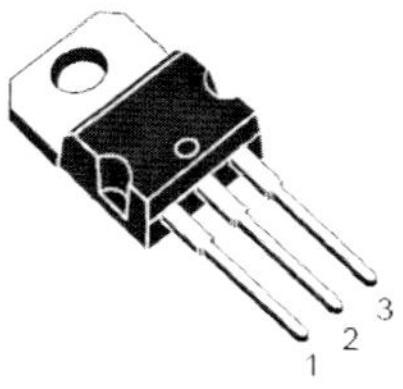 TO-220	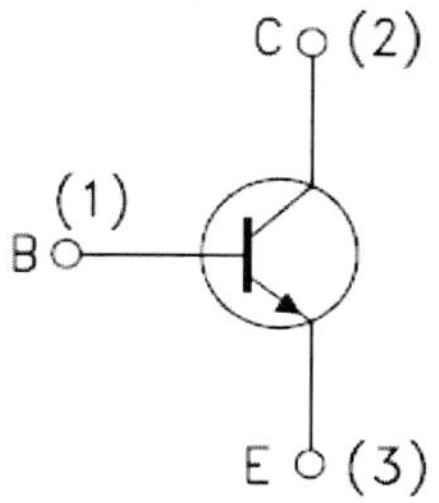	
Transistor kind	N-P-N, Bipolar	
Collector-base voltage ($I_E = 0$), V	V_{CBO}	-
Collector-emitter voltage ($I_B = 0$), V	V_{CEO}	700
Collector-emitter voltage ($U_{BE} = 0$)	V_{CES}	1600
Emitter-base voltage ($I_C = 0$), V	V_{EBO}	12
Collector current: continuous/peak, A	I_C	3/6
Base current, A	I_B	1
Total dissipation ($T_C = 25$°C), W	P_{tot}	100
Collector-emitter saturation voltage, V	$V_{CE\,(SAT)}$*	1.5
DC current gain	h_{FE}*	15 – 35
Dielectric strength (leads to case), V RMS	V_{ISOL}	-
Collector cut-off current ($V_{BE} = 0$), μA	I_{CES}	100
Max. operating junction temperature, °C	T_j	150
Thermal resistance: junction-case, °C/W	$R_{TH\ j\text{-}case}$	1.56

Transistor type	**BUX48C, BUV48C, BUV48CF1**
Manufacturer	**STMicroelectronics**

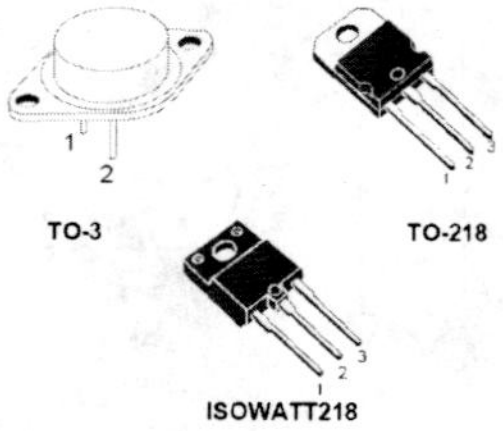

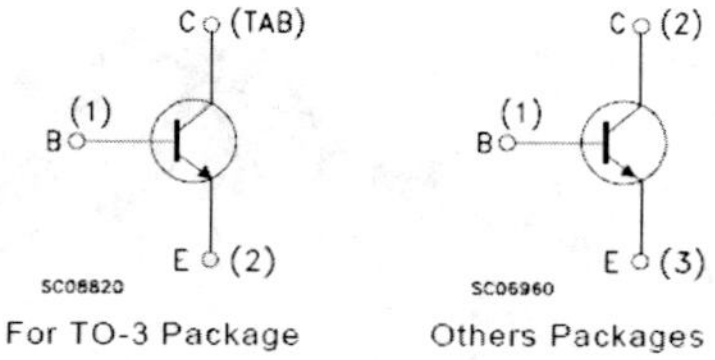

Transistor kind	N-P-N, Bipolar	
Collector-emitter voltage, V	V_{CEO}	700
Collector-emitter voltage, V	V_{CES}	1200
Emitter-base voltage ($I_C = 0$), V	V_{EBO}	7
Collector current: continuous/peak, A	I_C	15/30
Base current, A	I_B	4
Total dissipation (T_C = 25ºC), W	P_{tot}	175 (TO-3) 125 (TO-218) 55 (ISOWATT218)
Collector-emitter saturation voltage, V	$V_{CE\,(SAT)}$*	1.5 – 3
DC current gain	h_{FE}*	10 – 40
Collector cut-off current, μA	I_{CES}	500
Max. operating temperature, ºC	T_j	200 (TO-3) 150 (TO-218) 150 (ISOWATT218)

Transistor type	BUX98C	
Manufacturer	STMicroelectronics	
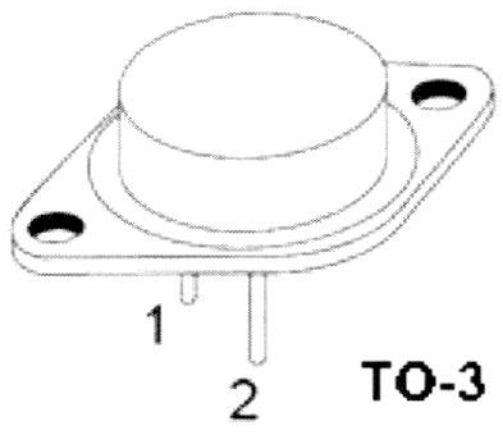	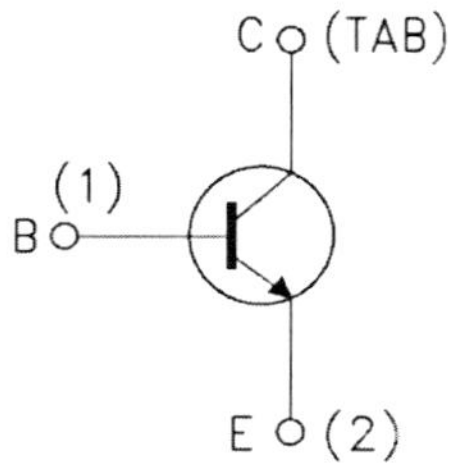	
Transistor kind	N-P-N, Bipolar	
Collector-base voltage ($I_E = 0$), V	V_{CBO}	-
Collector-emitter voltage ($I_B = 0$), V	V_{CEO}	700
Collector-emitter voltage ($U_{BE} = 0$)	V_{CES}	1200
Emitter-base voltage ($I_C = 0$), V	V_{EBO}	7
Collector current: continuous/peak, A	I_C	30/60
Base current, A	I_B	8
Total dissipation ($T_C = 25°C$), W	P_{tot}	250
Collector-emitter saturation voltage, V	$V_{CE\,(SAT)}$*	1.5 – 3
DC current gain	h_{FE}*	-
Dielectric strength (leads to case), V RMS	V_{ISOL}	-
Collector cut-off current ($V_{BE} = 0$), μA	I_{CES}	2000
Max. operating junction temperature, °C	T_j	200
Thermal resistance: junction-case, °C/W	$R_{TH\ j\text{-}case}$	0.7

Transistor type	**HD1750FX**
Manufacturer	**STMicroelectronics**

ISOWATT218FX

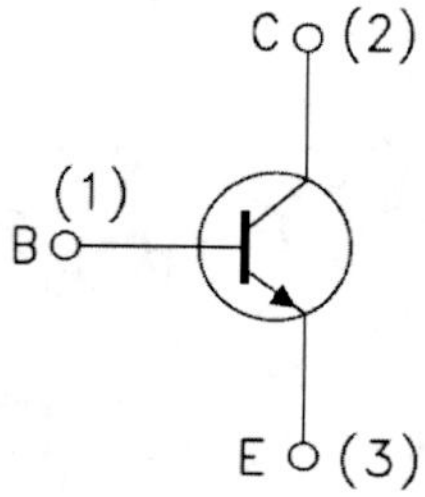

Transistor kind	N-P-N, Bipolar	
Collector-base voltage ($I_E = 0$), V	V_{CBO}	-
Collector-emitter voltage ($I_B = 0$), V	V_{CEO}	800
Collector-emitter voltage ($U_{BE} = 0$)	V_{CES}	1700
Emitter-base voltage ($I_C = 0$), V	V_{EBO}	10
Collector current: continuous/peak, A	I_C	24/36
Base current, A	I_B	12
Total dissipation ($T_C = 25°C$), W	P_{tot}	75
Collector-emitter saturation voltage, V	$V_{CE\,(SAT)}$*	3.0
DC current gain	h_{FE}*	6.5 – 9.5
Dielectric strength (leads to case), V RMS	V_{ISOL}	2500
Collector cut-off current ($V_{BE} = 0$), μA	I_{CES}	200
Max. operating junction temperature, °C	T_j	150
Thermal resistance: junction-case, °C/W	$R_{TH\ j\text{-}case}$	1.67

Transistor type	**HD1750FX (HD1750JL)**
Manufacturer	**STMicroelectronics**

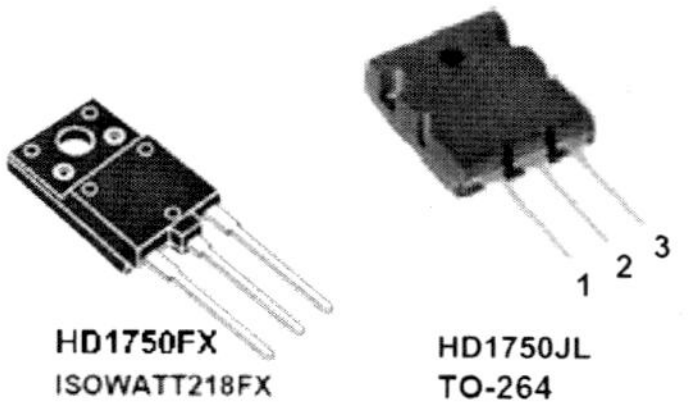

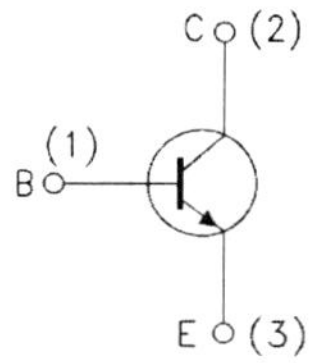

Transistor kind	-	N-P-N, Bipolar
Collector-base voltage (I_E = 0), V	V_{CBO}	-
Collector-emitter voltage (I_B = 0), V	V_{CEO}	800
Collector-emitter voltage (U_{BE} = 0)	V_{CES}	1700
Emitter-base voltage (I_C = 0), V	V_{EBO}	10
Collector current: continuous/peak, A	I_C	24/36
Base current, A	I_B	12
Total dissipation (T_C = 25°C), W	P_{tot}	75 (200)
Collector-emitter saturation voltage, V	$V_{CE(SAT)}$*	3.0
DC current gain	h_{FE}*	6.5 – 9.5
Dielectric strength (leads to case), V RMS	V_{ISOL}	2500
Collector cut-off current (V_{BE} = 0), μA	I_{CES}	200
Max. operating junction temperature, °C	T_j	150
Thermal resistance: junction-case, °C/W	$R_{TH\ j\text{-}case}$	1.67 (0.625)

Transistor type	**KT8121A2**	
Manufacturer	**OKB "Iskra" (Ulianovsk, Russia)**	
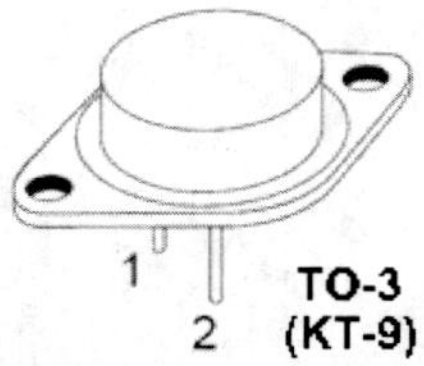	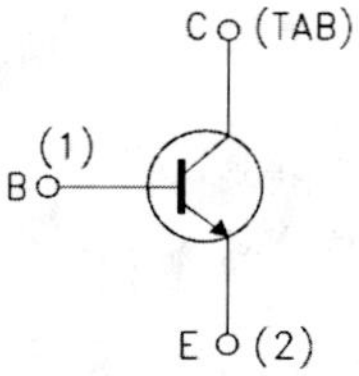	
Transistor kind	N-P-N, Bipolar	
Collector-base voltage ($I_E = 0$), V	V_{CBO}	1500
Collector-emitter voltage ($I_B = 0$), V	V_{CEO}	700
Collector-emitter voltage ($U_{BE} = 0$)	V_{CES}	-
Emitter-base voltage ($I_C = 0$), V	V_{EBO}	-
Collector current: continuous/peak, A	I_C	8/10
Base current, A	I_B	4
Total dissipation ($T_C = 25$°C), W	P_{tot}	100
Collector-emitter saturation voltage, V	$V_{CE\,(SAT)}$*	1
DC current gain	h_{FE}*	>1.7
Dielectric strength (leads to case), V RMS	V_{ISOL}	-
Collector cut-off current ($V_{BE} = 0$), μA	I_{CES}	-
Max. operating junction temperature, °C	T_j	150
Thermal resistance: junction-case, °C/W	$R_{TH\ j\text{-}case}$	-

Transistor type	**KT8157A**	
Manufacturer	**OKB "Iskra" (Ulianovsk, Russia)**	
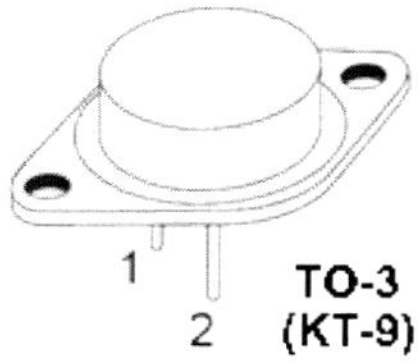 **TO-3 (KT-9)**	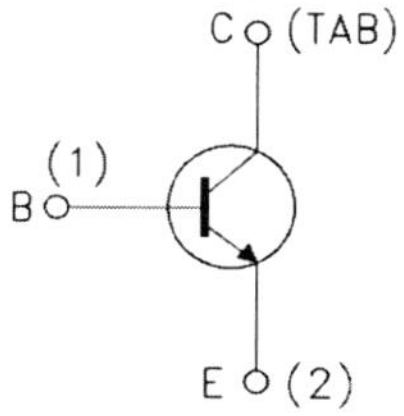	
Transistor kind	N-P-N, Bipolar	
Collector-base voltage ($I_E = 0$), V	V_{CBO}	1500
Collector-emitter voltage ($I_B = 0$), V	V_{CEO}	800
Collector-emitter voltage ($U_{BE} = 0$)	V_{CES}	-
Emitter-base voltage ($I_C = 0$), V	V_{EBO}	6
Collector current: continuous/peak, A	I_C	15/25
Base current, A	I_B	-
Total dissipation ($T_C = 25°C$), W	P_{tot}	150
Collector-emitter saturation voltage, V	$V_{CE(SAT)}$*	1.5
DC current gain	h_{FE}*	>8
Dielectric strength (leads to case), V RMS	V_{ISOL}	-
Collector cut-off current ($V_{BE} = 0$), μA	I_{CES}	-
Max. operating junction temperature, °C	T_j	150
Thermal resistance: junction-case, °C/W	$R_{TH\ j\text{-}case}$	-

Transistor type	**KT8192A**	
Manufacturer	**OKB "Iskra" (Ulianovsk, Russia)**	

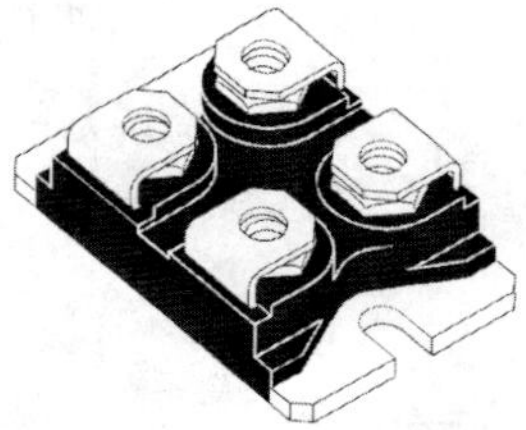

ISOTOP

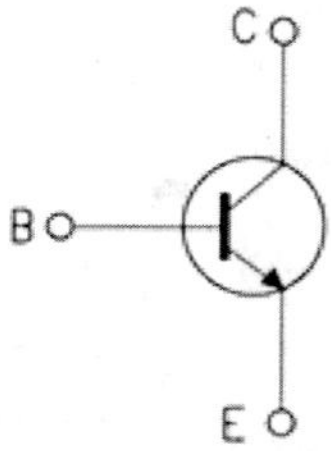

Transistor kind	N-P-N, Bipolar	
Collector-base voltage ($I_E = 0$), V	V_{CBO}	1500
Collector-emitter voltage ($I_B = 0$), V	V_{CEO}	700
Collector-emitter voltage ($U_{BE} = 0$)	V_{CES}	-
Emitter-base voltage ($I_C = 0$), V	V_{EBO}	7
Collector current: continuous/peak, A	I_C	30/50
Base current, A	I_B	15
Total dissipation ($T_C = 25$°C), W	P_{tot}	200
Collector-emitter saturation voltage, V	$V_{CE\,(SAT)}$*	1.5
DC current gain	h_{FE}*	> 5
Dielectric strength (leads to case), V RMS	V_{ISOL}	-
Collector cut-off current ($V_{BE} = 0$), μA	I_{CES}	5000
Max. operating junction temperature, °C	T_j	175
Thermal resistance: junction-case, °C/W	$R_{TH\ j\text{-}case}$	-

Transistor type	NTE2590	
Manufacturer	NTE	
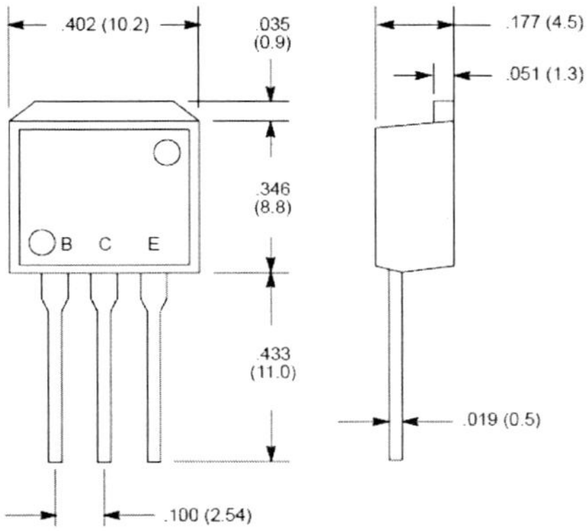	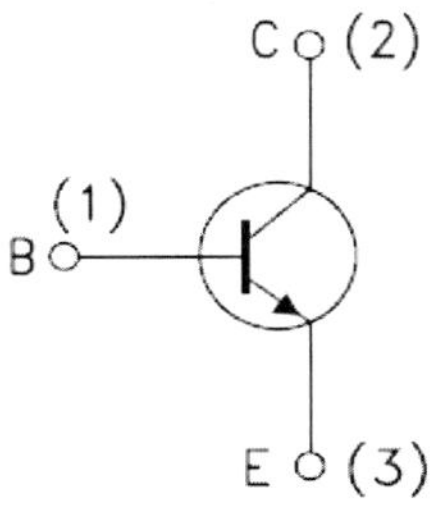	
Transistor kind	N-P-N, Bipolar	
Collector-base voltage ($I_E = 0$), V	V_{CBO}	1700
Collector-emitter voltage ($I_B = 0$), V	V_{CEO}	900
Emitter-base voltage ($I_C = 0$), V	V_{EBO}	5
Collector current: continuous/peak, A	I_C	0.05/0.15
Base current, A	I_B	-
Total dissipation (T_C = 25°C), W	P_{tot}	1.2
Collector-emitter saturation voltage, V	$V_{CE\,(SAT)}$*	5
DC current gain	h_{FE}*	20 – 120
Collector cut-off current ($V_{BE} = 0$), μA	I_{CES}	1
Max. operating junction temperature, °C	T_j	150
Thermal resistance: junction-case, °C/W	$R_{TH\ j\text{-}case}$	-

Transistor type	**ST8812FX**	
Manufacturer	**STMicroelectronics**	
 ISOWATT218FX	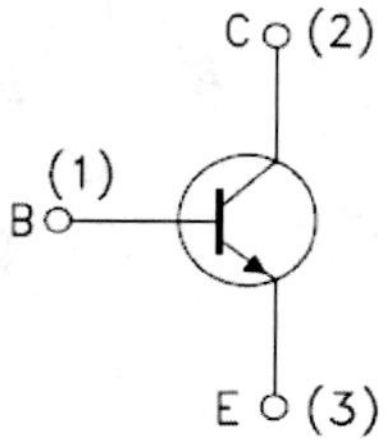	
Transistor kind	N-P-N, Bipolar	
Collector-base voltage (I_E = 0), V	V_{CBO}	1150
Collector-emitter voltage (I_B = 0), V	V_{CEO}	600
Collector-emitter voltage (U_{BE} = 0)	V_{CES}	-
Emitter-base voltage (I_C = 0), V	V_{EBO}	15
Collector current: continuous/peak, A	I_C	7/12
Base current, A	I_B	4
Total dissipation (T_C = 25°C), W	P_{tot}	50
Collector-emitter saturation voltage, V	$V_{CE\,(SAT)}$*	1.5 – 3
DC current gain	h_{FE}*	4.5 – 9
Dielectric strength (leads to case), V RMS	V_{ISOL}	2500
Collector cut-off current (V_{BE} = 0), μA	I_{CES}	1000
Max. operating junction temperature, °C	T_j	150
Thermal resistance: junction-case, °C/W	$R_{TH\ j\text{-}case}$	2.5

Transistor type	STI100 (STI1000)	
Manufacturer	Semiconductor Technology, Inc.	
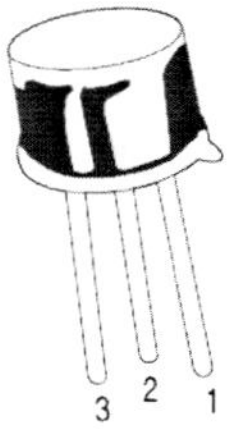 **TO-205AD** **TO39** PIN CONFIGURATION 1. EMITTER 2. BASE 3. COLLECTOR	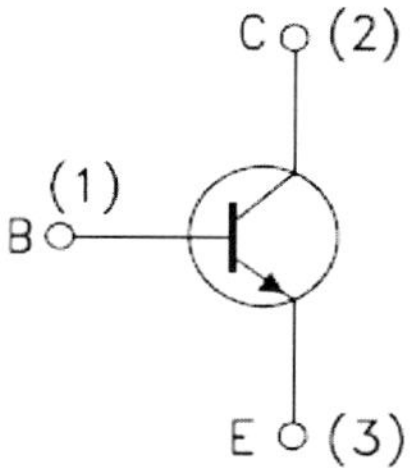	
Transistor kind	N-P-N, Bipolar	
Collector-base voltage (I_E = 0), V	V_{CBO}	1000
Collector-emitter voltage (I_B = 0), V	V_{CEO}	-
Collector-emitter voltage (U_{BE} = 0)	V_{CES}	1000
Emitter-base voltage (I_C = 0), V	V_{EBO}	-
Collector current: continuous, A	I_C	0.02
Base current, A	I_B	-
Total dissipation (T_C = 25°C), W	P_{tot}	1
Collector-emitter saturation voltage, V	$V_{CE(SAT)}$*	1.5
DC current gain	h_{FE}*	>30
Dielectric strength (leads to case), V RMS	V_{ISOL}	-
Collector cut-off current (V_{BE} = 0), μA	I_{CES}	-
Max. operating junction temperature, °C	T_j	-
Thermal resistance: junction-case, °C/W	$R_{TH\ j\text{-}case}$	-

Transistor type	**TIPL755A**	
Manufacturer	**Semiconductor Technology, Inc.**	
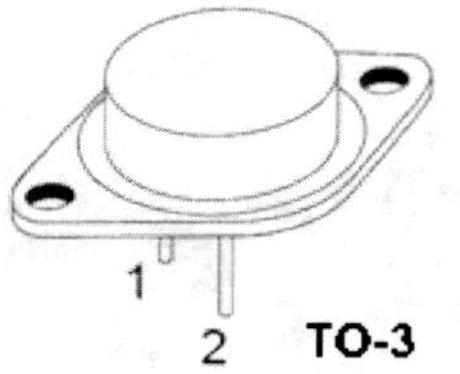	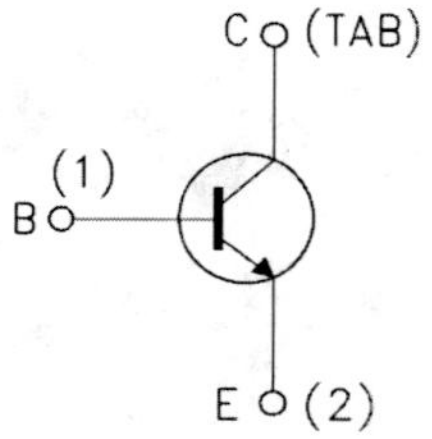	
Transistor kind	N-P-N, Bipolar	
Collector-base voltage (I_E = 0), V	V_{CBO}	-
Collector-emitter voltage (I_B = 0), V	V_{CEO}	420
Collector-emitter voltage (U_{BE} = 0)	V_{CES}	1000
Emitter-base voltage (I_C = 0), V	V_{EBO}	-
Collector current: continuous/peak, A	I_C	10
Base current, A	I_B	-
Total dissipation (T_C = 25ºC), W	P_{tot}	180
Collector-emitter saturation voltage, V	$V_{CE\,(SAT)}$*	2.5
DC current gain	h_{FE}*	>15
Dielectric strength (leads to case), V RMS	V_{ISOL}	-
Collector cut-off current (V_{BE} = 0), μA	I_{CES}	-
Max. operating junction temperature, ºC	T_j	150
Thermal resistance: junction-case, ºC/W	$R_{TH\ j\text{-}case}$	-

Transistor type	**TIPL761A**	
Manufacturer	**Power Innovative Ltd, Transys Electronics Ltd**	
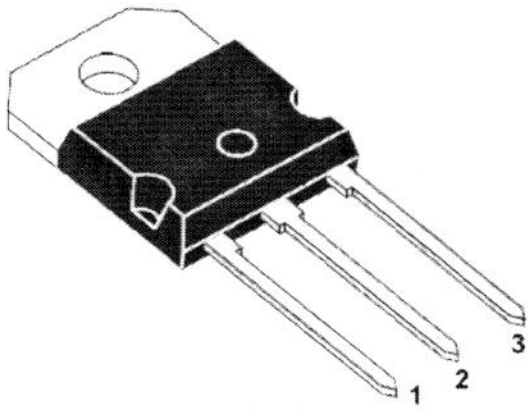	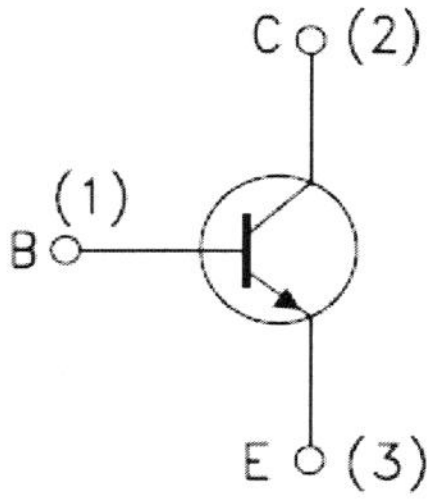	
Transistor kind	N-P-N, Bipolar	
Collector-base voltage ($I_E = 0$), V	V_{CBO}	1000
Collector-emitter voltage ($I_B = 0$), V	V_{CEO}	450
Collector-emitter voltage ($U_{BE} = 0$)	V_{CES}	1000
Emitter-base voltage ($I_C = 0$), V	V_{EBO}	10
Collector current: continuous/peak, A	I_C	4/8
Base current, A	I_B	-
Total dissipation ($T_C = 25°C$), W	P_{tot}	100
Collector-emitter saturation voltage, V	$V_{CE\,(SAT)}$*	2.5
DC current gain	h_{FE}*	20 – 60
Dielectric strength (leads to case), V RMS	V_{ISOL}	-
Collector cut-off current ($V_{BE} = 0$), μA	I_{CES}	1000
Max. operating junction temperature, °C	T_j	150
Thermal resistance: junction-case, °C/W	$R_{TH\ j\text{-}case}$	-

Transistor type	**TIPL763A**	
Manufacturer	**Semiconductor Technology, Inc.**	

TO-218

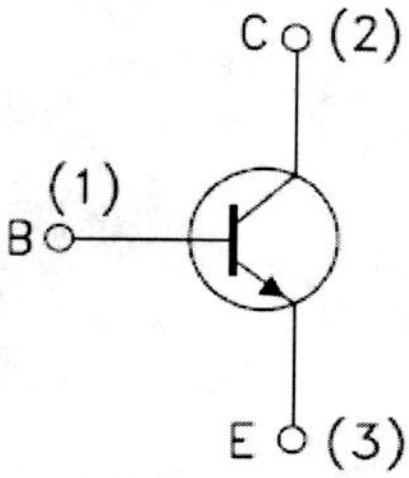

Transistor kind	N-P-N, Bipolar	
Collector-base voltage ($I_E = 0$), V	V_{CBO}	-
Collector-emitter voltage ($I_B = 0$), V	V_{CEO}	400
Collector-emitter voltage ($U_{BE} = 0$)	V_{CES}	1000
Emitter-base voltage ($I_C = 0$), V	V_{EBO}	-
Collector current: continuous/peak, A	I_C	8
Base current, A	I_B	-
Total dissipation ($T_C = 25$°C), W	P_{tot}	120
Collector-emitter saturation voltage, V	$V_{CE\,(SAT)}$*	2.5
DC current gain	h_{FE}*	>15
Dielectric strength (leads to case), V RMS	V_{ISOL}	-
Collector cut-off current ($V_{BE} = 0$), μA	I_{CES}	-
Max. operating junction temperature, °C	T_j	150
Thermal resistance: junction-case, °C/W	$R_{TH\ j\text{-}case}$	-

Transistor type	**TIPL765A**	
Manufacturer	**Power Innovative Ltd, Transys Electronics Ltd**	
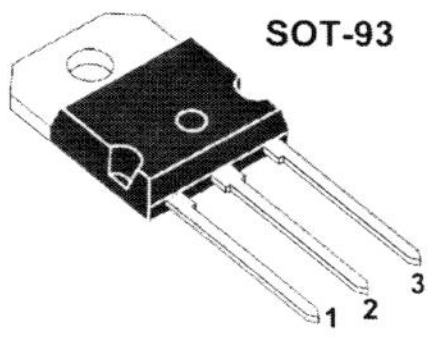	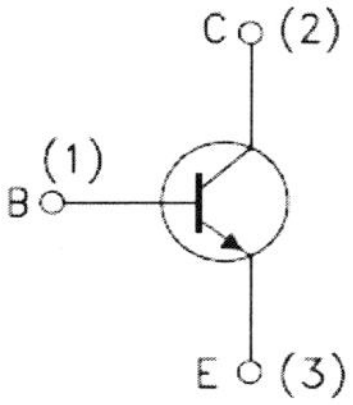	
Transistor kind	N-P-N, Bipolar	
Collector-base voltage ($I_E = 0$), V	V_{CBO}	1000
Collector-emitter voltage ($I_B = 0$), V	V_{CEO}	450
Collector-emitter voltage ($U_{BE} = 0$)	V_{CES}	1000
Emitter-base voltage ($I_C = 0$), V	V_{EBO}	10
Collector current: continuous/peak, A	I_C	10/15
Base current, A	I_B	-
Total dissipation ($T_C = 25°C$), W	P_{tot}	125
Collector-emitter saturation voltage, V	$V_{CE\,(SAT)}$*	2.5
DC current gain	h_{FE}*	15 – 60
Dielectric strength (leads to case), V RMS	V_{ISOL}	-
Collector cut-off current ($V_{BE} = 0$), μA	I_{CES}	200
Max. operating junction temperature, °C	T_j	150
Thermal resistance: junction-case, °C/W	$R_{TH\ j\text{-}case}$	-

Appendix B2: High-Voltage Darlington Transistors

Transistor type	BU808DFI	
Manufacturer	STMicroelectronics	
ISOWATT218	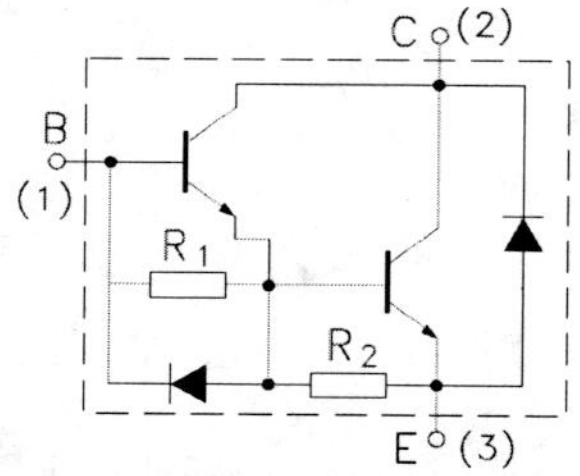	
Transistor kind	HV Darlington	
Collector-base voltage ($I_E = 0$), V	V_{CBO}	1400
Collector-emitter voltage ($I_B = 0$), V	V_{CEO}	700
Emitter-base voltage ($I_C = 0$), V	V_{EBO}	5
Collector current: continuous/peak, A	I_C	8/10
Base current, A	I_B	3
Total dissipation ($T_C = 25$°C), W	P_{tot}	52
Collector-emitter saturation voltage, V	$V_{CE\,(SAT)}$*	1.6
DC current gain	h_{FE}*	60 – 230
Dielectric strength , V RMS	V_{ISOL}	2500
Collector cut-off current, μA	I_{CES}	400
Max. operating temperature, °C	T_j	150
Thermal resistance: junction-case, °C/W	$R_{TH\ j\text{-}case}$	2.4

Transistor type	**ESM5045DV**	
Manufacturer	**STMicroelectronics**	

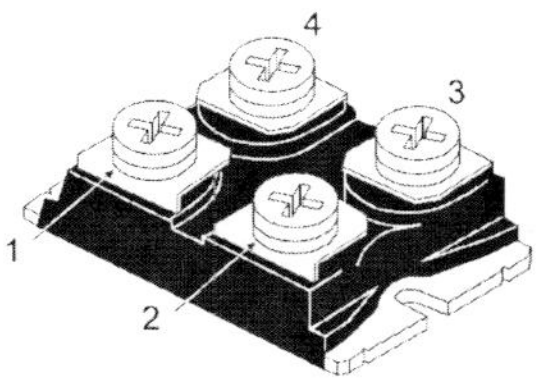

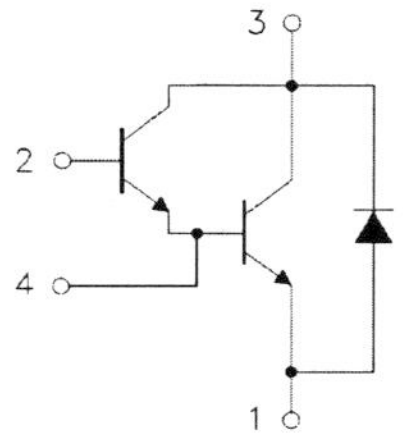

Transistor kind	Single Darlington transistor module	
Collector-base voltage (I_E = 0), V	V_{CBO}	-
Collector-emitter voltage (I_B = 0), V	V_{CEO}	450
Collector-emitter voltage (U_{BE} = -5V)	V_{CES}	600
Emitter-base voltage (I_C = 0), V	V_{EBO}	7
Collector current: continuous/peak, A	I_C	60/90
Base current, A	I_B	6
Total dissipation (T_C = 25°C), W	P_{tot}	175
Collector-emitter saturation voltage, V	$V_{CE\,(SAT)}$*	1.2 – 1.6
DC current gain	h_{FE}*	150
Dielectric strength (leads to case), V RMS	V_{ISOL}	2500
Collector cut-off current (V_{BE} = 0), μA	I_{CES}	1500
Max. operating junction temperature, °C	T_j	150
Thermal resistance: junction-case, °C/W	$R_{TH\,j\text{-}case}$	0.71

Transistor type	**KD7212A2**
Manufacturer	**Powerex, Inc**

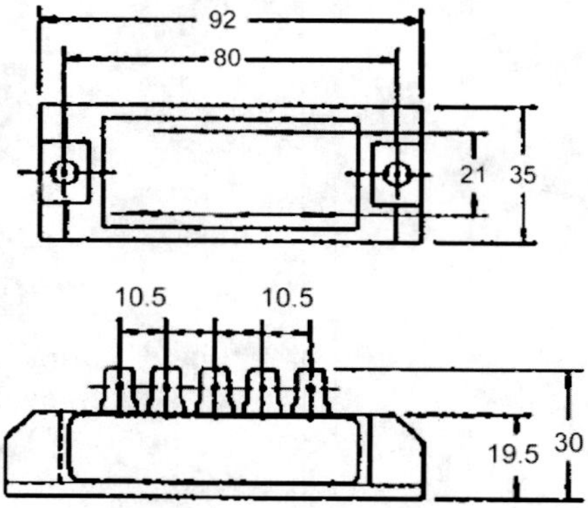

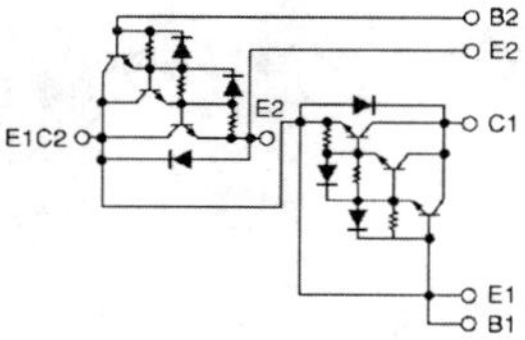

Transistor kind	Dual Darlington transistor module	
Collector-base voltage (I_E = 0), V	V_{CBO}	1200
Collector-emitter voltage (I_B = 0), V	V_{CEO}	950
Collector-emitter voltage (U_{BE} = -2V)	V_{CES}	1200
Emitter-base voltage (I_C = 0), V	V_{EBO}	7
Collector current: continuous/peak, A	I_C	25
Base current, A	I_B	1.5
Total dissipation (T_C = 25ºC), W	P_{tot}	208
Collector-emitter saturation voltage, V	$V_{CE\,(SAT)}$*	3
DC current gain	h_{FE}*	>75
Dielectric strength (leads to case), V RMS	V_{ISOL}	2500
Collector cut-off current (V_{BE} = 0), µA	I_{CES}	1000

Transistor type	**QCA50AA100**	
Manufacturer	**SanRex**	
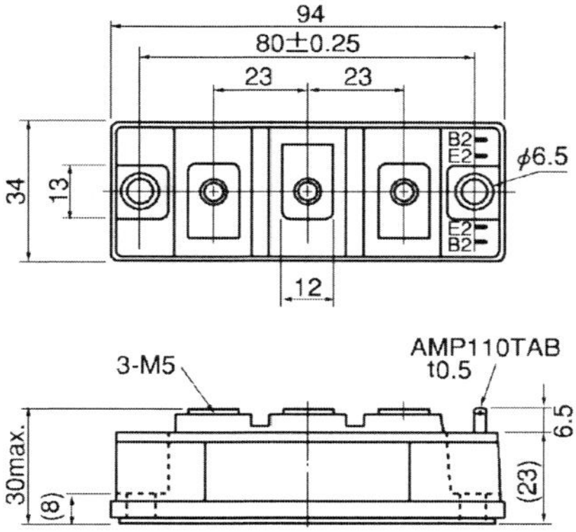	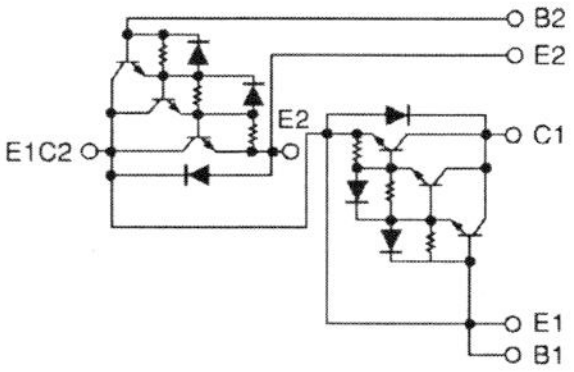	
Transistor kind	Dual Darlington transistor module	
Collector-base voltage (I_E = 0), V	V_{CBO}	1000
Collector-emitter voltage (U_{BE} = -2V)	V_{CES}	1000
Emitter-base voltage (I_C = 0), V	V_{EBO}	7
Collector current: continuous/peak, A	I_C	50
Base current, A	I_B	3
Total dissipation (T_C = 25ºC), W	P_{tot}	400
Collector-emitter saturation voltage, V	$V_{CE\ (SAT)}$*	2.5
DC current gain	h_{FE}*	>75
Dielectric strength (leads to case), V	V_{ISOL}	2500
Collector cut-off current (V_{BE} = 0), μA	I_{CES}	1000
Max. operating junction temperature, ºC	T_j	150

Transistor type	**QCA75AA120**
Manufacturer	**SanRex**

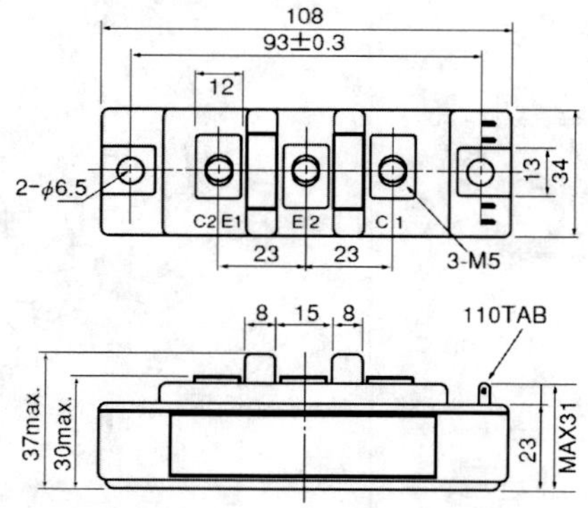

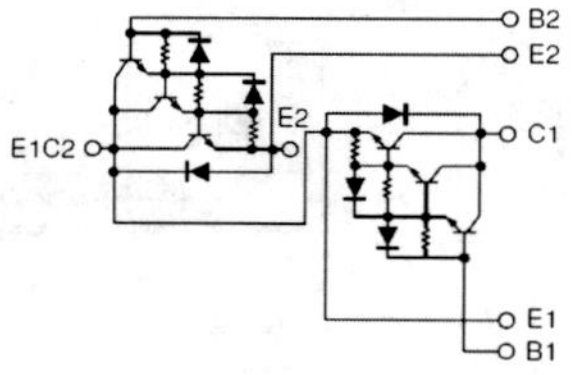

Transistor kind	Dual Darlington transistor module	
Collector-base voltage ($I_E = 0$), V	V_{CBO}	1200
Collector-emitter voltage (U_{BE} = -2V)	V_{CES}	1200
Emitter-base voltage ($I_C = 0$), V	V_{EBO}	10
Collector current: continuous, A	I_C	75
Base current, A	I_B	4
Total dissipation (T_C = 25°C), W	P_{tot}	500
Collector-emitter saturation voltage, V	$V_{CE\,(SAT)}$*	3.0
DC current gain	h_{FE}*	>75
Dielectric strength (leads to case), V RMS	V_{ISOL}	2500
Collector cut-off current ($V_{BE} = 0$), μA	I_{CES}	1000
Max. operating junction temperature, °C	T_j	-40 +150

Transistor type	QCA150AA120	
Manufacturer	SanRex	
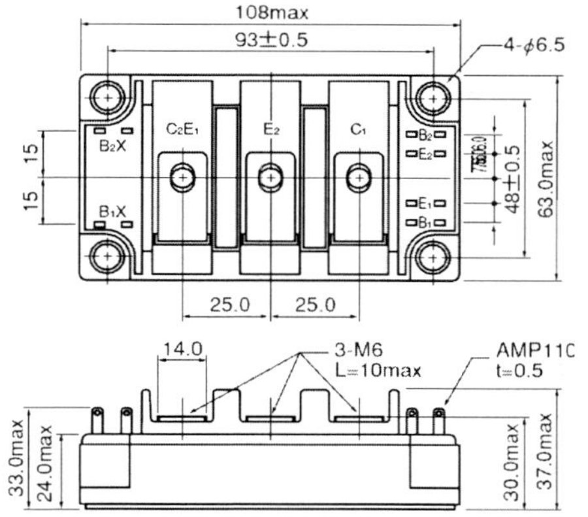	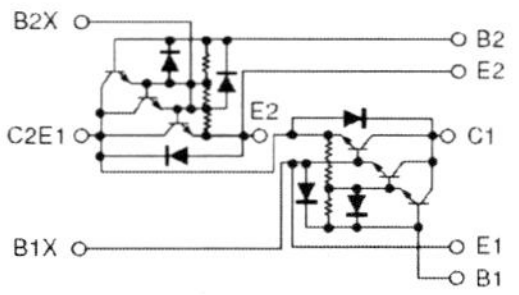	
Transistor kind	Dual Darlington transistor module	
Collector-base voltage (I_E = 0), V	V_{CBO}	1200
Collector-emitter voltage (U_{BE} = -2V)	V_{CES}	1200
Emitter-base voltage (I_C = 0), V	V_{EBO}	10
Collector current: continuous, A	I_C	150
Base current, A	I_B	8
Total dissipation (T_C = 25°C), W	P_{tot}	1000
Collector-emitter saturation voltage, V	$V_{CE\ (SAT)}$*	3.0
DC current gain	h_{FE}*	>75
Dielectric strength (leads to case), V RMS	V_{ISOL}	2500
Collector cut-off current (V_{BE} = 0), μA	I_{CES}	2000
Max. operating junction temperature, °C	T_j	-40 +125

Transistor type	SK50DAL100D	
Manufacturer	Semikron	
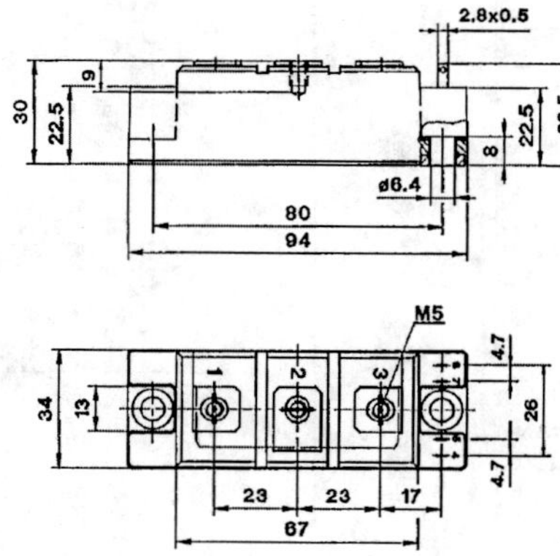	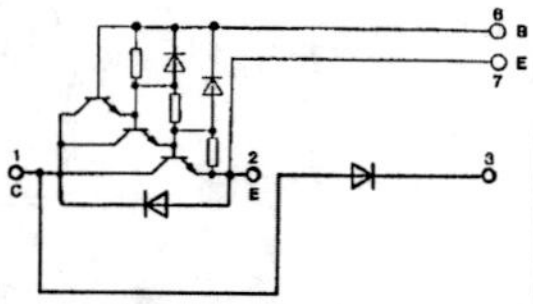	
Transistor kind	Single Darlington transistor module	
Collector-base voltage ($I_E = 0$), V	V_{CBO}	1000
Collector-emitter voltage ($I_B = 0$), V	V_{CEO}	1000
Collector-emitter voltage ($U_{BE} = -2V$)	V_{CES}	1000
Emitter-base voltage ($I_C = 0$), V	V_{EBO}	7
Collector current: continuous/peak, A	I_C	50
Base current, A	I_B	3
Total dissipation ($T_C = 25°C$), W	P_{tot}	400
Collector-emitter saturation voltage, V	$V_{CE (SAT)}$*	2.5
DC current gain	h_{FE}*	>75
Dielectric strength (leads to case), V RMS	V_{ISOL}	2500
Collector cut-off current ($V_{BE} = 0$), μA	I_{CES}	1000

Appendix B3: High-Voltage FET Transistors

Transistor type	**2SK1794**
Manufacturer	**NEC**

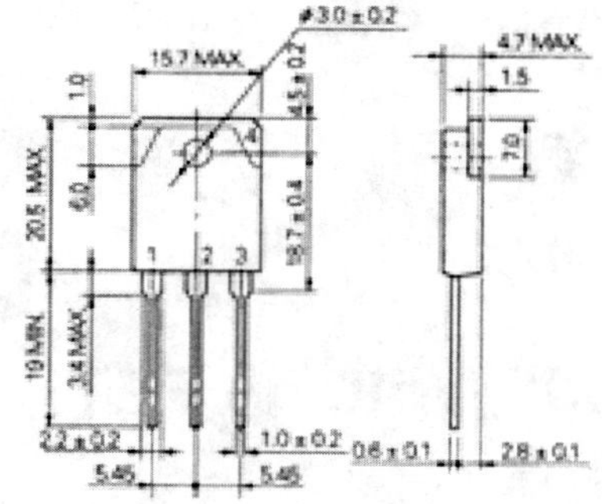

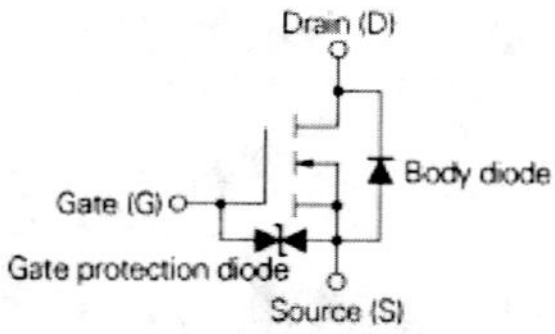

Transistor kind	Power MOSFET	
Drain-source voltage, V	V_{DS}	900
Static drain-source ON resistance, Ω	$R_{DS\ (ON)}$	2.8
Drain current continuous, A	I_{DM}	6
Drain current pulsed, A	$I_{DM\ (*)}$	12
Total power dissipation (T_C = 25°C), W	P_{tot}	100
Gate threshold voltage, V	V_{GS}	3 – 5
Turn-on delay time, ns	$t_{d(on)}$	20
Turn-off delay time, ns	$t_{d(off)}$	85
Input capacitance, pF	C_{ies}	1000
Drain-source breakdown voltage, V	$V_{(BR)\ DSS}$	-
Max. operating junction temperature, °C	T_j	150
Thermal resistance: junction-case, °C/W	R_{QJC}	-

Transistor type	**IXTP 05N100**	
Manufacturer	**IXYS**	
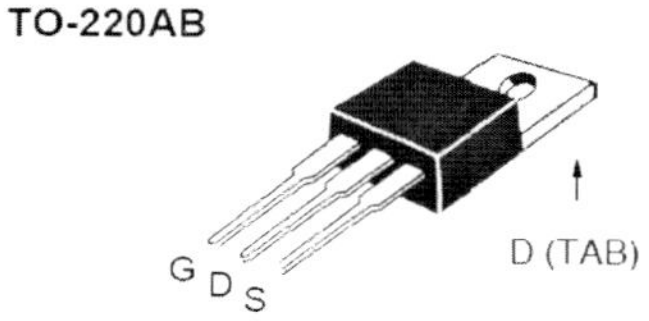	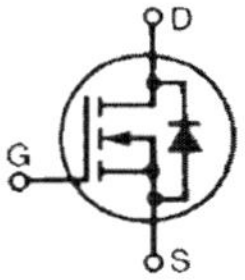	
Transistor kind	Power MOSFET	
Drain-source voltage, V	V_{DS}	1000
Drain-gate voltage, V	V_{DGR}	1000
Static drain-source ON resistance, Ω	$R_{DS\ (ON)}$	17
Drain current continuous, A (at 25°C)	I_{DM}	0.75
Drain current pulsed, A	$I_{DM\ (*)}$	3.0
Total power dissipation (T_C = 25°C), W	P_{tot}	40
Gate threshold voltage, V	V_{GS}	2.5 – 4.5
Gate-body leakage current, nA	I_{GSS}	100
Turn-on delay time, ns	$t_{d(on)}$	11
Turn-off delay time, ns	$t_{d(off)}$	40
Input capacitance, pF	C_{ies}	240
Drain-source breakdown voltage, V (Min)	$V_{(BR)\ DSS}$	-
Max. operating junction temperature, °C	T_j	150
Thermal resistance: junction-case, °C/W	R_{QJC}	-

Transistor type	**STFV4N150**	
Manufacturer	**STMicroelectronics**	
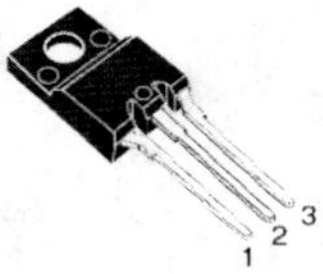 TO-220FH	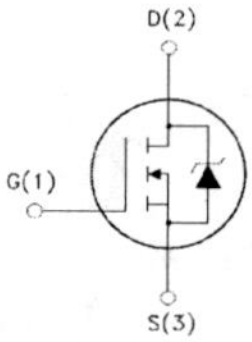	
Transistor kind	Power MOSFET	
Drain-source voltage, V	V_{DS}	1500
Drain-gate voltage, V	V_{DGR}	1500
Static drain-source ON resistance, Ω	$R_{DS\ (ON)}$	7.0
Drain current continuous, A (at 25ºC/100ºC)	I_{DM}	4/2.5
Drain current pulsed, A	$I_{DM\ (*)}$	12
Total power dissipation (T_C = 25ºC), W	P_{tot}	40
Gate threshold voltage, V	V_{GS}	3 – 5
Gate body leakage current, nA	I_{GSS}	100
Turn-on delay time, ns	$t_{d(on)}$	35
Turn-off delay time, ns	$t_{d(off)}$	45
Input capacitance, pF	C_{ies}	1300
Drain-source breakdown voltage, V	$V_{(BR)\ DSS}$	1500
Max. operating junction temperature, ºC	T_j	150
Thermal resistance: junction-case, ºC/W	R_{QJC}	3.12

Transistor type	**STP4N150, STW4N150**	
Manufacturer	**STMicroelectronics**	

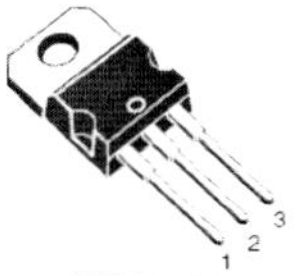

STP4N150
TO-220

STPW4N150
TO-247

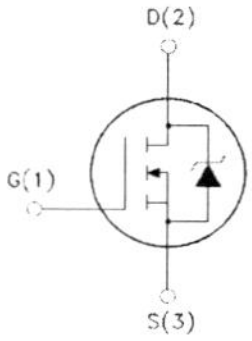

Transistor kind	Power MOSFET	
Drain-source voltage, V	V_{DS}	1500
Drain-gate voltage, V	V_{DGR}	1500
Static drain-source ON resistance, Ω	$R_{DS\,(ON)}$	7.0
Drain current continuous, A (at 25°C/100°C)	I_{DM}	4/2.5
Drain current pulsed, A	$I_{DM\,(*)}$	12
Total power dissipation (T_C = 25°C), W	P_{tot}	160
Gate threshold voltage, V	V_{GS}	3 – 5
Gate-body leakage current, nA	I_{GSS}	100
Turn-on delay time, ns	$t_{d(on)}$	35
Turn-off delay time, ns	$t_{d(off)}$	45
Input capacitance, pF	C_{ies}	1300
Drain-source breakdown voltage, V (Min)	$V_{(BR)\,DSS}$	1500
Max. operating junction temperature, °C	T_j	150
Thermal resistance: junction-case, °C/W	R_{QJC}	0.78

Transistor type	**STP5NB100, STP5NB100FP**
Manufacturer	**STMicroelectronics**

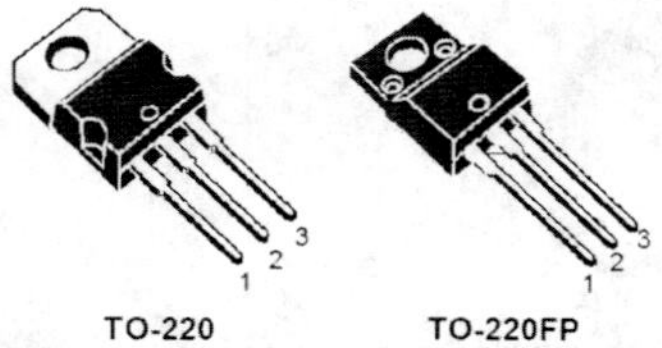

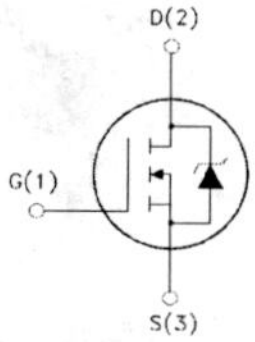

Transistor kind	Power MOSFET	
Drain-source voltage, V	V_{DS}	1000
Drain-gate voltage, V	V_{DGR}	1000
Static drain-source ON resistance, Ω	$R_{DS\ (ON)}$	2.7
Drain current continuous, A (at 25°C/100°C)	I_{DM}	5/3.1
Drain current pulsed, A	$I_{DM\ (*)}$	15
Total power dissipation (T_C = 25°C), W	P_{tot}	40
Gate threshold voltage, V	V_{GS}	3 – 5
Gate body leakage current, nA	I_{GSS}	100
Turn-on delay time, ns	$t_{d(on)}$	24
Turn-off delay time, ns	$t_{d(off)}$	-
Input capacitance, pF	C_{ies}	1500
Drain-source breakdown voltage, V	$V_{(BR)DSS}$	1000
Max. operating junction temperature, °C	T_j	150
Thermal resistance: junction-case, °C/W	R_{QJC}	0.32; 1.08

Transistor type	**STP5NK100Z, STF5NK100Z, STW5NK100Z**	
Manufacturer	**STMicroelectronics**	

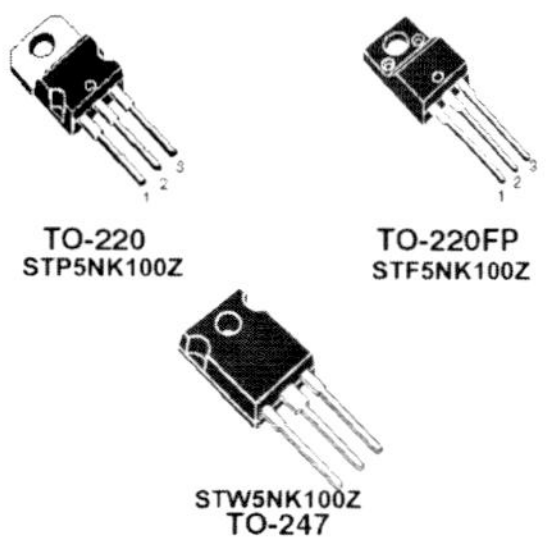

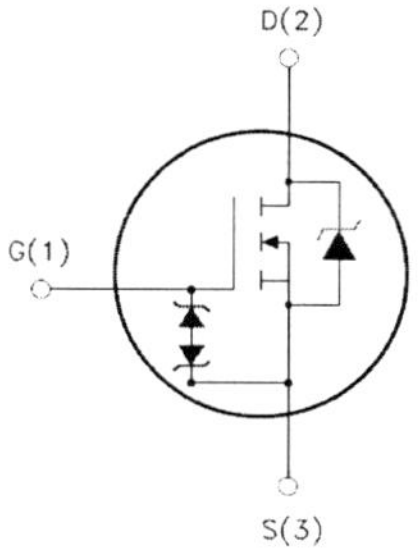

Transistor kind	Power MOSFET	
Drain-source voltage, V	V_{DS}	1000
Drain-gate voltage, V	V_{DGR}	1000
Static drain-source ON resistance, Ω	$R_{DS\,(ON)}$	3.7
Drain current continuous, A (at 25°C/100°C)	I_{DM}	3.5/2.2
Drain current pulsed, A	$I_{DM\,(*)}$	14
Total power dissipation (T_C = 25°C), W	P_{tot}	125 30 (STF)
Gate threshold voltage, V	V_{GS}	3 – 4.5
Turn-on delay time, ns	$t_{d(on)}$	22
Turn-off delay time, ns	$t_{d(off)}$	52
Input capacitance, pF	C_{ies}	1154
Drain-source breakdown voltage, V	$V_{(BR)\,DSS}$	1000
Max. operating junction temperature, °C	T_j	150

Transistor type	**STW11NK100Z**	
Manufacturer	**STMicroelectronics**	
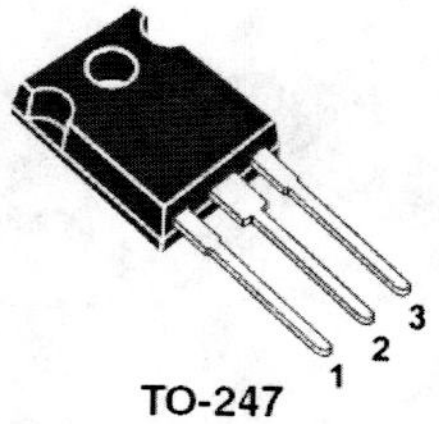 TO-247	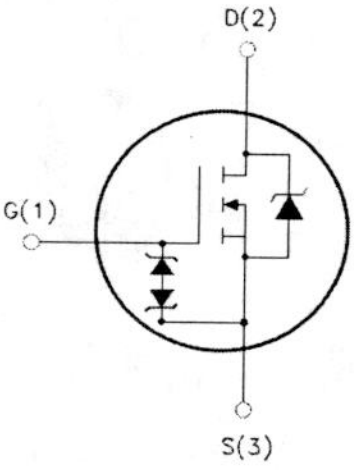	
Transistor kind	Power MOSFET	
Drain-source voltage, V	V_{DS}	1000
Drain-gate voltage, V	V_{DGR}	1000
Static drain-source ON resistance, Ω	$R_{DS\,(ON)}$	1.38
Drain current continuous, A (at 25°C/100°C)	I_{DM}	8.3/5.2
Drain current pulsed, A	$I_{DM\,(*)}$	33
Total power dissipation (T_C = 25°C), W	P_{tot}	230
Gate threshold voltage, V	V_{GS}	3 – 4.5
Gate-body leakage current, μA	I_{GSS}	10
Turn-on delay time, ns	$t_{d(on)}$	27
Turn-off delay time, ns	$t_{d(off)}$	98
Input capacitance, pF	C_{ies}	3500
Drain-source breakdown voltage, V (Min)	$V_{(BR)\,DSS}$	1000
Max. operating junction temperature, °C	T_j	150
Thermal resistance: junction-case, °C/W	R_{QJC}	0.54

Transistor type	STW12NK95Z	
Manufacturer	STMicroelectronics	
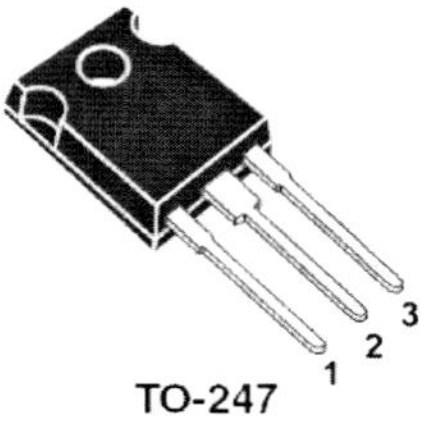 TO-247	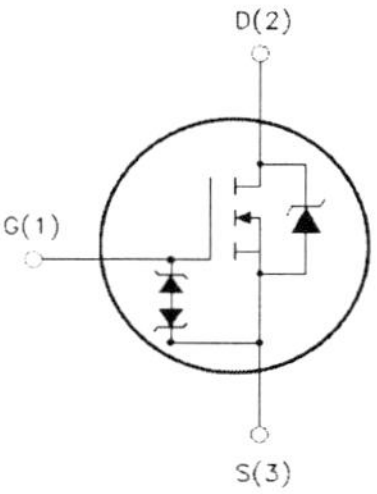	
Transistor kind	Power MOSFET	
Drain-source voltage, V	V_{DS}	950
Drain-gate voltage, V	V_{DGR}	950
Static drain-source ON resistance, Ω	$R_{DS\,(ON)}$	0.69
Drain current continuous, A (at 25°C/100°C)	I_{DM}	10/6.3
Drain current pulsed, A	$I_{DM\,(*)}$	40
Total power dissipation (T_C = 25°C), W	P_{tot}	230
Gate threshold voltage, V	V_{GS}	3 – 4.5
Gate-body leakage current, μA	I_{GSS}	10
Turn-on delay time, ns	$t_{d(on)}$	31
Turn-off delay time, ns	$t_{d(off)}$	88
Input capacitance, pF	C_{ies}	3500
Drain-source breakdown voltage, V (Min)	$V_{(BR)\,DSS}$	950
Max. operating junction temperature, °C	T_j	150
Thermal resistance: junction-case, °C/W	R_{QJC}	0.54

Transistor type	**STW13NK100Z**	
Manufacturer	**STMicroelectronics**	
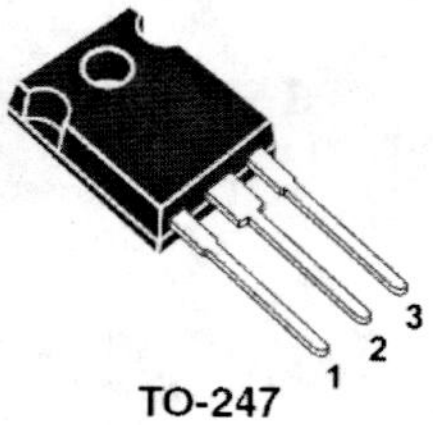 TO-247	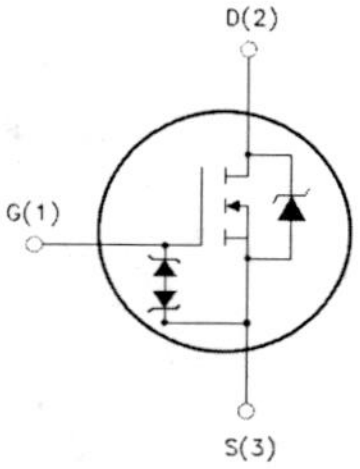	
Transistor kind	Power MOSFET	
Drain-source voltage, V	V_{DS}	1000
Drain-gate voltage, V	V_{DGR}	1000
Static drain-source ON resistance, Ω	$R_{DS\,(ON)}$	0.56
Drain current continuous, A (at 25°C/100°C)	I_{DM}	13/8.2
Drain current pulsed, A	$I_{DM\,(*)}$	52
Total power dissipation (T_C = 25°C), W	P_{tot}	350
Gate threshold voltage, V	V_{GS}	3 – 4.5
Gate-body leakage current, μA	I_{GSS}	10
Turn-on delay time, ns	$t_{d(on)}$	45
Turn-off delay time, ns	$t_{d(off)}$	145
Input capacitance, pF	C_{ies}	6000
Drain-source breakdown voltage, V (Min)	$V_{(BR)\,DSS}$	1000
Max. operating junction temperature, °C	T_j	150
Thermal resistance: junction-case, °C/W	R_{QJC}	0.36

Appendix B4: High-Voltage IGBT Transistors

Transistor type	**APT15GN120BDQ1**	
Manufacturer	**Advanced Power Technology**	
	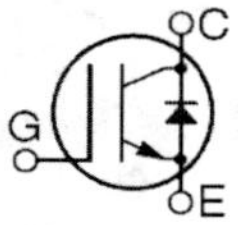	
Transistor kind	Insulated Gate Bipolar Transistor	
Collector-emitter voltage, V	V_{CES}	1200
Gate-emitter voltage, V	V_{GE}	±30
Collector current: continuous/peak, A	I_C	45/45
Total power dissipation (T_C = 25°C), W	P_{tot}	195
Gate threshold voltage (V_{CE} = V_{GE}), V	V_{GE}	5 – 6.5
Collector cut-off current (V_{GE} = 0), μA	I_{CES}	200
Collector-emitter ON voltage, V	$V_{CE\,(ON)}$*	1.4 – 2.1
Gate-emitter leakage current (V_{GE} = 0), nA	I_{GES}	120
Turn-on delay time, (inductive switching), ns	$t_{d(on)}$	10
Turn-off delay time, (inductive switching), ns	$t_{d(off)}$	150
Input capacitance, pF	C_{ies}	1200
Max. operating junction temperature, °C	T_j	150
Thermal resistance: junction-case, °C/W	R_{QJC}	0.64

Transistor type	**APT35GN120B**	
Manufacturer	**Advanced Power Technology**	
 TO-247	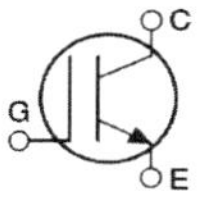	
Transistor kind	HV Insulated Gate Bipolar Transistor	
Collector-emitter voltage, V	V_{CES}	1200
Gate-emitter voltage, V	V_{GE}	±30
Collector current: continuous/peak, A	I_C	94/105
Total power dissipation (T_C = 25°C), W	P_{tot}	379
Gate threshold voltage (V_{CE} = V_{GE}), V	V_{GE}	5.8 – 6.5
Collector cut-off current (V_{GE} = 0), μA	I_{CES}	100
Collector-emitter saturation voltage, V	$V_{CE\,(SAT)}$*	2.5 – 4.7
Gate-emitter leakage current (V_{GE} = 0), nA	I_{GES}	600
Turn-on delay time, (inductive switching), ns	$t_{d(on)}$	24
Turn-off delay time, (inductive load), ns	$t_{d(off)}$	300
Input capacitance, pF	C_{ies}	2500
Max. operating junction temperature, °C	T_j	150
Thermal resistance: junction-case, °C/W	R_{QJC}	0.33
Integrated gate resistor, Ohm	$R_{G(inp)}$	6

Transistor type	**APT100GN120J**	
Manufacturer	**Advanced Power Technology**	
 **ISOTOP**	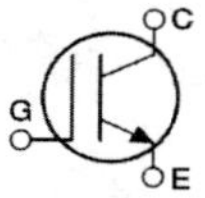	
Transistor kind	IGBT	
Collector-emitter voltage, V	V_{CES}	1200
Gate-emitter voltage, V	V_{GE}	±30
Collector current: continuous/peak, A	I_C	153/300
Total power dissipation (T_C = 25°C), W	P_{tot}	446
Gate threshold voltage ($V_{CE} = V_{GE}$), V	V_{GE}	5 – 6.5
Collector cut-off current (V_{GE} = 0), μA	I_{CES}	100
Collector-emitter ON voltage, V	$V_{CE\,(ON)}$*	1.4 – 2.1
Gate-emitter leakage current (V_{GE} = 0), nA	I_{GES}	600
Turn-on delay time, (inductive switching), ns	$t_{d(on)}$	50
Turn-off delay time, (inductive switching), ns	$t_{d(off)}$	615
Input capacitance, pF	C_{ies}	6500
Dielectric strength (leads to case), V RMS	V_{ISOL}	2500
Max. operating junction temperature, ºC	T_j	150
Thermal resistance: junction-case, °C/W	R_{QJC}	0.28
Integrated gate resistor, Ohm	$R_{G(inp)}$	7.5

Transistor type	APT150GN120J	
Manufacturer	Advanced Power Technology	

ISOTOP

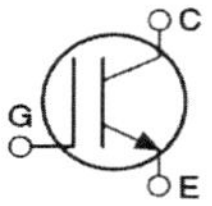

Transistor kind	IGBT	
Collector-emitter voltage, V	V_{CES}	1200
Gate-emitter voltage, V	V_{GE}	±30
Collector current: continuous/peak, A	I_C	215/450
Total power dissipation (T_C = 25°C), W	P_{tot}	625
Gate threshold voltage ($V_{CE} = V_{GE}$), V	V_{GE}	5 – 6.5
Collector cut-off current (V_{GE} = 0), µA	I_{CES}	100
Collector-emitter ON voltage, V	$V_{CE\,(ON)}$*	1.4 – 2.1
Gate-emitter leakage current (V_{GE} = 0), nA	I_{GES}	600
Turn-on delay time, (inductive load), ns	$t_{d(on)}$	55
Turn-off delay time, (inductive load), ns	$t_{d(off)}$	675
Input capacitance, pF	C_{ies}	9500
Dielectric strength (leads to case), V RMS	V_{ISOL}	2500
Max. operating junction temperature, °C	T_j	150
Thermal resistance: junction-case, °C/W	R_{QJC}	0.20
Integrated gate resistor, Ohm	$R_{G(inp)}$	5

Transistor type	FIO 50-12BD	
Manufacturer	IXYS	
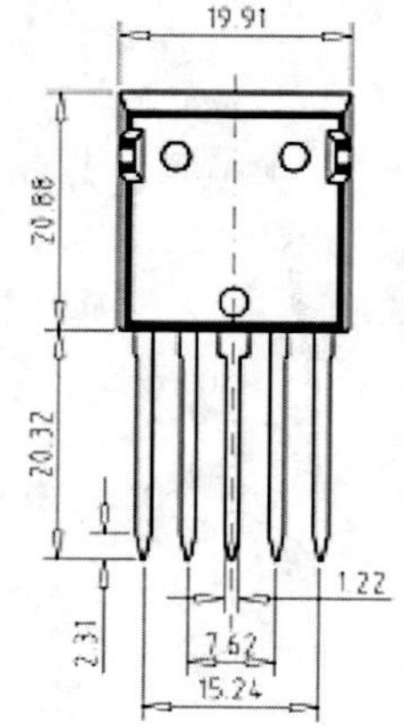 	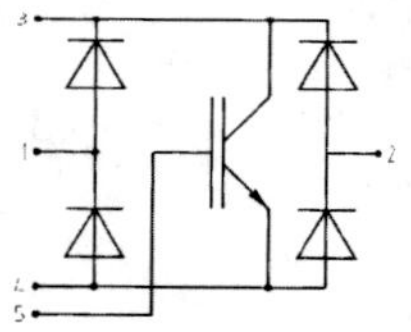	
Transistor kind	HV IGBT	
Collector-emitter voltage, V	V_{CES}	1200
Gate-emitter voltage, V	V_{GE}	±20
Collector current: continuous, A	I_C	50
Total power dissipation (T_C = 25°C), W	P_{tot}	200
Gate threshold voltage ($V_{CE} = V_{GE}$), V	$V_{GE(TH)}$	4.5 – 6.5
Collector cut-off current (V_{GE} = 0), μA	I_{CES}	400
Collector-emitter saturation voltage, V	$V_{CE\ (SAT)}$*	2 – 2.6
Gate-emitter leakage current (V_{GE} = 0), nA	I_{GES}	200
Turn-on delay time, (inductive switching), ns	$t_{d(on)}$	150
Turn-off delay time, (inductive switching), ns	$t_{d(off)}$	700
Input capacitance, pF	C_{ies}	2000

Transistor type	**GT25Q301**
Manufacturer	**Toshiba**

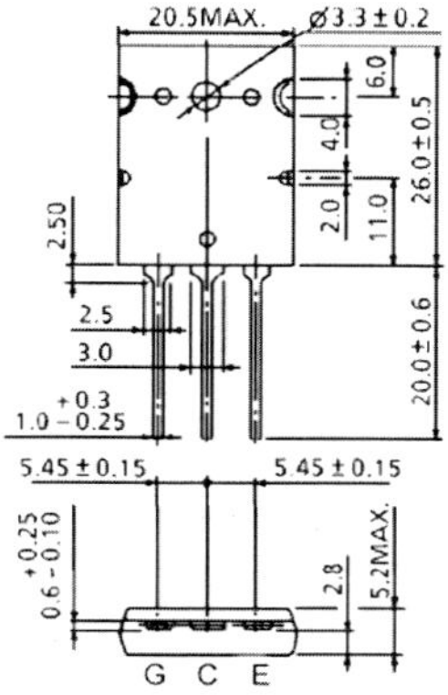

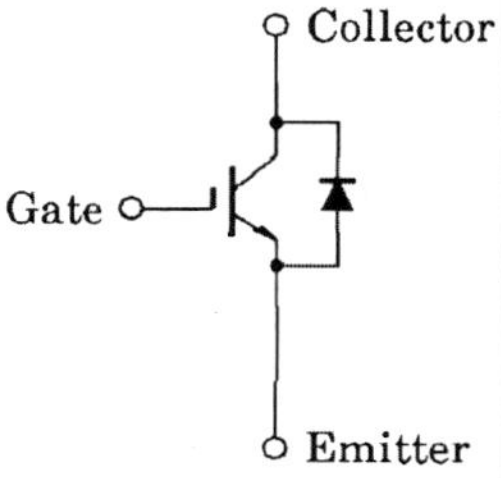

Transistor kind	Insulated Gate Bipolar Transistor	
Collector-emitter voltage, V	V_{CES}	1200
Gate-emitter voltage, V	V_{GE}	±20
Collector current: continuous/peak, A	I_C	25/50
Total power dissipation (T_C = 25°C), W	P_{tot}	200
Collector cut-off current (V_{GE} = 0), μA	I_{CES}	1000
Collector-emitter saturation voltage, V	$V_{CE\,(ON)}$*	2.1 – 2.7
Gate-emitter leakage current (V_{GE} = 0), nA	I_{GES}	500
Turn-on delay time, (inductive load), μs	$t_{d(on)}$	0.3
Turn-off delay time, (inductive load), μs	$t_{d(off)}$	0.68
Max. operating junction temperature, °C	T_j	150
Thermal resistance: junction-case, °C/W	R_{QJC}	0.625

Transistor type	IRG4PH20K	
Manufacturer	IXYS	

TO-247AC

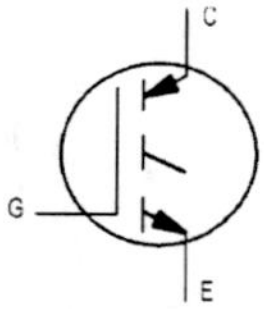

Transistor kind	Insulated Gate Bipolar Transistor	
Collector-emitter voltage, V	V_{CES}	1200
Gate-emitter voltage, V	V_{GE}	±20
Collector current: continuous/peak, A	I_C	11/22
Total power dissipation (T_C = 25°C), W	P_{tot}	60
Gate threshold voltage ($V_{CE} = V_{GE}$), V	V_{GE}	3.5 – 6.5
Collector cut-off current (V_{GE} = 0), μA	I_{CES}	250
Collector-emitter saturation voltage, V	$V_{CE\ (SAT)}$*	2.8 – 4.3
Gate-emitter leakage current (V_{GE} = 0), nA	I_{GES}	100
Turn-on delay time, (inductive load), ns	$t_{d(on)}$	23
Turn-off delay time, (inductive load), ns	$t_{d(off)}$	93 – 140
Input capacitance, pF	C_{ies}	435
Dielectric strength (leads to case), V RMS	V_{ISOL}	-
Max. operating junction temperature, °C	T_j	150
Thermal resistance: junction-case, °C/W	R_{QJC}	2.1

Transistor type	**IXDH20N120**	
Manufacturer	**IXYS**	
	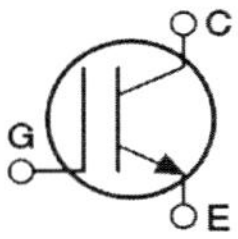	
Transistor kind	Insulated Gate Bipolar Transistor	
Collector-emitter voltage, V	V_{CES}	1200
Gate-emitter voltage, V	V_{GE}	±20
Collector current: continuous/peak, A	I_C	38/50
Total power dissipation (T_C = 25°C), W	P_{tot}	200
Gate threshold voltage ($V_{CE} = V_{GE}$), V	V_{GE}	4.5 – 6.5
Collector cut-off current (V_{GE} = 0), μA	I_{CES}	1000
Collector-emitter saturation voltage, V	$V_{CE\ (SAT)}$*	2.4 – 3.0
Gate-emitter leakage current (V_{GE} = 0), nA	I_{GES}	500
Turn-on delay time, (inductive switching), ns	$t_{d(on)}$	100
Turn-off delay time, (inductive switching), ns	$t_{d(off)}$	500
Input capacitance, pF	C_{ies}	1000
Dielectric strength (leads to case), V RMS	V_{ISOL}	-
Max. operating junction temperature, °C	T_j	150
Thermal resistance: junction-case, °K/W	R_{QJC}	1.6

Transistor type	**IXDN75N120**	
Manufacturer	**IXYS**	
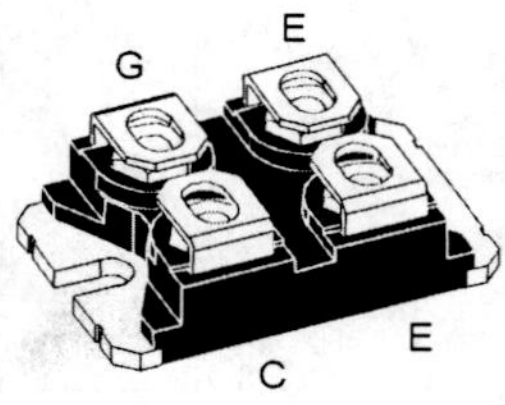	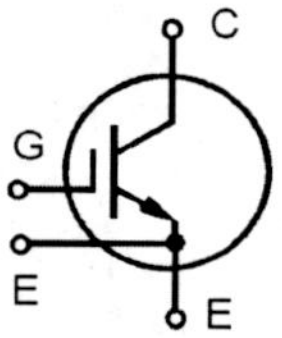	
Transistor kind	Insulated Gate Bipolar Transistor	
Collector-emitter voltage, V	V_{CES}	1200
Gate-emitter voltage, V	V_{GE}	±20
Collector current: continuous/peak, A	I_C	150/190
Total power dissipation (T_C = 25°C), W	P_{tot}	660
Gate threshold voltage ($V_{CE} = V_{GE}$), V	V_{GE}	4.5 – 6.5
Collector cut-off current (V_{GE} = 0), μA	I_{CES}	600
Collector-emitter saturation voltage, V	$V_{CE\,(SAT)}$*	2.2 – 2.7
Gate-emitter leakage current (V_{GE} = 0), nA	I_{GES}	500
Turn-on delay time, (inductive sload), ns	$t_{d(on)}$	100
Turn-off delay time, (inductive load), ns	$t_{d(off)}$	650
Input capacitance, pF	C_{ies}	5500
Dielectric strength (leads to case), V RMS	V_{ISOL}	2500
Max. operating junction temperature, °C	T_j	150
Thermal resistance: junction-case, °K/W	R_{QJC}	0.19

Transistor type	**IXEL40N400**	
Manufacturer	**IXYS**	
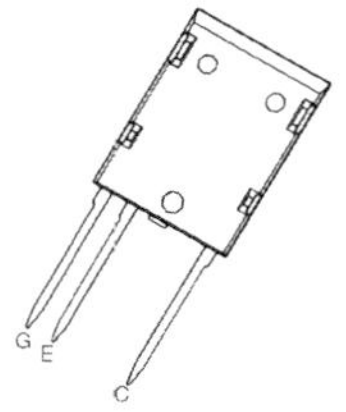	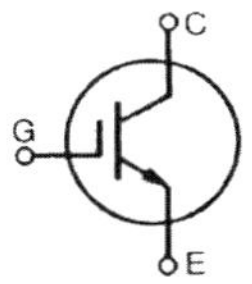	
Transistor kind	Very HV IGBT	
Collector-emitter voltage, V	V_{CES}	4000
Gate-emitter voltage, V	V_{GE}	±20
Collector current: continuous/peak, A	I_C	40/170
Total power dissipation (T_C = 25°C), W	P_{tot}	380
Gate threshold voltage (V_{CE} = V_{GE}), V	V_{GE}	6 – 7.5
Collector cut-off current (V_{GE} = 0), μA	I_{CES}	200
Collector-emitter saturation voltage, V	$V_{CE\,(SAT)}$*	3.0 – 4.0
Gate-emitter leakage current (V_{GE} = 0), nA	I_{GES}	500
Turn-on delay time, (inductive switching), ns	$t_{d(on)}$	170
Turn-off delay time, (inductive switching), ns	$t_{d(off)}$	660
Input capacitance, pF	C_{ies}	7450
Dielectric strength (leads to case), V RMS	V_{ISOL}	2500
Max. operating junction temperature, °C	T_j	125
Thermal resistance: junction-case, °K/W	R_{QJC}	0.33

Transistor type	IXGH16N160
Manufacturer	IXYS

TO-247

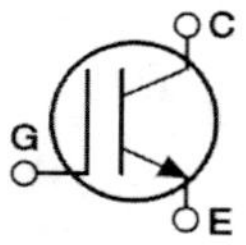

Transistor kind	HV Insulated Gate Bipolar Transistor	
Collector-emitter voltage, V	V_{CES}	1700
Gate-emitter voltage, V	V_{GE}	±20
Collector current: continuous/peak, A	I_C	32/80
Total power dissipation (T_C = 25°C), W	P_{tot}	190
Gate threshold voltage ($V_{CE} = V_{GE}$), V	V_{GE}	3 – 5
Collector cut-off current (V_{GE} = 0), µA	I_{CES}	500
Collector-emitter saturation voltage, V	$V_{CE\,(SAT)}$*	2.7 – 3.5
Gate-emitter leakage current (V_{GE} = 0), nA	I_{GES}	100
Turn-on delay time, (inductive switching), ns	$t_{d(on)}$	48
Turn-off delay time, (inductive switching), ns	$t_{d(off)}$	200
Input capacitance, pF	C_{ies}	1700
Dielectric strength (leads to case), V RMS	V_{ISOL}	-
Max. operating junction temperature, °C	T_j	150
Thermal resistance: junction-case, °K/W	R_{QJC}	0.65

Transistor type	**IXGH25N160**	
Manufacturer	**IXYS**	
 **TO-247**	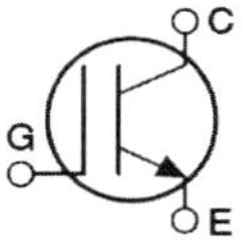 	
Transistor kind	HV Insulated Gate Bipolar Transistor	
Collector-emitter voltage, V	V_{CES}	1600
Gate-emitter voltage, V	V_{GE}	±20
Collector current: continuous/peak, A	I_C	75/200
Total power dissipation (T_C = 25°C), W	P_{tot}	300
Gate threshold voltage ($V_{CE} = V_{GE}$), V	V_{GE}	3 – 5
Collector cut-off current (V_{GE} = 0), μA	I_{CES}	1000
Collector-emitter saturation voltage, V	$V_{CE\,(SAT)}$*	2.5 – 4.7
Gate-emitter leakage current (V_{GE} = 0), nA	I_{GES}	600
Turn-on delay time, (inductive switching), ns	$t_{d(on)}$	47
Turn-off delay time, (inductive switching), ns	$t_{d(off)}$	86
Input capacitance, pF	C_{ies}	2090
Dielectric strength (leads to case), V RMS	V_{ISOL}	-
Max. operating junction temperature, °C	T_j	150
Thermal resistance: junction-case, °K/W	R_{QJC}	0.42

Transistor type	**IXLF19N250A**	
Manufacturer	**IXYS**	
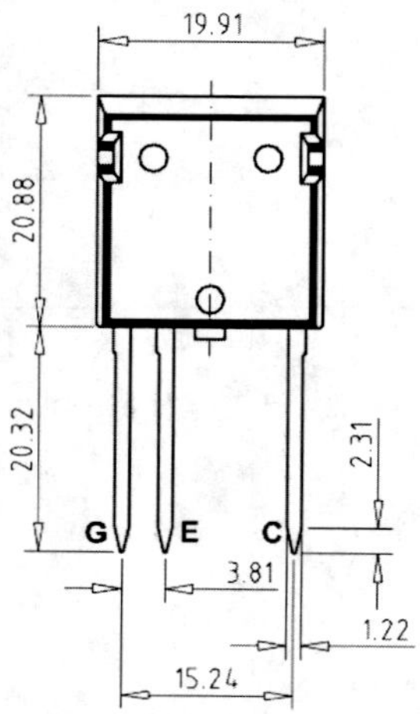	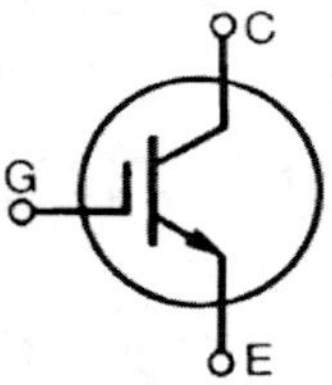	
Transistor kind	HV Insulated Gate Bipolar Transistor	
Collector-emitter voltage, V	V_{CES}	2500
Gate-emitter voltage, V	V_{GE}	±20
Collector current: continuous, A	I_C	32
Total power dissipation (T_C = 25°C), W	P_{tot}	250
Collector cut-off current (V_{GE} = 0), mA	I_{CES}	0.15
Collector-emitter saturation voltage, V	$V_{CE\,(SAT)}$*	3.2 – 4.0
Gate-emitter leakage current (V_{GE} = 0), nA	I_{GES}	500
Turn-on delay time, (inductive switching), ns	$t_{d(on)}$	100
Turn-off delay time, (inductive switching), ns	$t_{d(off)}$	600
Input capacitance, nF	C_{ies}	2.28

Transistor type	IXSK35N120AU1	
Manufacturer	IXYS	
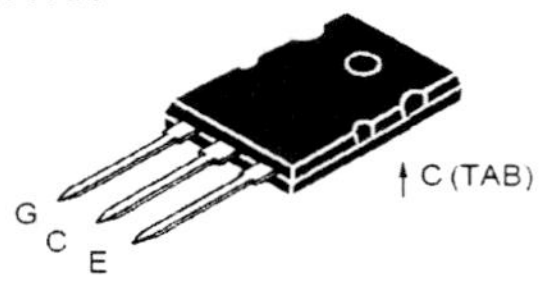	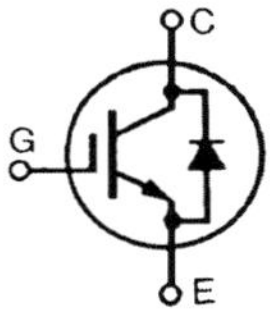	
Transistor kind	Insulated Gate Bipolar Transistor	
Collector-emitter voltage, V	V_{CES}	1200
Gate-emitter voltage, V	V_{GE}	±20
Collector current: continuous/peak, A	I_C	35/140
Total power dissipation (T_C = 25°C), W	P_{tot}	300
Gate threshold voltage (V_{CE} = V_{GE}), V	V_{GE}	4 – 8
Collector cut-off current (V_{GE} = 0), μA	I_{CES}	750 – 1500
Collector-emitter saturation voltage, V	$V_{CE\,(SAT)}$*	4.0
Gate-emitter leakage current (V_{GE} = 0), nA	I_{GES}	100
Turn-on delay time, (inductive load), ns	$t_{d(on)}$	80
Turn-off delay time, (inductive load), ns	$t_{d(off)}$	900
Input capacitance, pF	C_{ies}	3900
Max. operating junction temperature, °C	T_j	150
Thermal resistance: junction-case, °K/W	R_{QJC}	0.15

Transistor type	**SCP-4979**	
Manufacturer	**Sensitron Semiconductor**	
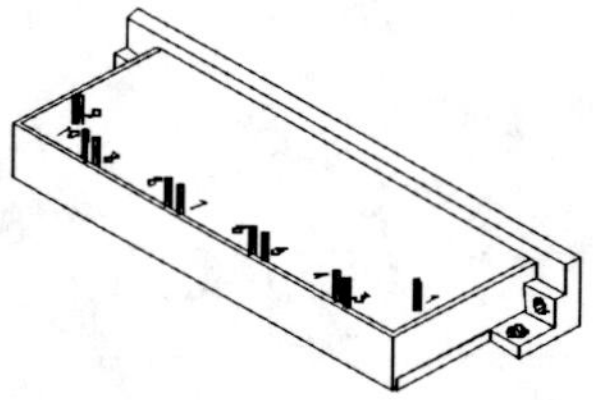	-	
Transistor kind	HV IGBT Solid-State Switch	
Collector-emitter voltage, V	V_{CES}	2900
Gate-emitter voltage, V	V_{GE}	-
Collector current: continuous/peak, A	I_C	1/100
Total power dissipation (T_C = 25°C), W	P_{tot}	-
Gate threshold voltage ($V_{CE} = V_{GE}$), V	V_{GE}	5 – 7.5
Collector cut-off current (V_{GE} = 0), μA	I_{CES}	-
Collector-emitter ON voltage, V	$V_{CE\,(ON)}$*	0.8
Gate-emitter leakage current (V_{GE} = 0), μA	I_{GES}	500
Turn-on delay time, (inductive switching), ns	$t_{d(on)}$	-
Turn-off delay time, (inductive switching), ns	$t_{d(off)}$	-
Input capacitance, pF	C_{ies}	12.500
Dielectric strength (leads to case), V RMS	V_{ISOL}	10.000
Max. operating junction temperature, °C	T_j	100
Thermal resistance: junction-case, °C/W	R_{QJC}	1.0

Appendix C: High-Voltage Thyristors

Thyristor type	**25TTS12, 25TTS16**	
Manufacturer	**International Rectifier**	
 TO-220AC	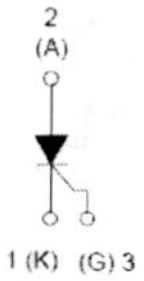	
Thyristor kind	Phase Control SCR	
Max. repetitive off-state peak voltage (sine wave), V	V_{DRM}, V_{RRM}	1200, 1600
Max. on-state RMS current (for 85°C), A	$I_{T\,(RMS)}$	25
Max. average on-state current, A	$I_{T\,(AV)}$	16
Peak, 1/2 cycle surge current, A	I_{TSM}	300
Max. peak forward on-state voltage, V	V_{TM}	1.25
Max. required DC gate voltage to trigger, V	V_{GT}	2.5
Max. required DC gate current, mA	I_{GT}	20 – 60
Max. peak gate current, A	I_{GM}	1.5
Max. off-state leakage current, mA	I_{DRM}, I_{RRM}	10
Critical rate of rise off-state voltage, V/μs	dV/dt	500
Critical rate of rise on-state current, A/μs	dI/dt	150
Max. holding current (gate open), mA	I_H	150
Max. latching current, mA	I_L	200
Turn-On time, μs	t_{gd}	0.9
Turn-Off time, μs	t_q	110
Max. junction temperature range, °C	T_j	-40 +125

Thyristor type	**30TPS12, 30TPS16**	
Manufacturer	**International Rectifier**	

Thyristor kind	Phase Control SCR	
Max. repetitive off-state peak voltage, V	V_{DRM}, V_{RRM}	1200, 1600
Max. on-state RMS current (for 85ºC), A	$I_{T\,(RMS)}$	30
Max. average on-state current, A	$I_{T\,(AV)}$	20
Peak, 1/2 cycle surge current, A	I_{TSM}	250
Max. peak forward on-state voltage, V	V_{TM}	1.85
Max. required DC gate voltage, V	V_{GT}	2.5
Max. required DC gate current, mA	I_{GT}	60
Max. peak gate current, A	I_{GM}	2.5
Max. off-state leakage current, mA	I_{DRM}, I_{RRM}	10
Critical rate of rise off-state voltage, V/µs	dV/dt	500
Critical rate of rise on-state current, A/µs	dI/dt	150
Max. holding current (gate open), mA	I_H	100
Max. latching current, mA	I_L	200
Turn-On time, µs	t_{gd}	0.9
Turn-Off time, µs	t_q	110

Thyristor type	40TPS12	
Manufacturer	**International Rectifier**	
 TO-247AC	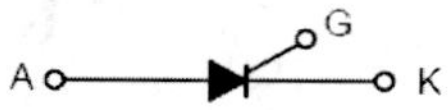	
Thyristor kind	Phase Control SCR	
Max. repetitive off-state peak voltage, V	V_{DRM}, V_{RRM}	1200
Max. on-state RMS current (for 85°C), A	$I_{T\,(RMS)}$	55
Max. average on-state current, A	$I_{T\,(AV)}$	35
Peak, 1/2 cycle surge current, A	I_{TSM}	500
Max. peak forward on-state voltage, V	V_{TM}	1.85
Max. required DC gate voltage, V	V_{GT}	4
Max. required DC gate current, mA	I_{GT}	250
Max. peak gate current, A	I_{GM}	2.5
Max. off-state leakage current, mA	I_{DRM}, I_{RRM}	10
Critical rate of rise off-state voltage, V/µs	dV/dt	1000
Critical rate of rise on-state current, A/µs	dI/dt	100
Max. holding current (gate open), mA	I_H	150
Max. latching current, mA	I_L	300
Turn-On time, µs	t_{gd}	-
Max. junction operating temperature, °C	T_j	-40 +125

Thyristor type	AK55HB120, AK55HB160 AK90HB120, AK90HB160
Manufacturer	**Sanrex**

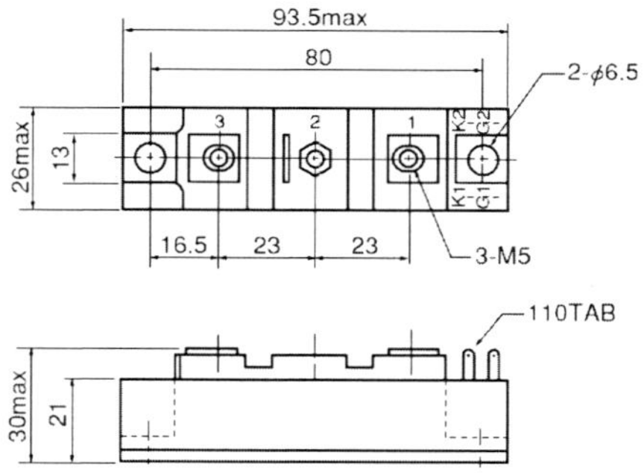

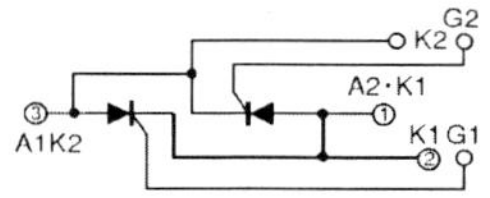

Thyristor kind	Power Thyristor Module	
Max. off-state peak voltage, V	V_{DRM}, V_{RRM}	1200, 1600
Max. on-state RMS current, A	$I_{T\,(RMS)}$	122 (AK55), 200 (AK90)
Max. average on-state current, A	$I_{T\,(AV)}$	55 (AK55), 90 (AK90)
Peak, 1/2 cycle surge current, A	I_{TSM}	1000 (AK55), 1650 (AK90)
Max. peak forward on-state, V	V_{TM}	1.5
Max. DC gate current to trigger, mA	I_{GT}	100
Max. off-state leakage current, mA	I_{DRM}, I_{RRM}	20 – 30
Critical rate of rise of voltage, V/μs	dV/dt	500
Critical rate of rise of current, A/μs	dI/dt	150 (AK55) 200 (AK90)
Max. holding current, mA	I_H	50
Max. latching current, mA	I_L	100
Turn-On time, μs	t_{gd}	10

Thyristor type	**BTW67-1200, BTW69-1200**	
Manufacturer	**STMicroelectronics**	
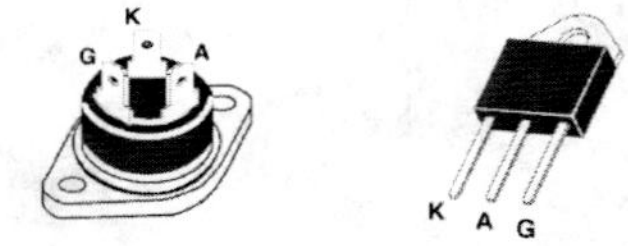 RD91 (BTW67) TOP3 (BTW69)		
Thyristor kind	Phase Control SCR	
Max. off-state peak voltage (sine wave), V	V_{DRM}, V_{RRM}	1200
Max. on-state RMS current (for 85°C), A	$I_{T\,(RMS)}$	50
Max. average on-state current, A	$I_{T\,(AV)}$	32
Peak, 1/2 cycle surge current, A	I_{TSM}	580
Max. peak forward on-state voltage, V	V_{TM}	1.9
Max. required DC gate voltage to trigger, V	V_{GT}	1.3
Max. required DC gate current to trigger, mA	I_{GT}	80
Max. peak gate current, A	I_{GM}	8
Max. off-state leakage current, mA	I_{DRM}, I_{RRM}	5
Critical rate of rise of off-state voltage, V/µs	dV/dt	1000
Critical rate of rise of on-state current, A/µs	dI/dt	50
Max. holding current (gate open), mA	I_H	150
Max. latching current, mA	I_L	200
Turn-On time, µs	t_{gd}	-
Max. junction operating temperature range, °C	T_j	-40 +125

Thyristor type	**C106M**	
Manufacturer	**ON Semiconductor**	
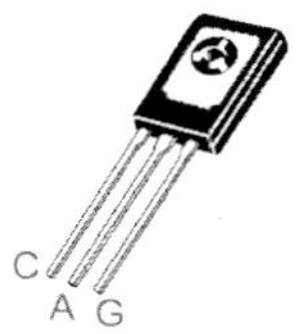 TO-225AA CASE 077 STYLE 2	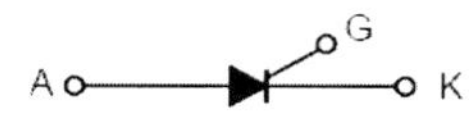	
Thyristor kind	Sensitive Gate Miniature SCR	
Max. repetitive off-state peak voltage, V	V_{DRM}, V_{RRM}	600
Max. on-state RMS current (for 85°C), A	$I_{T(RMS)}$	4
Max. average on-state current, A	$I_{T(AV)}$	2.5
Peak, 1/2 cycle (sine) surge current, A	I_{TSM}	20
Max. peak forward on-state voltage, V	V_{TM}	2.2
Max. required DC gate voltage to trigger, V	V_{GT}	0.8
Max. required DC gate current to trigger, mA	I_{GT}	0.2
Max. peak gate current, A	I_{GM}	0.2
Max. off-state leakage current, mA	I_{DRM}, I_{RRM}	0.1
Critical rate of rise of off-state voltage, V/μs	dV/dt	8
Critical rate of rise of on-state current, A/μs	dI/dt	-
Max. holding current (gate open), mA	I_H	3.0
Max. latching current, mA	I_L	5.0
Turn-On time, μs	t_{gd}	-
Max. junction operating temperature range, °C	T_j	-40 +110

Thyristor type	**CS 29-12io1C**	
Manufacturer	**IXYS**	
ISOPLUS 220™ 	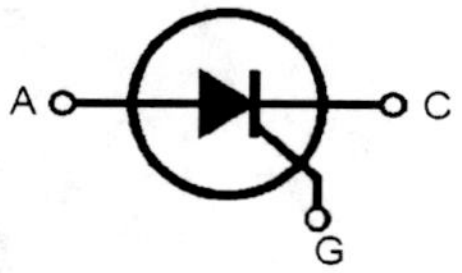	
Thyristor kind	Phase Control Thyristor with Electrically Isolated Back Surface	
Max. repetitive off-state peak voltage, V	V_{DRM} , V_{RRM}	1200
Max. on-state RMS current (for 85°C), A	$I_{T\,(RMS)}$	35
Max. average on-state current, A	$I_{T\,(AV)}$	23
Peak, 1/2 cycle non-repetitive surge current, A	I_{TSM}	200
Max. peak forward on-state voltage, V	V_{TM}	1.5
Max. required DC gate voltage to trigger, V	V_{GT}	1.2
Max. required DC gate current to trigger, mA	I_{GT}	80
Max. off-state leakage current, mA	I_{DRM} , I_{RRM}	2
Critical rate of rise of off-state voltage, V/μs	dV/dt	1000
Critical rate of rise of on-state current, A/μs	dI/dt	150
Max. holding current (gate open), mA	I_H	50
Max. latching current, mA	I_L	150
Turn-On time, μs	t_{gd}	2
Max. junction temperature range, °C	T_j	-40 +125

Thyristor type	**CS 60-12io1, CS 60-14io1, CS 60-16io1**	
Manufacturer	**IXYS**	
PLUS247 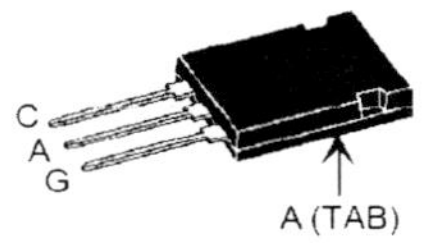	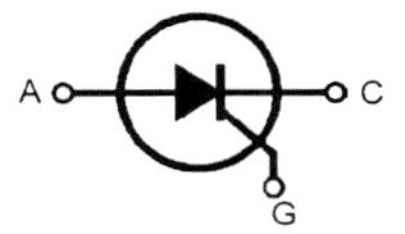	
Thyristor kind	Phase Control Thyristor	
Max. repetitive off-state peak voltage, V	V_{DRM}, V_{RRM}	1200, 1400, 1600
Max. on-state RMS current (85°C), A	$I_{T\,(RMS)}$	75
Max. average on-state current, A	$I_{T\,(AV)}$	48
Peak, 1/2 cycle non-repetitive surge current, A	I_{TSM}	1500
Max. peak forward on-state voltage, V	V_{TM}	1.4
Max. required DC gate voltage to trigger, V	V_{GT}	1.5
Max. required DC gate current to trigger, mA	I_{GT}	200
Max. off-state leakage current, mA	I_{DRM}, I_{RRM}	10
Critical rate of rise of off-state voltage, V/μs	dV/dt	1000
Critical rate of rise of on-state current, A/μs	dI/dt	150
Max. holding current (gate open), mA	I_H	200
Max. latching current, mA	I_L	450
Turn-On time, μs	t_{gd}	2
Max. junction temperature range, °C	T_j	-40 +125

Thyristor type	**MSS50-1200**
Manufacturer	**STMicroelectronics**

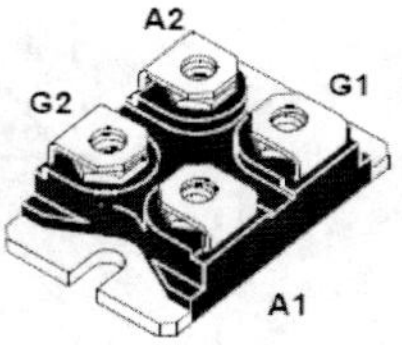

ISOTOP®

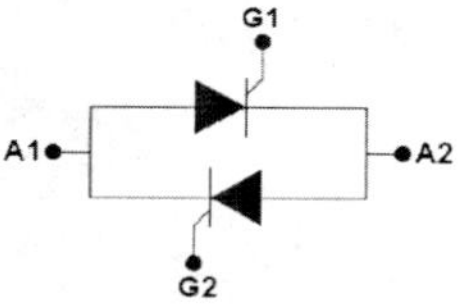

Thyristor kind	Back to Back SCR Module	
Max. repetitive off-state peak voltage, V	V_{DRM}, V_{RRM}	1200
Max. on-state RMS current (for 85°C), A	$I_{T\,(RMS)}$	70
Max. average on-state current, A	$I_{T\,(AV)}$	-
Peak, 1 cycle non-repetitive surge current, A	I_{TSM}	600
Max. peak forward on-state voltage, V	V_{TM}	1.7
Max. required DC gate voltage to trigger, V	V_{GT}	1.3
Max. required DC gate current to trigger, mA	I_{GT}	50
Max. peak gate current, A	I_{GM}	4
Max. off-state leakage current, mA	I_{DRM}, I_{RRM}	10
Critical rate of rise of off-state voltage, V/μs	dV/dt	1000
Critical rate of rise of on-state current, A/μs	dI/dt	50
Max. holding current (gate open), mA	I_H	80
Max. latching current, mA	I_L	120
Max. junction temperature range, °C	T_j	-40 +125

Thyristor type	**PK25F120, PK25F160**
Manufacturer	**SanRex**

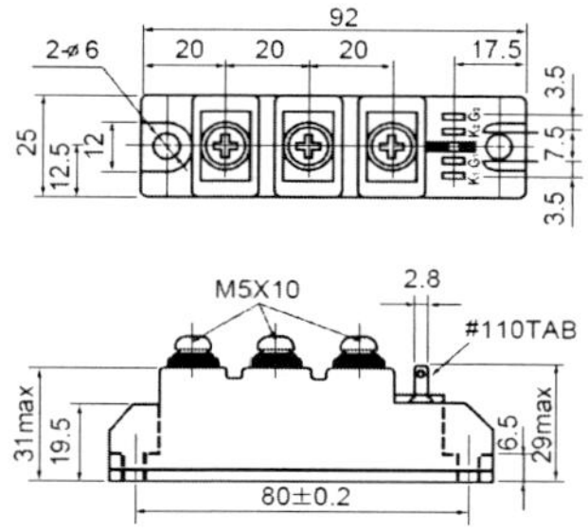

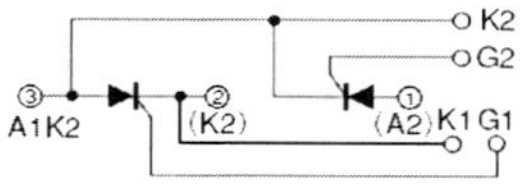

Thyristor kind	Power thyristor module	
Max. repetitive off-state peak voltage, V	V_{DRM}, V_{RRM}	1200, 1600
Max. on-state RMS current (for 85°C), A	$I_{T\,(RMS)}$	39
Max. average on-state current, A	$I_{T\,(AV)}$	25
Peak, 1/2 cycle surge current, A	I_{TSM}	530
Max. peak forward on-state voltage, V	V_{TM}	1.55
Max. required DC gate current, mA	I_{GT}	50
Max. peak gate current, A	I_{GM}	3.0
Max. off-state leakage current, mA	I_{DRM}, I_{RRM}	10
Critical rate of rise off-state voltage, V/μs	dV/dt	500
Critical rate of rise on-state current, A/μs	dI/dt	100
Max. holding current (gate open), mA	I_H	50
Max. latching current, mA	I_L	100
Turn-On time, μs	t_{gd}	10
Max. operating temperature range, °C	T_j	-40 +125

Thyristor type	**PK110F120, PK110F160**
Manufacturer	**SanRex**

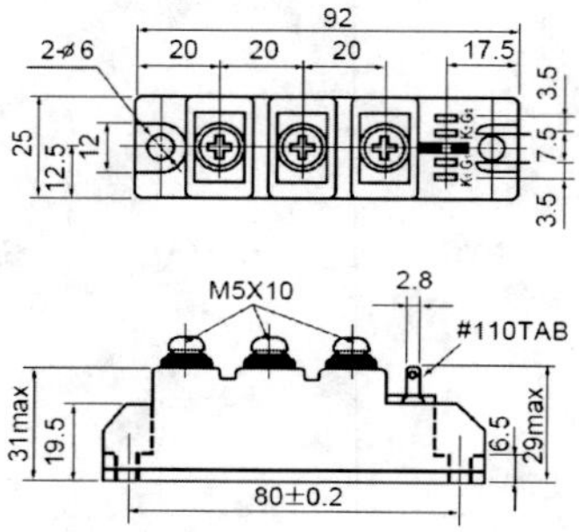

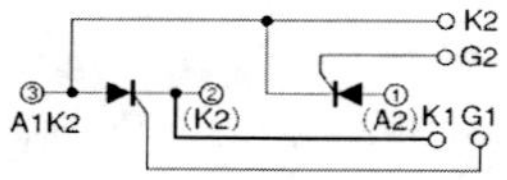

Thyristor kind	Power thyristor module	
Max. repetitive off-state peak voltage, V	V_{DRM}, V_{RRM}	1200, 1600
Max. on-state RMS current (for 85°C), A	$I_{T\,(RMS)}$	172
Max. average on-state current, A	$I_{T\,(AV)}$	110
Peak, 1/2 cycle surge current, A	I_{TSM}	2300
Max. peak forward on-state voltage, V	V_{TM}	1.45
Max. required DC gate current, mA	I_{GT}	100
Max. peak gate current, A	I_{GM}	3.0
Max. off-state leakage current, mA	I_{DRM}, I_{RRM}	20
Critical rate of rise of off-state voltage, V/μs	dV/dt	500
Critical rate of rise of on-state current, A/μs	dI/dt	200
Max. holding current (gate open), mA	I_H	50
Max. latching current, mA	I_L	100
Turn-On time, μs	t_{gd}	10

Thyristor type	**SKKT 122/12E, SKKT 122/14E, SKKT 122/16E, SKKT 122/18E,**	
Manufacturer	**Semikron**	
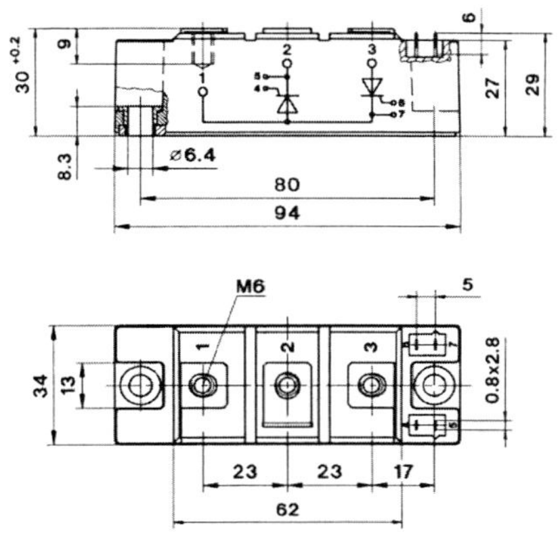	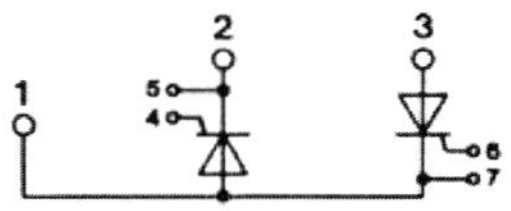	
Max. repetitive off-state peak voltage, V	V_{DRM}, V_{RRM}	1200, 1400, 1600, 1800
Max. on-state RMS current (for 85°C), A	$I_{T\,(RMS)}$	160
Max. average on-state current, A	$I_{T\,(AV)}$	129
Peak, 1/2 cycle surge current, A	I_{TSM}	3200
Max. peak forward on-state voltage, V	V_{TM}	1.55
Max. required DC gate current, mA	I_{GT}	150
Max. off-state leakage current, mA	I_{DRM}, I_{RRM}	40
Critical rate of rise of off-state voltage, V/μs	dV/dt	1000
Critical rate of rise of on-state current, A/μs	dI/dt	200
Max. holding current (gate open), mA	I_H	300
Max. latching current, mA	I_L	500
Turn-On time, μs	t_{gd}	1

Thyristor type	SKT10/12E	
Manufacturer	Semikron	
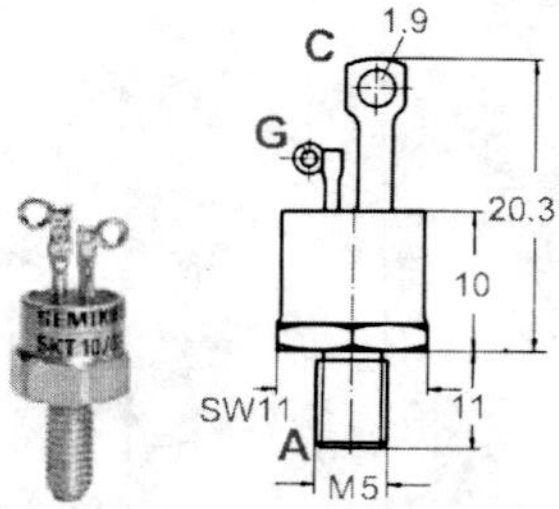	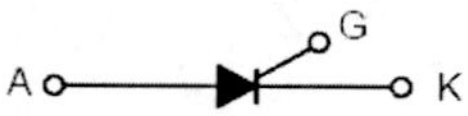	
Thyristor kind	Line Thyristor with Hermetic Metal Case	
Max. repetitive off-state peak voltage, V	V_{DRM}, V_{RRM}	1200
Max. on-state RMS current (for 85°C), A	$I_{T\,(RMS)}$	30
Max. average on-state current, A	$I_{T\,(AV)}$	10
Peak, 1/2 cycle surge current, A	I_{TSM}	210
Max. peak forward on-state voltage, V	V_{TM}	1.6
Max. required DC gate current, mA	I_{GT}	100
Max. off-state leakage current, mA	I_{DRM}, I_{RRM}	4
Critical rate of rise off-state voltage, V/μs	dV/dt	1000
Critical rate of rise on-state current, A/μs	dI/dt	50
Max. holding current (gate open), mA	I_H	80 – 150
Max. latching current, mA	I_L	150 – 300
Turn-On time, μs	t_{gd}	1
Turn-Off time, μs	t_q	80
Max. operating temperature range, °C	T_j	-40 +130

Thyristor type	**SKT 24/12E, SKT 24/14E, SKT 24/16E, SKT 24/18E**	
Manufacturer	**Semikron**	
 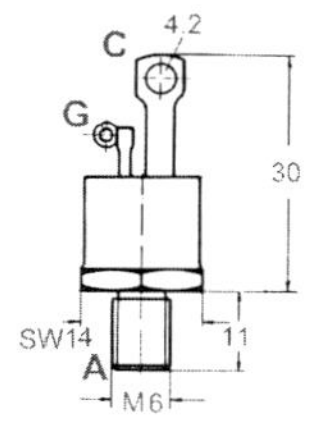		
Thyristor kind	Line Thyristor with Hermetic Metal Case	
Max. repetitive off-state peak voltage, V	V_{DRM}, V_{RRM}	1200, 1400, 1600, 1800
Max. on-state RMS current (for 85°C), A	$I_{T\,(RMS)}$	50
Max. average on-state current, A	$I_{T\,(AV)}$	24
Peak, 1/2 cycle surge current, A	I_{TSM}	380
Max. peak forward on-state voltage, V	V_{TM}	1.9
Max. required DC gate current, mA	I_{GT}	100
Max. off-state leakage current, mA	I_{DRM}, I_{RRM}	8
Critical rate of rise off-state voltage, V/µs	dV/dt	1000
Critical rate of rise on-state current, A/µs	dI/dt	50
Max. holding current (gate open), mA	I_H	80 – 150
Max. latching current, mA	I_L	150 – 300
Turn-On time, µs	t_{gd}	1
Turn-Off time, µs	t_q	80
Max. junction operating temperature, °C	T_j	-40 +130

Thyristor type	**SKT 50/12E, SKT 50/14E, SKT 50/16E, SKT 50/18E**	
Manufacturer	**Semikron**	
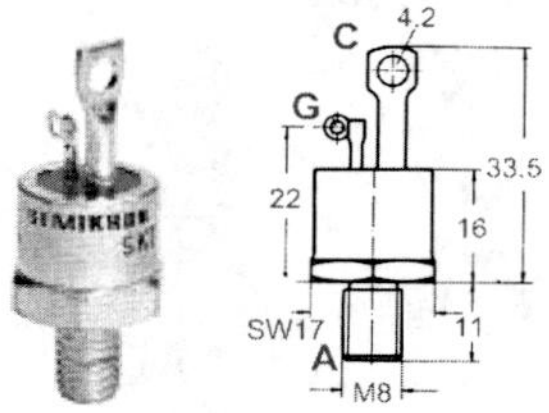	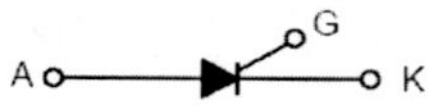	
Thyristor kind	Line Thyristor with Hermetic Metal Case	
Max. repetitive off-state peak voltage, V	V_{DRM}, V_{RRM}	1200, 1400, 1600, 1800
Max. on-state RMS current (for 85°C), A	$I_{T\,(RMS)}$	75
Max. average on-state current, A	$I_{T\,(AV)}$	45
Peak, 1/2 cycle surge current, A	I_{TSM}	900
Max. peak forward on-state voltage, V	V_{TM}	1.8
Max. required DC gate current, mA	I_{GT}	150
Max. off-state leakage current, mA	I_{DRM}, I_{RRM}	8
Critical rate of rise off-state voltage, V/μs	dV/dt	1000
Critical rate of rise on-state current, A/μs	dI/dt	50
Max. holding current (gate open), mA	I_H	200
Max. latching current, mA	I_L	400
Turn-On time, μs	t_{gd}	1.5
Turn-Off time, μs	t_q	100
Max. junction operating temperature, °C	T_j	-40 +130

Thyristor type	**T50RIA120, T70RIA120, T90RIA120**	
Manufacturer	**International Rectifier**	
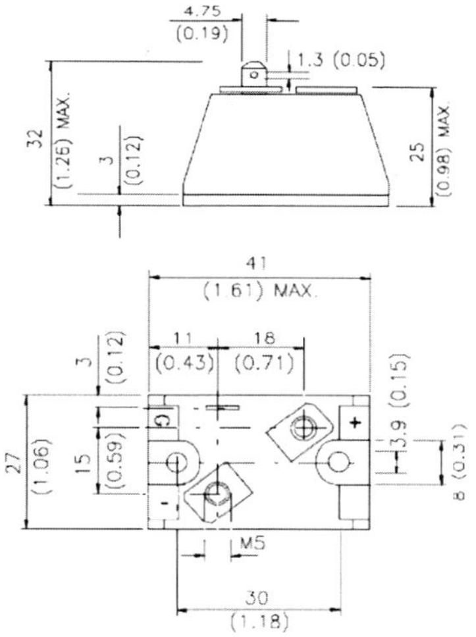	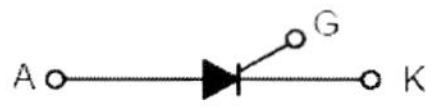	
Max. repetitive off-state peak voltage, V	V_{DRM}, V_{RRM}	1200
Max. on-state RMS current (for 85°C), A	$I_{T(RMS)}$	80, 110, 141
Max. average on-state current, A	$I_{T(AV)}$	50, 70, 90
Peak, 1/2 cycle surge current, kA	I_{TSM}	1.1; 1.4; 1.5
Max. peak forward on-state voltage, V	V_{TM}	1.6
Max. required DC gate current, mA	I_{GT}	120
Max. off-state leakage current, mA	I_{DRM}, I_{RRM}	15
Critical rate of rise off-state voltage, V/µs	dV/dt	500
Critical rate of rise on-state current, A/µs	dI/dt	150 – 200
Max. latching current, mA	I_L	400
Turn-On time, µs	t_{gd}	0.9

Thyristor type	VCO 132-12io7, VCO 132-14io7, VCO 132-16io7, VCO 132-18io7	
Manufacturer	IXYS	
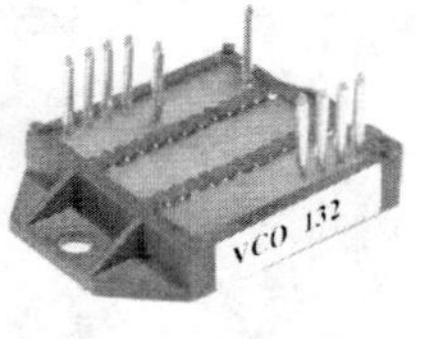 ECO-PAC2	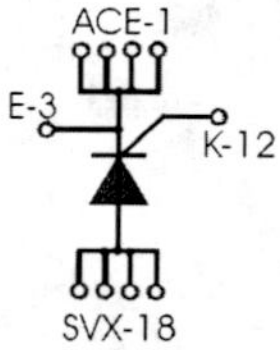	
Thyristor kind	Thyristor Module with Leads Suitable for PC Board	
Max. repetitive off-state peak voltage, V	V_{DRM}, V_{RRM}	1200, 1400, 1600, 1800
Max. on-state RMS current (for 85°C), A	$I_{T\,(RMS)}$	200
Max. average on-state current, A	$I_{T\,(AV)}$	130
Peak, 1/2 cycle surge current, A	I_{TSM}	3600
Max. peak forward on-state voltage, V	V_{TM}	1.3
Max. required DC gate voltage, V	V_{GT}	1.6
Max. required DC gate current, mA	I_{GT}	400
Max. off-state leakage current, mA	I_{DRM}, I_{RRM}	10
Critical rate of rise off-state voltage, V/μs	dV/dt	1000
Critical rate of rise on-state current, A/μs	dI/dt	150
Max. holding current (gate open), mA	I_H	200
Max. latching current, mA	I_L	450
Turn-On time, μs	t_{gd}	2
Max. junction operating temperature, °C	T_j	-40 +125

Thyristor type	**VWO 140-12io1; VWO 140-14io1; VWO 140-16io1**
Manufacturer	**IXYS**

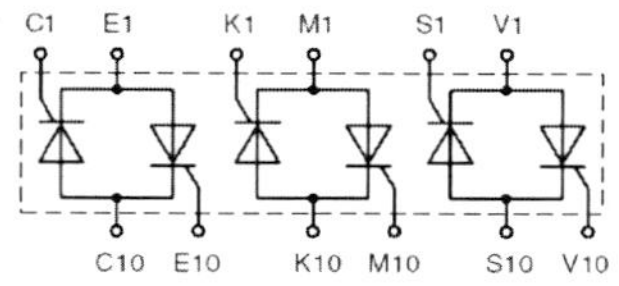

Thyristor kind	Three Phase AC Controlled Modules	
Max. repetitive off-state peak voltage, V	V_{DRM}, V_{RRM}	1200, 1400, 1600
Max. on-state RMS current (for 85°C), A	$I_{T\,(RMS)}$	101 (per phase)
Max. average on-state current, A	$I_{T\,(AV)}$	46
Peak, 1/2 cycle surge current, A	I_{TSM}	1040
Max. peak forward on-state voltage, V	V_{TM}	1.5
Max. required DC gate current, mA	I_{GT}	100
Max. off-state leakage current, mA	I_{DRM}, I_{RRM}	5
Critical rate of rise off-state voltage, V/µs	dV/dt	1000
Critical rate of rise on-state current, A/µs	dI/dt	150
Max. holding current (gate open), mA	I_H	200
Max. latching current, mA	I_L	450
Turn-On time, µs	t_{gd}	2
Turn-Off time, µs	t_q	150
Max. junction operating temperature, °C	T_j	125

Thyristor type	70TPS12, 70TPS16	
Manufacturer	International Rectifier	
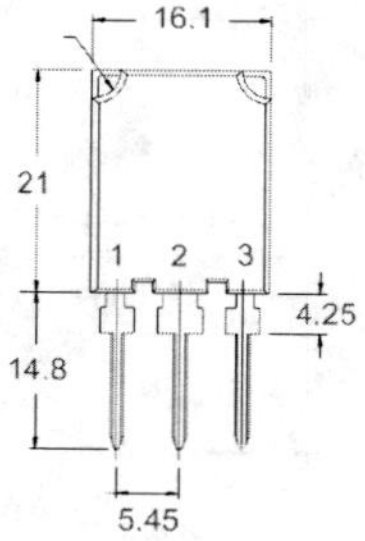	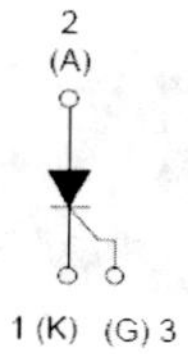	
Thyristor kind	Phase Control SCR	
Max. repetitive off-state peak voltage, V	V_{DRM}, V_{RRM}	1200, 1600
Max. on-state RMS current (for 85ºC), A	$I_{T\,(RMS)}$	75
Max. average on-state current, A	$I_{T\,(AV)}$	70
Peak, 1/2 cycle surge current, A	I_{TSM}	1200
Max. peak forward on-state voltage, V	V_{TM}	1.4
Max. required DC gate voltage, V	V_{GT}	4.0
Max. required DC gate current, mA	I_{GT}	80 – 270
Max. peak gate current, A	I_{GM}	2.5
Max. off-state leakage current, mA	I_{DRM}, I_{RRM}	15
Critical rate of rise off-state voltage, V/µs	dV/dt	500
Critical rate of rise on-state current, A/µs	dI/dt	150
Max. holding current (gate open), mA	I_H	200
Max. latching current, mA	I_L	400
Turn-On time, µs	t_{gd}	-

Thyristor type	**IRKT.105/12, IRKT.105/14, IRKT.105/16,**	
Manufacturer	**International Rectifier**	
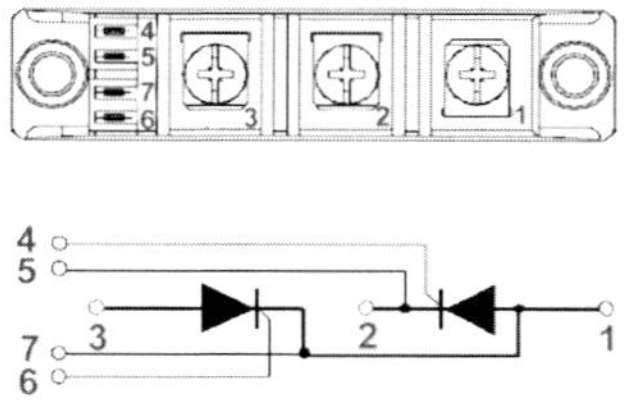	**TO-240AA**	
Thyristor kind	Thyristor Module	
Max. repetitive off-state peak voltage, V	V_{DRM}, V_{RRM}	1200, 1400, 1600
Max. on-state RMS current (for 85°C), A	$I_{T(RMS)}$	235
Max. average on-state current, A	$I_{T(AV)}$	105
Peak, 1/2 cycle non-repetitive surge current, A	I_{TSM}	1785
Max. peak forward on-state voltage, V	V_{TM}	1.64
Max. required DC gate voltage to trigger, V	V_{GT}	4
Max. required DC gate current to trigger, mA	I_{GT}	80 – 270
Max. off-state leakage current, mA	I_{DRM}, I_{RRM}	20
Critical rate of rise of off-state voltage, V/μs	dV/dt	500
Critical rate of rise of on-state current, A/μs	dI/dt	150
Max. holding current (gate open), mA	I_H	240
Max. latching current, mA	I_L	400
Max. junction temperature range, °C	T_j	-40 +130

Appendix D: High-Voltage Triacs

Triac type	**ACST8-8CFP, ACST8-8CT**	
Manufacturer	**STMicroelectronics**	
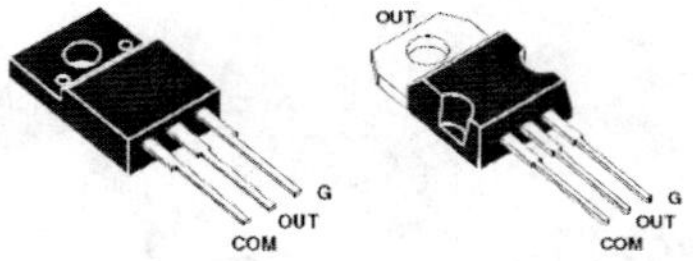 TO-220FPAB ACST8-8CFP TO-220AB ACST8-8CT	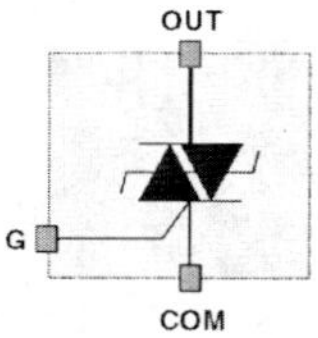	
Triac kind	Alternistor triac. Overvoltage protected.	
Max. repetitive off-state peak voltage, V	V_{DRM}	800
Max. on-state RMS current (for 85°C), A	$I_{T\,(RMS)}$	8
Peak, 1cycle (sine) surge current, A	I_{TSM}	80
Max. peak forward on-state voltage, V	V_{TM}	1.5
Max. required DC gate voltage, V	V_{GT}	1.5
Max. required DC gate current, mA	I_{GT}	30
Max. peak gate current, A	I_{GM}	1
Max. off-state leakage current, mA	I_{DRM}, I_{RRM}	10 – 1000
Critical rate of rise off-state voltage, V/μs	dV/dt	750
Critical rate of rise on-state current, A/μs	dI/dt	100
Max. holding current (gate open), mA	I_H	40
Max. latching current, mA	I_L	70
Turn-On time, μs	t_{gd}	-
Turn-Off time, μs	t_q	-
Max. junction operating temperature, °C	T_j	-

Triac type	**BTA208X-1000B**	
Manufacturer	**Phillips Semiconductors**	
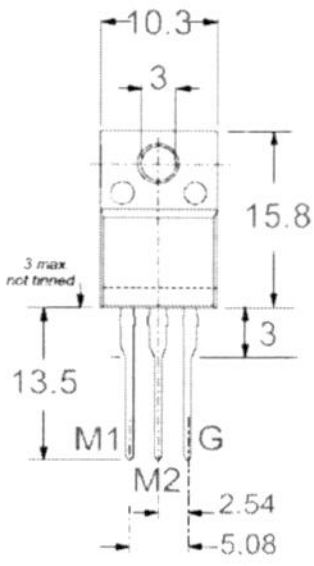 SOT186A, isolated	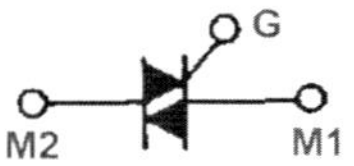	
Triac kind	Alternistor triac (Bidirectional triac)	
Max. repetitive off-state peak voltage, V	V_{DRM}	1000
Max. on-state RMS current (for 85°C), A	$I_{T\,(RMS)}$	8
Peak, 1 cycle surge current, A	I_{TSM}	65
Max. peak forward on-state voltage, V	V_{TM}	1.65
Max. required DC gate voltage, V	V_{GT}	1.5
Max. required DC gate current, mA	I_{GT}	2 – 50
Max. peak gate current, A	I_{GM}	2
Max. off-state leakage current, mA	I_{DRM}, I_{RRM}	0.5
Critical rate of rise off-state voltage, V/µs	dV/dt	4000
Critical rate of rise on-state current, A/µs	dI/dt	38
Max. holding current (gate open), mA	I_H	60
Max. latching current, mA	I_L	60
Turn-On time, µs	t_{gd}	2
Max. junction operating temperature, °C	T_j	-40 + 125

Triac type	**QK006LH4, QK006RH4**	
Manufacturer	**Teccor Electronics (Littelfuse Inc.)**	
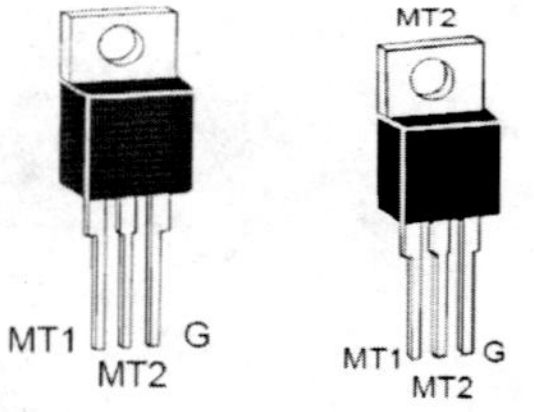 TO-220 TO-220	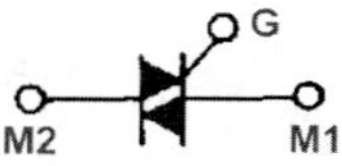	
Max. repetitive off-state peak voltage, V	V_{DRM}	1000
Max. on-state RMS current (for 85°C), A	$I_{T\,(RMS)}$	6
Max. average on-state current, A	$I_{T\,(AV)}$	-
Peak, 1/2 cycle surge current, A	I_{TSM}	80
Max. peak forward on-state voltage, V	V_{TM}	1.6
Max. required DC gate voltage, V	V_{GT}	1.3
Max. required DC gate current, mA	I_{GT}	35
Max. peak gate current, A	I_{GM}	1.6
Max. off-state leakage current, mA	I_{DRM}, I_{RRM}	3
Critical rate of rise off-state voltage, V/μs	dV/dt	150
Critical rate of rise on-state current, A/μs	dI/dt	70
Max. holding current (gate open), mA	I_H	35
Max. latching current, mA	I_L	-
Turn-On time, μs	t_{gd}	4
Max. junction operating temperature, °C	T_j	-

Triac type	QK008L5, QK008R5	
Manufacturer	Teccor Electronics (Littelfuse Inc.)	
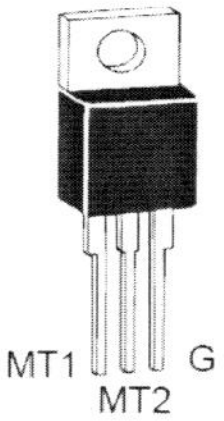	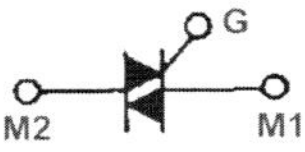	
Max. repetitive off-state peak voltage, V	V_{DRM}	1000
Max. on-state RMS current (for 85°C), A	$I_{T\,(RMS)}$	8
Max. average on-state current, A	$I_{T\,(AV)}$	
Peak, 1/2 cycle surge current, A	I_{TSM}	83
Max. peak forward on-state voltage, V	V_{TM}	1.6
Max. required DC gate voltage, V	V_{GT}	2.5
Max. required DC gate current, mA	I_{GT}	50 – 75
Max. peak gate current, A	I_{GM}	1.8
Max. off-state leakage current, mA	I_{DRM}, I_{RRM}	3
Critical rate of rise off-state voltage, V/µs	dV/dt	100
Critical rate of rise on-state current, A/µs	dI/dt	70
Max. holding current (gate open), mA	I_H	50
Turn-On time, µs	t_{gd}	3
Turn-Off time, µs	t_q	-
Max. junction operating temperature, °C	T_j	-

Triac type	QK008VH4	
Manufacturer	Teccor Electronics (Littelfuse Inc.)	
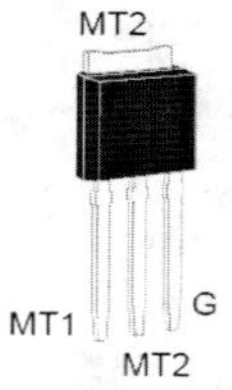 TO-251	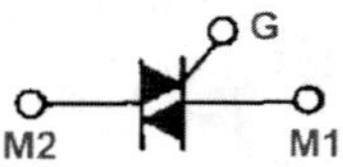	
Max. repetitive off-state peak voltage, V	V_{DRM}	1000
Max. on-state RMS current (for 85°C), A	$I_{T\,(RMS)}$	8
Max. average on-state current, A	$I_{T\,(AV)}$	-
Peak, 1/2 cycle surge current, A	I_{TSM}	80
Max. peak forward on-state voltage, V	V_{TM}	1.6
Max. required DC gate voltage, V	V_{GT}	1.3
Max. required DC gate current, mA	I_{GT}	35
Max. peak gate current, A	I_{GM}	1.6
Max. off-state leakage current, mA	I_{DRM}, I_{RRM}	2
Critical rate of rise off-state voltage, V/μs	dV/dt	150
Critical rate of rise on-state current, A/μs	dI/dt	70
Max. holding current (gate open), mA	I_H	35
Turn-On time, μs	t_{gd}	4
Turn-Off time, μs	t_q	-
Max. junction operating temperature, °C	T_j	-

Triac type	**QK012LH5, QK012RH5**	
Manufacturer	**Teccor Electronics (Littelfuse Inc.)**	
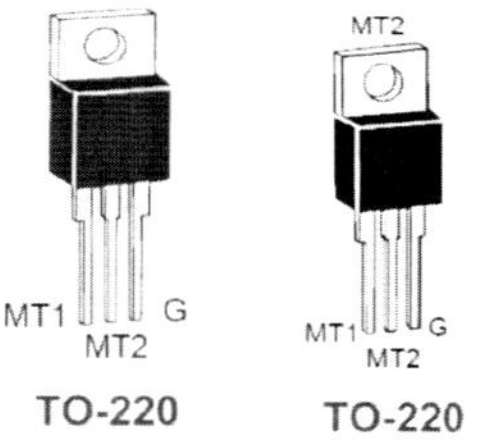	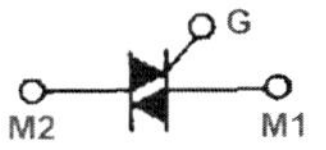	
Triac kind	Alternistor triac (Bidirectional triac)	
Max. repetitive off-state peak voltage, V	V_{DRM}	1000
Max. on-state RMS current (for 85°C), A	$I_{T\,(RMS)}$	12
Max. average on-state current, A	$I_{T\,(AV)}$	-
Peak, 1/2 cycle surge current, A	I_{TSM}	110
Max. peak forward on-state voltage, V	V_{TM}	1.6
Max. required DC gate voltage, V	V_{GT}	1.3
Max. required DC gate current, mA	I_{GT}	50
Max. peak gate current, A	I_{GM}	2
Max. off-state leakage current, mA	I_{DRM}, I_{RRM}	3
Critical rate of rise off-state voltage, V/μs	dV/dt	300
Critical rate of rise on-state current, A/μs	dI/dt	70
Max. holding current (gate open), mA	I_H	50
Max. latching current, mA	I_L	-
Turn-On time, μs	t_{gd}	4

Triac type	**QK025L6, QK025R6**	
Manufacturer	**Teccor Electronics (Littelfuse Inc.)**	

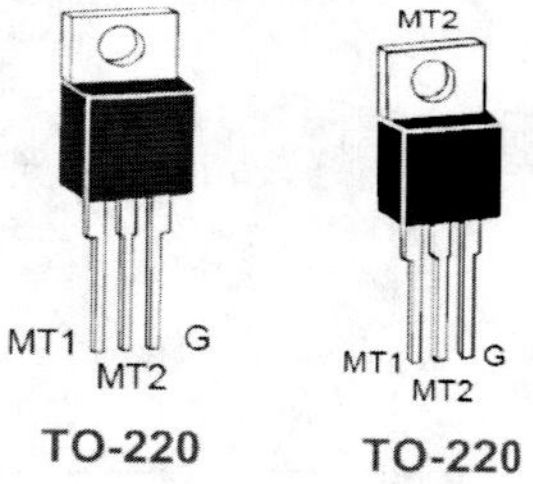

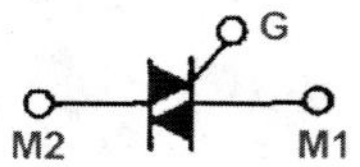

Triac kind	Alternistor triac (Bidirectional triac)	
Max. repetitive off-state peak voltage, V	V_{DRM}	1000
Max. on-state RMS current (for 85°C), A	$I_{T\,(RMS)}$	25
Peak, 1/2 cycle surge current, A	I_{TSM}	208
Max. peak forward on-state voltage, V	V_{TM}	1.8
Max. required DC gate voltage, V	V_{GT}	2.5
Max. required DC gate current, mA	I_{GT}	80
Max. peak gate current, A	I_{GM}	2
Max. off-state leakage current, mA	I_{DRM}, I_{RRM}	3
Critical rate of rise off-state voltage, V/µs	dV/dt	400
Critical rate of rise on-state current, A/µs	dI/dt	100
Max. holding current (gate open), mA	I_H	100
Turn-On time, µs	t_{gd}	5
Max. junction operating temperature, °C	T_j	-

Triac type	QK040K6	
Manufacturer	Teccor Electronics (Littelfuse Inc.)	
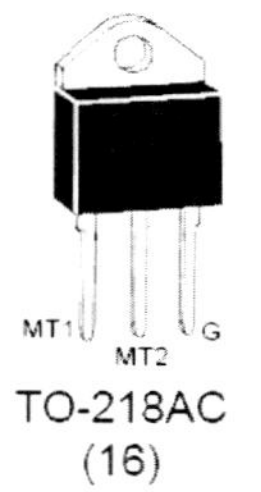	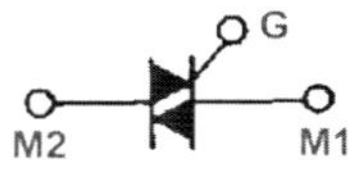	
Triac kind	Alternistor triac (Bidirectional triac)	
Max. repetitive off-state peak voltage, V	V_{DRM}	1000
Max. on-state RMS current (for 85°C), A	$I_{T(RMS)}$	40
Max. average on-state current, A	$I_{T(AV)}$	-
Peak, 1/2 cycle surge current, A	I_{TSM}	335
Max. peak forward on-state voltage, V	V_{TM}	1.8
Max. required DC gate voltage, V	V_{GT}	2.5
Max. required DC gate current, mA	I_{GT}	100
Max. peak gate current, A	I_{GM}	4
Max. off-state leakage current, mA	I_{DRM}, I_{RRM}	5
Critical rate of rise off-state voltage, V/µs	dV/dt	500
Critical rate of rise on-state current, A/µs	dI/dt	150
Max. holding current (gate open), mA	I_H	120
Max. latching current, mA	I_L	-
Turn-On time, µs	t_{gd}	5

Triac type	SSG25C100, SSG25C120
Manufacturer	SanRex

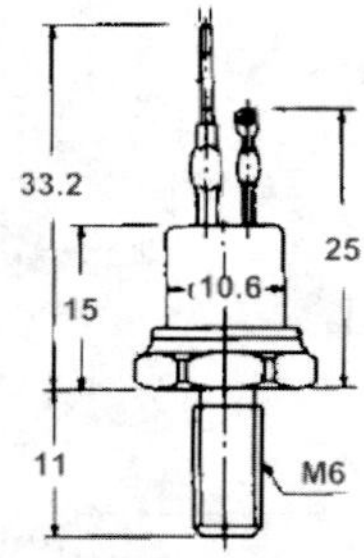

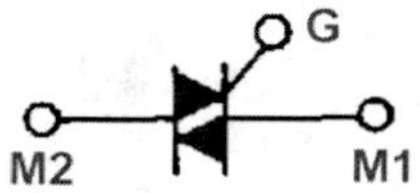

Triac kind	Stud type package	
Max. repetitive off-state peak voltage, V	V_{DRM}	1000, 1200
Max. on-state RMS current (for 85ºC), A	$I_{T\ (RMS)}$	25
Peak, 1 cycle surge current, A	I_{TSM}	220
Max. peak forward on-state voltage, V	V_{TM}	1.6
Max. required DC gate voltage, V	V_{GT}	3
Max. required DC gate current, mA	I_{GT}	70
Max. peak gate current, A	I_{GM}	3
Max. off-state leakage current, mA	I_{DRM}, I_{RRM}	3
Critical rate of rise off-state voltage, V/μs	dV/dt	100
Critical rate of rise on-state current, A/μs	dI/dt	50
Max. holding current (gate open), mA	I_H	30
Turn-On time, μs	t_{gd}	10
Max. junction operating temperature, ºC	T_j	-30 +125

Triac type	**T50AC100A, T50AC120A 50AC100A, 50AC120A**	
Manufacturer	**International Rectifier**	
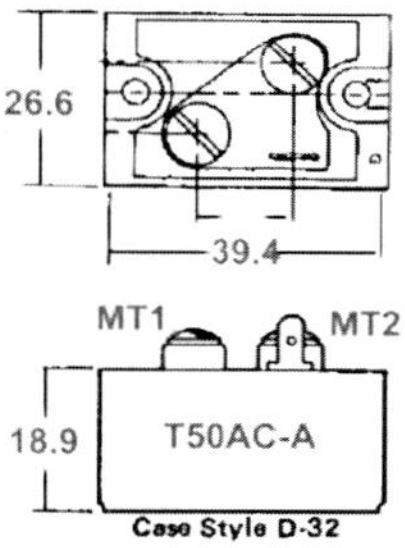 Case Style D-32 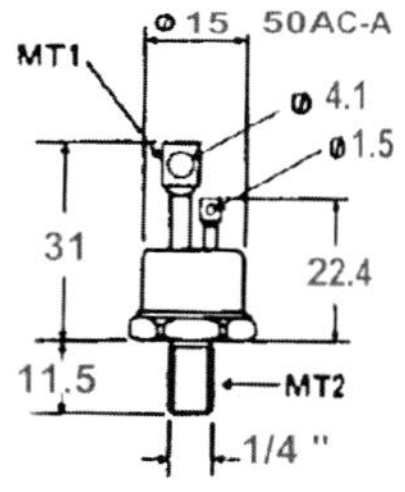 TO-208AC (TO65)	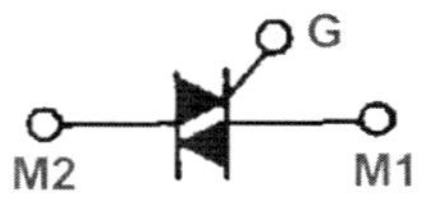	
Max. repetitive off-state peak voltage, V	V_{DRM}	1000, 1200
Max. on-state RMS current (for 85°C), A	$I_{T\,(RMS)}$	50
Max. average on-state current, A	$I_{T\,(AV)}$	-
Peak, 1/2 cycle surge current, A	I_{TSM}	520
Max. peak forward on-state voltage, V	V_{TM}	-
Max. required DC gate voltage, V	V_{GT}	2.5
Max. required DC gate current, mA	I_{GT}	200
Max. peak gate current, A	I_{GM}	3
Max. off-state leakage current, mA	I_{DRM}, I_{RRM}	10
Critical rate of rise off-state voltage, V/µs	dV/dt	200
Critical rate of rise on-state current, A/µs	dI/dt	100
Max. holding current (gate open), mA	I_H	90
Max. junction operating temperature, °C	T_j	-40 +125

Triac type	**TPDV 625-1025, TPDV 625-1225**	
Manufacturer	**SGS-Thomson Microelectronics**	
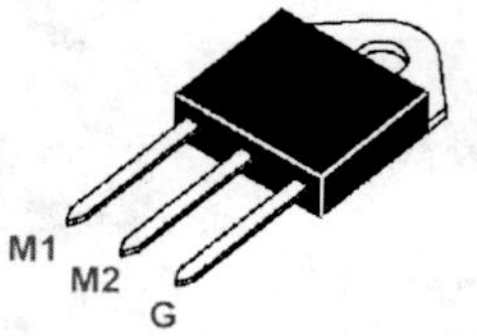 TOP 3 (Plastic)	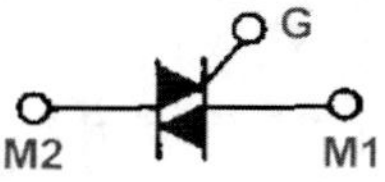	
Triac kind	Alternistor triac (Bidirectional triac)	
Max. repetitive off-state peak voltage, V	V_{DRM}	1000, 1200
Max. on-state RMS current (for 85°C), A	$I_{T\,(RMS)}$	25
Peak, 1/2 cycle surge current, A	I_{TSM}	230
Max. peak forward on-state voltage, V	V_{TM}	1.8
Max. required DC gate voltage, V	V_{GT}	1.5
Max. required DC gate current, mA	I_{GT}	150
Max. peak gate current, A	I_{GM}	-
Max. off-state leakage current, mA	I_{DRM}, I_{RRM}	8
Critical rate of rise off-state voltage, V/µs	dV/dt	500
Critical rate of rise on-state current, A/µs	dI/dt	100
Max. holding current (gate open), mA	I_H	50
Max. latching current, mA	I_L	200
Turn-On time, µs	t_{gd}	2/5
Max. junction operating temperature, °C	T_j	-40 +125

Appendix E: Bilateral Voltage-Trigger Switches

Device type	**BS08D-112, DS08D-T112**	
Manufacturer	**Powerex**	
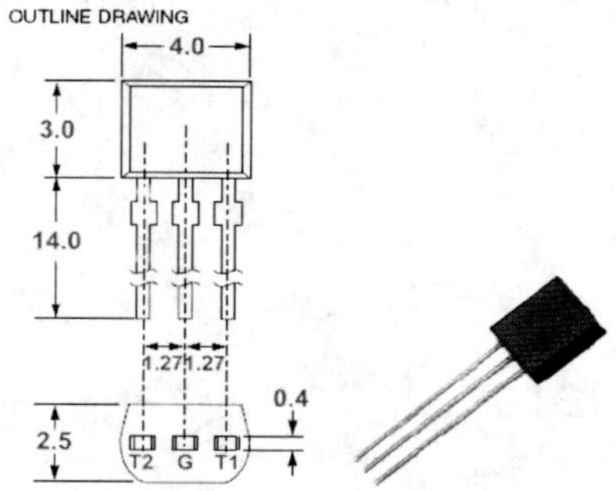	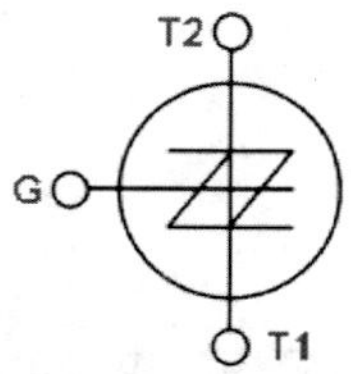	
Device king	Bilateral trigger thyristor	
Max. repetitive on-state peak current (for 10 μs), A	I_{TRM}	1.0
Breakover voltage (forward and reverse), V	V_{BO}	7 – 9
Dynamic breakover voltage (forward and reverse), V	ΔV	1.4
Breakover voltage symmetry, V $\Delta V_{BO} = \Delta V_{BO1} - \Delta V_{BO2}$	ΔV_{BO}	0.5
Forward gate current to trigger (max.), μA	I_G	200
Rise time, μs	t_r	-
Leakage current, μA	I_R	1.0
Peak current, A	I_P	0.175
Holding current, mA	I_H	1.5
Operating junction temperature range, °C	T_j	-55 +125

Device type	**DB3, DB4**	
Manufacturer	**STMicroelectronics**	
 DO-35		
Device king	Bilateral trigger diac	
Max. repetitive on-state peak current (for 10 μs), A	I_{TRM}	2
Breakover voltage (forward and reverse), V	V_{BO}	28 – 36 (DB3) 35 – 45 (DB4)
Dynamic breakover voltage (forward and reverse), V	ΔV	>5
Breakover voltage symmetry, V $\Delta V_{BO} = \Delta V_{BO1} - \Delta V_{BO2}$	ΔV_{BO}	3
Breakover current at breakover voltage, μA	I_{BO}	50
Rise time, μs	t_r	2
Leakage current, μA	I_R	10
Peak current, A	I_P	0.3
Operating junction temperature range, °C	T_j	-40 +125

Device type	**HT-32, HT-32A, HT-32B, HT-34, HT35, HT-36A, HT-36B, HT-40, HT-60**	
Manufacturer	**Teccor Electronics**	
 DO-35		
Sidac or Diac king	Bilateral trigger diac	
Max. repetitive on-state peak current (for 10 μs), A	I_{TRM}	2.0 (1.5 for HT-60)
Breakover voltage (forward and reverse), V	V_{BO}	27 – 37 (HT-32) 28 – 36 (HT-32A) 30 – 34 (HT-32B) 32 – 36 (HT-34B) 30 – 40 (HT-35) 32 – 40 (HT-36A) 34 – 38 (HT-36B) 35 – 45 (HT-40) 56 – 70 (HT-60)
Dynamic breakover voltage (forward and reverse), V	ΔV	7 – 10 (20 for HT-60)
Breakover voltage symmetry, V $\Delta V_{BO} = \Delta V_{BO1} - \Delta V_{BO2}$	ΔV_{BO}	2 – 3 (4 for HT-60)
Breakover current at breakover voltage, μA	I_{BO}	50
Peak current, A	I_P	-
Operating junction temperature, °C	T_j	-40 +125

Device type	**K1V22, KIV24, KIV26**	
Manufacturer	**Shindengen America, Inc.**	
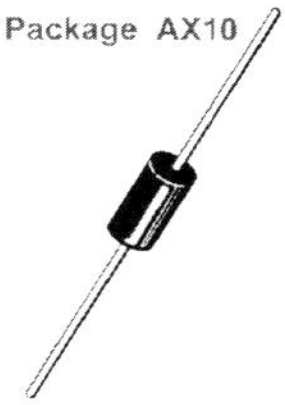		
Device king	High Voltage Silicon Diode for Alternating Current (SIDAC)	
On-state rms current (50 Hz sine wave), A	$I_{T(RMS)}$	1.0
Breakover voltage (forward and reverse), V	V_{BO}	200 – 230 | KIV22 220 – 250 | KIV24 240 – 270 | KIV26
Repetitive peak off-state voltage, V	V_{DRM}	±180
Peak on-state voltage, V	V_{TM}	1.5
Critical rite-of-rise of On-state current, A/ μs	dI/dt	80
Switching resistance, R_S	kΩ	0.1
Leakage current, μA	I_{DRM}	10
Peak non-repetitive surge current (1 cycle sine), A	I_{TSM}	20
Holding current, mA	I_H	20
Operating junction temperature range, °C	T_j	-40 +125

Device type	**K1V33(W), KIV34(W), KIV36(W), KIV38(W)**	
Manufacturer	**Shindengen America, Inc.**	
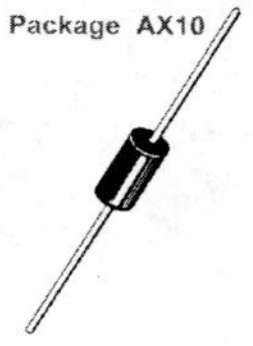 **AX10**		
Device king	High Voltage Silicon Diode for Alternating Current (SIDAC)	
On-state rms current (50 Hz sine wave), A	$I_{T(RMS)}$	1.0
Breakover voltage (forward and reverse), V	V_{BO}	309 – 355 KIV33(W) 320 – 360 KIV34(W) 340 – 380 KIV36(W) 360 – 400 KIV38(W)
Repetitive peak off-state voltage, V	V_{DRM}	±270
Peak on-state voltage, V	V_{TM}	3.0
Critical rite-of-rise of On-state current, A/ μs	dI/dt	50
Switching resistance, R_S	kΩ	0.1
Leakage current, μA	I_{DRM}	10
Peak surge current (1 cycle), A	I_{TSM}	13
Holding current, mA	I_H	50
Operating temperature, °C	T_j	-40 +125

Device type	K105, K110, K120, K130, K140, K150, K195, K200, K220, K240, K250, K300		
Manufacturer	Yixing City Huanzhou Microelectron Co. Wuxi Xuyang Electronics Co.		
DO-15  dia 3.5 x 6.75 mm	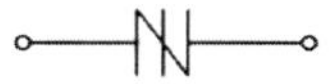		
On-state rms current, A	$I_{T(RMS)}$	1.0	
Breakover voltage (forward and reverse), V	V_{BO}	95 – 113	K105
		104 – 118	K110
		110 – 125	K120
		120 – 138	K130
		130 – 146	K140
		140 – 170	K150
		165 – 190	K195
		190 – 215	K200
		205 – 230	K220
		220 – 250	K240
		240 – 280	K250
		270 – 330	K300
Repetitive peak (off-state) voltage, V	V_{DRM}	75	K105
		85	K110
		90	K120
		95	K130
		105	K140
		115	K150
		130	K195
		150	K200
		165	K220
		175	K240
		190	K250
		215	K300
Peak on-state voltage, V	V_{TM}	1.5	
Continuous on-state DC or RMS current, A	I_T	1.0	
Critical rate-of-rise of on-state current, A/μs	dI/dt	150	
Leakage current (off-state), μA	I_{DRM}	5	
Peak surge current, A	I_{TSM}	16.7	
Holding current, mA	I_H	100	

Device type	**K1050, K1100, K1200, K1300, K1400, K1500 (K...E70 and K...G series)**
Manufacturer	**Teccor Electronics (Littelfuse Inc.)**

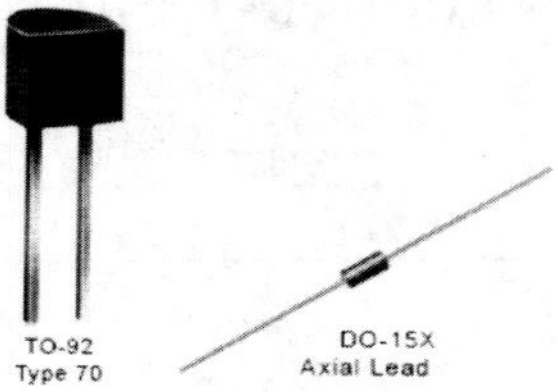

K....E70 K.....G

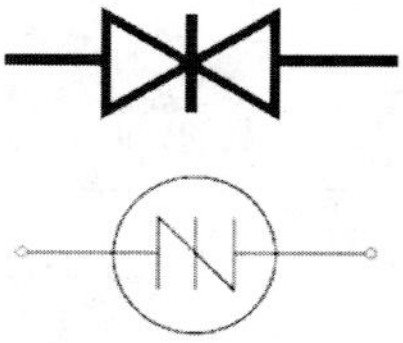

Device king	Bilateral voltage-trigger switch, Silicon Diode for Alternating Current (SIDAC)		
On-state rms current (50 Hz sine wave), A	$I_{T(RMS)}$	1.0	
Breakover voltage (forward and reverse), V	V_{BO}	95 – 113 104 – 118 110 – 125 120 – 138 130 – 136 140 – 170	K1050 K1100 K1200 K1300 K1400 K1500
Repetitive peak off-state voltage, V	V_{DRM}	±90	
Peak on-state voltage, V	V_{TM}	1.5	
Critical rite-of-rise on-state current, A/ μs	dI/dt	150	
Critical rite-of-rise of off-state voltage, V/ μs	dV/dt	1500	
Switching resistance, R_S	kΩ	0.1	
Leakage current, μA	I_{DRM}	5.0	
Peak switching current, A	I_S	0.15	
Holding current, mA	I_H	60 – 150	

Device type	**K2000, K2200, K2400, K2500 (K...E70 and K...G series)**
Manufacturer	**Teccor Electronics (Littelfuse Inc.)**

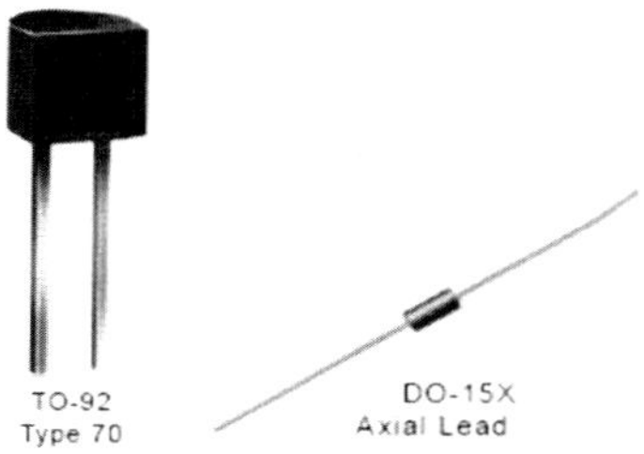

K....E70 K.....G

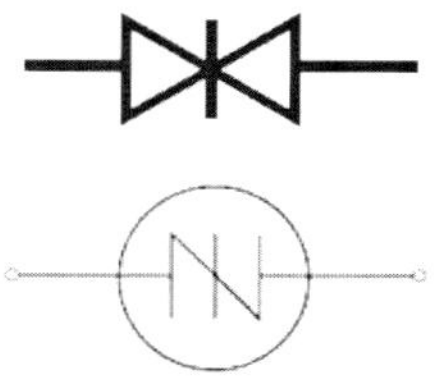

Device king	Bilateral voltage-trigger switch, Silicon Diode for Alternating Current (SIDAC)		
On-state rms current (50 Hz sine wave), A	$I_{T(RMS)}$	1.0	
Breakover voltage (forward and reverse), V	V_{BO}	95 – 113	K1050
		104 – 118	K1100
		110 – 125	K1200
		120 – 138	K1300
		130 – 136	K1400
		140 – 170	K1500
Repetitive peak off-state voltage, V	V_{DRM}	±90	
Critical rite-of-rise of on-state current, A/ µs	dI/dt	150	
Critical rite-of-rise of off-state voltage, V/ µs	dV/dt	1500	
Switching resistance, R_S	kΩ	0.1	
Leakage current, µA	I_{DRM}	5.0	
Peak switching current, A	I_S	0.15	
Holding current, mA	I_H	60 – 150	

Device type	**K2201G, K2401G**		
Manufacturer	**Teccor Electronics (Littelfuse Inc.)**		
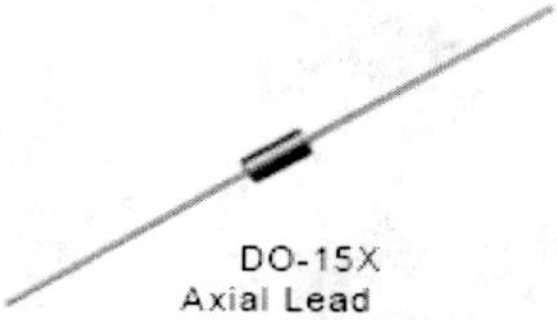	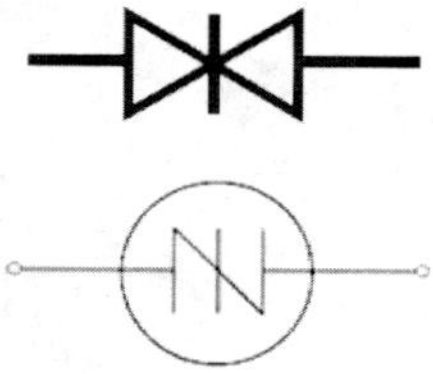		
Device king	Bilateral voltage-trigger switch, Silicon Diode for Alternating Current (SIDAC)		
On-state rms current (50 Hz sine wave), A	$I_{T(RMS)}$	10	
Breakover voltage (forward and reverse), V	V_{BO}	205 – 230 220 – 250	K2201G K2401G
Repetitive peak off-state voltage, V	V_{DRM}	±180	
Peak on-state voltage, V	V_{TM}	2.5	
Critical rite-of-rise of on-state current, A/ μs	dI/dt	150	
Critical rite-of-rise of off-state voltage, V/ μs	dV/dt	1500	
Switching resistance, R_S	kΩ	2.0	
Leakage current, μA	I_{DRM}	5.0	
Peak switching current, A	I_S	0.15	
Holding current, mA	I_H	60 – 150	

Device type	**MKP1V120, MKP1V130, MKP1V160, MKP1V240**		
Manufacturer	**ON Semiconductor**		
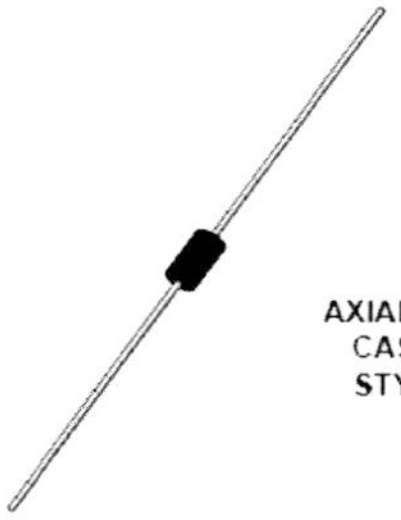 AXIAL LEAD CASE 59 STYLE 2			
Device king	High Voltage Silicon Diode for Alternating Current (SIDAC)		
On-state rms current (50 Hz sine wave), A	$I_{T(RMS)}$	1.0	
Breakover voltage (forward and reverse), V	V_{BO}	110 – 130 120 – 140 150 – 170 220 – 250	MKP1V120 MKP1V120 MKP1V160 MKP1V240
Repetitive peak off-state voltage, V	V_{DRM}	±90 ±180	MKP3V120-160 MKP1V240
Peak on-state voltage, V	V_{TM}	1.3 – 1.5	
Critical rite-of-rise of on-state current, A/µs	dI/dt	120	
Switching resistance, R_S	kΩ	0.1	
Leakage current, µA	I_{DRM}	5	
Peak non-repetitive surge current (1 cycle sine), A	I_{TSM}	4	
Holding current, mA	I_H	100	
Operating temperature, °C	T_j	-40 +125	

Device type	**MKP3V120, MKP3V240**		
Manufacturer	**ON Semiconductor**		
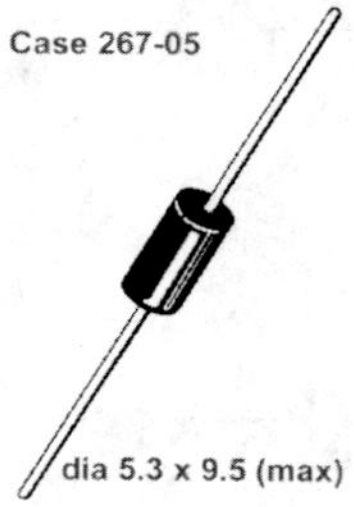 	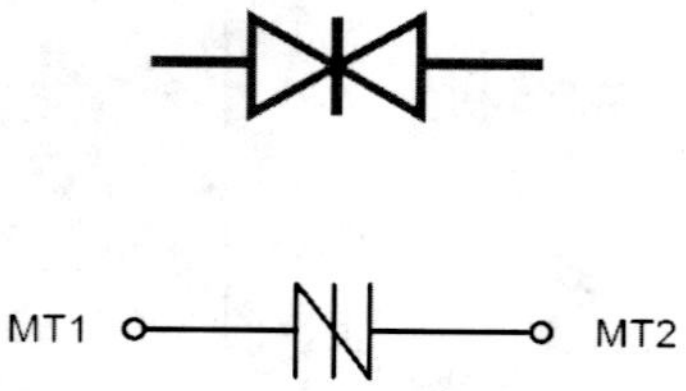 		
Device king	High Voltage Silicon Diode for Alternating Current (SIDAC)		
On-state rms current (50 Hz sine wave), A	$I_{T(RMS)}$	1.0	
Breakover voltage (forward and reverse), V	V_{BO}	110 – 130 220 – 250	MKP3V120 MKP3V240
Repetitive peak off-state voltage, V	V_{DRM}	±90 ±180	MKP3V120 MKP3V240
Peak on-state voltage, V	V_{TM}	1.1 – 1.5	
Critical rite-of-rise of on-state current, A/μs	dI/dt	120	
Switching resistance, R_S	kΩ	0.1	
Leakage current, μA	I_{DRM}	10	
Peak non-repetitive surge current (1 cycle sine), A	I_{TSM}	20	
Holding current, mA	I_H	100	
Oper. temperature, °C	T_j	-40 +125	

Device type	NTE6403	
Manufacturer	NTE Electronics	

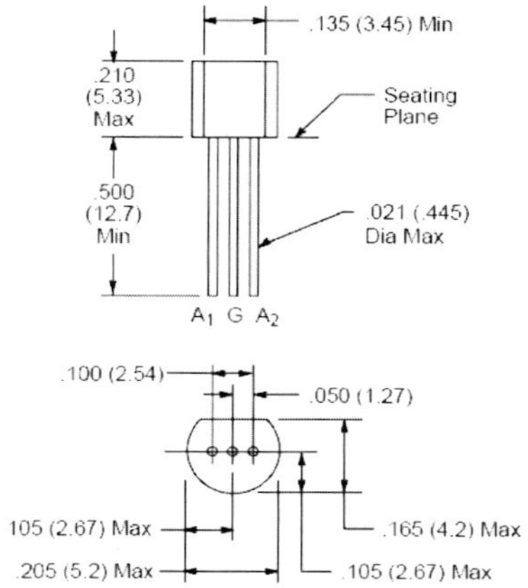

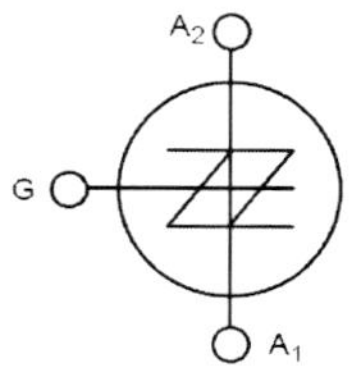

Device king	Bilateral trigger thyristor	
Max. on-state peak current (for 10 μs), A	I_{TRM}	1
Breakover voltage (forward and reverse), V	V_{BO}	7.5 – 9.0
Dynamic breakover voltage (forward and reverse), V	ΔV	3.5
Breakover voltage symmetry, V $\Delta V_{BO} = \Delta V_{BO1} - \Delta V_{BO2}$	ΔV_{BO}	0.2
Forward gate current to trigger (max.), μA	I_G	100
Rise time, μs	t_r	1
Leakage current, μA	I_R	10
Peak current, A	I_P	0.175
Holding current, mA	I_H	0.5

Index